Generalized Integral Transformations

Generalized Integral Transformations

A. H. ZEMANIAN

College of Engineering and Applied Sciences
State University of New York at Stony Brook

Dover Publications, Inc., New York

Published in Canada by General Publishing Company, Ltd., 30 Lesmill Road, Don Mills, Toronto, Ontario.
Published in the United Kingdom by Constable and Company, Ltd.

This Dover edition, first published in 1987, is an unabridged, corrected republication of the work first published by Interscience Publishers, New York, 1968, as Volume XVIII in the series "Pure and Applied Mathematics," edited by R. Courant, L. Bers and J. J. Stoker.

Manufactured in the United States of America
Dover Publications, Inc., 31 East 2nd Street, Mineola, N.Y. 11501

Library of Congress Cataloging-in-Publication Data

Zemanian, A. H. (Armen H.)
Generalized integral transformations.

Bibliography: p.
Includes indexes.
1. Integral transforms. 2. Distributions, Theory of (Functional analysis) I. Title.
QA432.Z45 1987 517′.5 87-551
ISBN 0-486-65375-7 (pbk.)

Peter
Thomas
Lewis
Susan

Preface

The subject of this book arises from the confluence of two mathematical disciplines, the theory of integral transformations and the theory of generalized functions. The former theory is a classical subject in mathematics whose literature can be traced back through at least 150 years. The latter subject, on the other hand, is of recent origin, its advent being the publication of Laurent Schwartz's works, which appeared from 1944 onward; the most notable is his two volume work, "Théorie des Distributions," published in 1950 and 1951. Some fragments of the theory appeared still earlier in the works of S. Bochner [1], around 1927, and S.L. Soboleff [1], around 1936. (The numbers in brackets indicate references to the bibliography at the end of the book.)

An important achievement was the extension to generalized functions of the Fourier transformation, which became thereby a remarkably powerful tool especially in the theory of partial differential equations. The theory and applications of the generalized Fourier transformation has been an active research area for the past fifteen years. In 1952, L. Schwartz [2] also extended the Laplace transformation to generalized functions, and research into this subject has since proceeded apace. Still another integral transformation, whose generalization has been investigated quite thoroughly, is the Hilbert transformation (see Beltrami and Wohlers [1], [2], Bremermann [1], Bremermann and Durand [1], Gelfand and Shilov [2], Griffith [2], Guttinger [1], Horvath [1], [2], Jones [1], [2], Lauwerier [1], Schwartz [3], and Tillman [1], [2].)

However, research into the extension to generalized functions of other integral transformations, remained dormant until quite recently despite the fact that the various types of such transformations are numerous. It is the purpose of this book to present an account of the recent generalizations of some of the simpler and more commonly encountered integral transformations, in particular, the Laplace, Mellin, Hankel, K, Weierstrass, and convolution transformations, as well as those that arise from a variety of orthogonal series expansions. The convolution transformation is especially interesting in that it encompasses as special cases a number of specific transformations such as the one-sided Laplace, Stieltjes, and K transformations (see Hirschman and Widder [1], pp. 65–79). We do not discuss either the generalized Fourier or generalized Hilbert transformations since their theories have already appeared in a number of books, and we have nothing new to add. On the other hand, we do discuss the Laplace transformation since the Mellin and Weierstrass transformations can be obtained from it by certain changes of variables; the theory presented here does not rely upon the Fourier transformation, and in this way it is distinct from (but equivalent to) Schwartz's approach to the Laplace transformation.

Actually, for any of the integral transformations mentioned above, it is not at all difficult to define a generalized transformation if one restricts sufficiently the type of generalized function on which it is to act. The difficulty arises in obtaining either an inversion formula or a uniqueness theorem, and this must be accomplished if the generalized integral transformation is to become significant as an analytical tool. Such results are obtained for every integral transformation considered herein.

As is the case in the theory of distributions and generalized functions, one meets in this book quite a variety of testing-function spaces and their duals. This may be disconcerting to the reader especially since the resulting notation becomes somewhat formidable. However, there is no avoiding this situation if one wishes to achieve the generality aimed for in this book. Each integral transformation requires a different testing-function space, which is tailored to suit certain properties of the kernel function of the transformation. However, there is one unifying concept for all cases: Let I be the open interval of integration for the conventional integral transformation under consideration, and let $\mathscr{E}'(I)$ be the space of distributions whose supports are compact subsets of I (see Sec. 2.3). Then, it turns out that the corresponding generalized integral transformation is always defined on the members of $\mathscr{E}'(I)$. Thus, we have in all cases this simple criterion under which the transformation may be applied.

Incidentally, even in the classical theory of integral transformations there is at least implicitly an abundance of function spaces. Indeed, every time one states a set of conditions under which a transformation can be applied to a function, he is indeed specifying a space of functions in the domain of the transformation. However, in the classical theory there is in general no need to denote these function spaces by symbols, in contrast to the generalized theory.

This book is based upon a graduate course given at the State University of New York at Stony Brook; the course is usually addressed to both mathematics and engineering students. This is reflected in the fact that a substantial part of the book is devoted to the applications of generalized integral transformations to various initial-value and boundary-value problems, as well as to certain problems in system theory. Nevertheless, the emphasis of this work is on the theory of these transformations.

It is presumed that the reader has had a first course in advanced calculus and is therefore familiar with the standard theorems on the interchange of limit processes. Some knowledge of Lebesgue integration including Fubini's theorem and of functions of a complex variable is also assumed. On the other hand, those results concerning topological linear spaces and generalized functions that will be needed are discussed in the first two chapters. On a few occasions we refer to Zemanian [1] for certain results concerning distributions. We freely use various properties of special functions, which appear in such standard reference works as Jahnke, Emde, and Losch [1] and Erdélyi (Ed.) [1]. We also use without proving a number of classical results from the conventional theory of integral transformations, in particular, the complex inversion formula for the Laplace transformation (Sec. 3.5), the inversion formula for the Hankel transformation (Sec. 5.1), orthonormal series expansions in the space $L_2(a,b)$ (Sec. 9.2), and the Riesz-Fischer theorem (Sec. 9.2). Since the proofs of these results appear in so many books, still another presentation seems hardly warranted. Finally, in Secs. 8.5 and 8.6 we present two special cases of the generalized convolution transformation, and, in doing so, we borrow several results from Hirschman and Widder's book on the convolution transformation.

A triple numbering system is used for all theorems, corollaries, lemmas, examples, and figures; the first two numbers indicate the sections in which they appear. For example, Lemma 1.8-1 and Theorem 1.8-1 are respectively the first lemma and first theorem appearing in Sec. 1.8. On the other hand, equations are single numbered starting with (1) in each section.

A. H. Zemanian

September 1968

Acknowledgments

Most of the results presented in this book were developed in research sponsored by the Air Force Cambridge Research Laboratories under contract AF19(628)-2981. The interest and encouragement of Dr. K. Haase of those Research Laboratories is gratefully acknowledged. The author is also indebted to V. K. Balakrishnan, L. D'Amato, E. Koh, D. S. Lee, R. Meidan, J. N. Pandey, and W. Queen for their critical appraisal of various portions of the manuscript. Finally, to the author's wife for her assistance in so many ways, thanks.

A. H. ZEMANIAN

Contents

CHAPTER I

Countably Multinormed Spaces, Countable-Union Spaces, and Their Duals

1.1. Introduction

The theory of generalized functions is founded on that of topological linear spaces. For our purposes, not much is needed, especially if one takes advantage of the simplifying concept of a sequential-convergence linear space. In this chapter we develop those results from the theories of these two types of spaces that will be needed for an understanding of our later discussions. However, we first pause in the next section to state those conventions of terminology and notation that we shall adhere to throughout this book.

1.2. Notations and Terminology

First of all, we call attention to the Index of Symbols at the end of the book; it contains the more important symbols used in the text.

$\mathscr{R}^n$ and $\mathscr{C}^n$ denote respectively the real and complex n-dimensional euclidean spaces. Thus, an arbitrary point t in $\mathscr{R}^n$ (in $\mathscr{C}^n$) is an ordered n-tuple of real (respectively, complex) numbers, $t = \{t_1, t_2, \ldots, t_n\}$, whose magnitude is

$$|t| = \left[\sum_{\nu=1}^{n} |t_\nu|^2\right]^{1/2} \tag{1}$$

The distance between any two such points t, τ is $|t - \tau|$, where we use componentwise subtraction.

A compact set in $\mathscr{R}^n$ or $\mathscr{C}^n$ is simply a closed bounded set. (However, in more general types of topological linear spaces a closed bounded set need not be compact.) If I is an open set in $\mathscr{R}^n$, if K is a compact set in $\mathscr{R}^n$, and if K is contained in I, we say that K is a compact subset of I.

A nonnegative element t of $\mathscr{R}^n$ is one whose components t_ν are all nonnegative; in this case we write $t \geq 0$. Moreover, for $x, t \in \mathscr{R}^n$, the notations $x \leq t$ and $x < t$ mean $x_\nu \leq t_\nu$ and respectively $x_\nu < t_\nu$ $(\nu = 1, 2, \ldots, n)$. An integer $k = \{k_1, k_2, \ldots, k_n\}$ in $\mathscr{R}^n$ is an element of $\mathscr{R}^n$ whose components k_ν are all integers.

If k is a nonnegative integer in $\mathscr{R}^n$, then a partial differenitation with respect to t is denoted by

$$D^k \triangleq D_t{}^k \triangleq \frac{\partial^{k_1+k_2+\cdots+k_n}}{\partial t_1{}^{k_1}\,\partial t_2{}^{k_2}\cdots\partial t_n{}^{k_n}} \tag{2}$$

(We shall use the symbol $\triangleq$ whenever we wish to emphasize that a particular equality is a definition.) We follow a standard notation in denoting the order $k_1 + k_2 + \cdots + k_n$ of the partial differentiation (2) by $|k|$. The context in which this symbol is used should clarify any possible confusion with the notation (1). The $|k|$th order ordinary or partial derivative of a function f will be denoted alternatively by $D^k f$, $D_t{}^k f(t)$, or $f^{(k)}(t)$. It will turn out in all of our subsequent discussions that the order of differentiation in (2) can be changed in any fashion.

By a *conventional function* we mean a function whose domain is contained in either $\mathscr{R}^n$ or $\mathscr{C}^n$ and whose range is in either $\mathscr{R}^1$ or $\mathscr{C}^1$ (not necessarily respectively). Here, we use the adjective "conventional" to distinguish this from the concept of a generalized function, which we shall introduce later on. We choose not to use the adjective "ordinary" for the following reason. Had we consistently used the terminology "ordinary function," "ordinary transformation," and so forth, to distinguish these concepts from the generalized notions (i.e., generalized function, generalized transformation, etc.), we would have eventually come across the disconcerting phrase "an ordinary partial derivative."

If the range of a conventional function is required to be in $\mathscr{R}^1$, we call it a *real-valued function*. On the other hand, a *complex-valued function* is the same thing as a conventional function and may therefore have an entirely real range. This is no misnomer since $\mathscr{R}^1$ can be identified with the real axis in $\mathscr{C}^1$.

Let I be an open set in $\mathscr{R}^n$. By a *locally integrable function on* I we mean a conventional function that is Lebesgue integrable on every open set J in $\mathscr{R}^n$ whose closure $\bar{J}$ is a compact subset of I. Let $p \in \mathscr{R}^1$ satisfy $1 \leq p < \infty$; as is customary, $L_p(I)$ denotes the collection of all (equivalence classes of) locally integrable functions f on I satisfying

$$\int_I |f(t)|^p \, dt < \infty$$

If I is the open interval $a < t < b$ in $\mathscr{R}^1$, we also use the notation $L_p(a, b)$. For $p = 2$, $f \in L_2(I)$ is called *quadratically integrable on* I.

A *function of rapid descent* is a conventional function $f(t)$ on $\mathscr{R}^n$ or $\mathscr{C}^n$ such that $|f(t)| = o(|t|^{-m})$ as $|t| \to \infty$ for every integer $m \in \mathscr{R}^1$. On the other hand, $f(t)$ is said to be a *function of slow growth* if it is a conventional function on $\mathscr{R}^n$ or $\mathscr{C}^n$ such that there exists an integer $k \in \mathscr{R}^1$ for which $|f(t)| = O(|t|^k)$ as $|t| \to \infty$. The same terminology is used when f is defined only on the integers or on some other unbounded subset of $\mathscr{R}^n$ or $\mathscr{C}^n$.

A conventional function is said to be *smooth* (in Zemanian [1] we called it *infinitely smooth*) if all its derivatives of all orders are continuous at all points of its domain. It is a standard result from calculus that the order of differentiation in any nth-order partial derivative ($|n| \geq 2$) of a smooth function may be changed in any fashion. The support of a continuous function $f(t)$ defined on some open set Ω in $\mathscr{R}^n$ is the closure with respect to Ω of the set of points t where $f(t) \neq 0$. It is denoted by *supp f*.

Let z be a variable in $\mathscr{C}^1$ and μ a fixed member of $\mathscr{C}^1$. Unless the contrary is specifically mentioned, the (in general) multivalued function z^μ is always restricted to its principal branch by requiring that $-\pi < \arg z \leq \pi$. Thus, z^μ assumes real positive values whenever z is real and positive. A similar convention is followed for other multivalued functions such as the Bessel function $J_\mu(z)$ of first kind and order μ and the modified Bessel function $K_\mu(z)$ of third kind and order μ. As a consequence, when μ is real, both $J_\mu(z)$ and $K_\mu(z)$ are real-valued when z is real and positive (Jahnke, Emde, and Losch [1], p. 135 and p. 210).

We will at times use the notation $\{\phi \colon P(\phi)\}$ to denote the set of all elements ϕ for which the proposition $P(\phi)$ concerning ϕ is true. Moreover, $\{\phi_\nu\}_{\nu \in A}$ is a collection of indexed elements where the index ν varies through some set A. On the other hand, a sequence is denoted by either $\{\phi_\nu\}_{\nu=1}^\infty$ or $\{\phi_1, \phi_2, \phi_3, \ldots\}$, whereas a directed set is denoted by $\{\phi_\nu\}_{\nu \to \infty}$. Finite collections are indicated by $\{\phi_1, \phi_2, \ldots, \phi_n\}$. We sometimes use the abbreviated notation $\{\phi_\nu\}$ when it is clear what type of collection we are dealing with.

1.3. Linear Spaces

A linear space (or, synonymously, a vector space) is an abstraction of the collection of vectors in a euclidean space. It is defined in such a manner that the concepts "addition of vectors" and "multiplication of a vector by a number" are preserved. Also, the "numbers" that are multiplied with vectors are taken to be the members of some unspecified field (Paige and Swift [1], pp. 4–5). In all the specific examples of linear spaces encountered in this book this field will always be $\mathscr{C}^1$, the space of complex numbers. Therefore, we use the following more restrictive definition of a linear space.

A collection $\mathscr{V}$ of elements ϕ, ψ, θ, ... is said to be a *linear space* if the following axioms are satisfied.

1. There is an operation $+$, called "addition," by which any pair of elements ϕ and ψ can be combined to yield a unique element $\phi + \psi$ in $\mathscr{V}$. Moreover, $+$ has the following properties:

1a. $\phi + \psi = \psi + \phi$ (commutativity)

1b. $(\phi + \psi) + \theta = \phi + (\psi + \theta)$ (associativity)

1c. There exists a unique element $\varnothing$ in $\mathscr{V}$ such that $\phi + \varnothing = \phi$ for every $\phi \in \mathscr{V}$.

1d. For every $\phi \in \mathscr{V}$ there exists a unique element $-\phi$ in $\mathscr{V}$ such that $\phi + (-\phi) = \varnothing$.

2. There is an operation, called "multiplication by a complex number," by which any complex number α and any $\phi \in \mathscr{V}$ can be combined to yield a unique element $\alpha\phi$ in $\mathscr{V}$. Moreover, the following properties are satisfied for every choice of $\phi \in \mathscr{V}$ and the complex numbers α and β:

2a. $\alpha(\beta\phi) = (\alpha\beta)\phi$

2b. $1\phi = \phi$ (1 denotes the number one.)

3. In addition, the following distributive laws must be fulfilled:

3a. $\alpha(\phi + \psi) = \alpha\phi + \alpha\psi$

3b. $(\alpha + \beta)\phi = \alpha\phi + \beta\phi$

This ends the definition of a linear space.

The customary rules for the addition of conventional functions and their multiplication by numbers follow from these axioms. (See Problem 1.3-1.) Subtraction is defined by $\phi - \psi = \phi + (-\psi)$. Moreover, $-\psi$ is called the *negative of* ψ. Finally, $\varnothing$ is called the *zero element* or the *origin* in $\mathscr{V}$; in most cases we denote it simply by 0.

A subset $\mathscr{U}$ of a linear space $\mathscr{V}$ is called a *linear subspace* (or simply a *subspace*) of $\mathscr{V}$ if for every ϕ and ψ in $\mathscr{U}$ and for every complex number α, $\phi + \psi$ and $\alpha\phi$ are both in $\mathscr{U}$.

EXAMPLE 1.3-1. Let K be a compact subset of $\mathscr{R}^n$. $\mathscr{D}_K$ is the set of all complex-valued smooth functions on $\mathscr{R}^n$ that vanish at all points outside K. $\mathscr{D}_K$ is a linear space under the customary definitions of addition and multiplication by a complex number, the zero element in $\mathscr{D}_K$ being the identically zero function. This space will be of importance to us later on.

A specific example of a function in $\mathscr{D}_K$ is the following: Assume that K contains the domain $\{t: t \in \mathscr{R}^n, |t| \leq 1\}$.
Set

$$\zeta(t) = \begin{cases} \exp \dfrac{1}{|t|^2 - 1} & |t| < 1 \\ 0 & |t| \geq 1 \end{cases} \tag{1}$$

$\zeta(t)$ is a member of $\mathscr{D}_K$; this assertion becomes obvious once it has been shown that all the partial derivatives of $\zeta(t)$ are continuous at those points t for which $|t| = 1$.

EXAMPLE 1.3-2. $\mathscr{D}$ is the union of all $\mathscr{D}_K$ where K varies through all possible compact subsets of $\mathscr{R}^n$. Thus, $\phi(t)$ is in $\mathscr{D}$ if and only if it is a complex-valued smooth function whose support is a compact set. $\mathscr{D}$ is also a linear space under the usual definitions of addition and multiplication by a complex number. The zero element is again the identically zero function.

PROBLEM 1.3-1. Prove the following rules using the axioms of a linear space.

a. $\phi + \psi = \phi + \theta$ implies $\psi = \theta$.
b. $\alpha\varnothing = \varnothing$
c. $0\phi = \varnothing$ (Here, 0 denotes the number zero.)
d. $(-1)\phi = -\phi$
e. If $\alpha\phi = \beta\phi$ and $\phi \neq \varnothing$, then $\alpha = \beta$.
f. If $\alpha\phi = \alpha\psi$ and $\alpha \neq 0$, then $\phi = \psi$.

PROBLEM 1.3-2. Verify that a linear subspace $\mathscr{U}$ of a linear space $\mathscr{V}$ is also a linear space under the same rules for addition and multiplication by a complex number.

PROBLEM 1.3-3. Show that the intersection $\mathscr{I}$ of any collection of linear subspaces of a linear space $\mathscr{V}$ is also a linear subspace of $\mathscr{V}$.

PROBLEM 1.3-4. Show that the function $\zeta(t)$ given by (1) is truly in $\mathscr{D}_K$.

1.4. Sequential-Convergence Spaces

We turn now to the concepts of a sequential-convergence space and a sequential-convergence linear space. (See, for example, Dudley [1].)

Let $\mathscr{V}$ be some set. A sequence $\{\phi_\nu\}$ is said to be a *sequence in* $\mathscr{V}$ if all its members are in $\mathscr{V}$. $\mathscr{V}$ is called a *sequential-convergence space* if there is assigned to it a rule that singles out certain sequences in $\mathscr{V}$, called the *convergent sequences in* $\mathscr{V}$, and assigns to each such sequence some element in $\mathscr{V}$, called the *limit* of the sequence, and if in addition the first three axioms listed below are satisfied. If $\{\phi_\nu\}$ is a convergent sequence in $\mathscr{V}$ and ϕ is its limit, we say that "$\{\phi_\nu\}$ converges in $\mathscr{V}$ to ϕ," and we write "$\phi_\nu \to \phi$ in $\mathscr{V}$ as $\nu \to \infty$" or "$\lim_{\nu\to\infty} \phi_\nu = \phi$ in $\mathscr{V}$."

1. For each $\phi \in \mathscr{V}$, the sequence $\{\phi, \phi, \phi, \ldots\}$ converges in $\mathscr{V}$ to ϕ.

2. If $\{\phi_\nu\}$ converges in $\mathscr{V}$ to ϕ, then every subsequence $\{\phi_{\nu_k}\}_{k=1}^{\infty}$ of $\{\phi_\nu\}$ also converges in $\mathscr{V}$ to ϕ.

3. The limit of every convergent sequence is unique. That is, if $\phi_\nu \to \phi$ in $\mathscr{V}$ and $\phi_\nu \to \psi$ in $\mathscr{V}$, then $\phi = \psi$.

The concept of convergence can be extended to a directed set $\{\phi_\nu\}_{\nu\to a}$, where the numerical index ν tends to some limit a, as follows. We say that $\{\phi_\nu\}_{\nu\to a}$ converges in $\mathscr{V}$ to a limit ϕ if and only if every sequence $\{\phi_{\nu_k}\}_{k=1}^{\infty}$ with $\nu_k \to a$ that is contained in $\{\phi_\nu\}_{\nu\to a}$ converges in $\mathscr{V}$ to ϕ.

The collection $\mathscr{V}$ is called a *sequential-convergence linear space* if it is simultaneously a linear space and a sequential-convergence space and if the rule defining the convergent sequences satisfies the following additional two axioms.

4. If $\{\phi_\nu\}$ converges in $\mathscr{V}$ to ϕ and $\{\psi_\nu\}$ converges in $\mathscr{V}$ to ψ, then $\{\phi_\nu + \psi_\nu\}$ converges in $\mathscr{V}$ to $\phi + \psi$.

5. If $\{\phi_\nu\}$ converges in $\mathscr{V}$ to ϕ and the sequence $\{\alpha_\nu\}$ of complex numbers converges in the conventional sense to the complex number α, then $\{\alpha_\nu \phi_\nu\}$ converges in $\mathscr{V}$ to $\alpha\phi$.

Finally, $\mathscr{V}$ will be called a *sequential-convergence* linear space* if in addition to the above five axioms the following one is also satisfied.

6. If it is false that $\phi_\nu \to \phi$ in $\mathscr{V}$ as $\nu \to \infty$, then there exists a subsequence $\{\psi_\mu\}$ of $\{\phi_\nu\}$ such that for every subsequence $\{\theta_\lambda\}$ of $\{\psi_\mu\}$ it is false that $\theta_\lambda \to \phi$ in $\mathscr{V}$ as $\lambda \to \infty$.

EXAMPLE 1.4-1. Consider the linear space $\mathscr{D}$ which was defined in Example 1.3-2. Define the sequence $\{\phi_\nu\}_{\nu=1}^{\infty}$ as being convergent in $\mathscr{D}$ if all the ϕ_ν are in $\mathscr{D}$ and have their supports contained in some fixed compact subset K of $\mathscr{R}^n$ and if, for each choice of the nonnegative integer $k \in \mathscr{R}^n$,

$$|D^k\phi_\nu(t) - D^k\phi_\mu(t)| \to 0$$

uniformly on $\mathscr{R}^n$ as ν and μ tend independently to infinity. It follows from a standard theorem of calculus (Apostol [1], p. 395 and p. 402) that there exists a unique smooth function ϕ on $\mathscr{R}^n$, whose support is contained in K, such that, for each k,

$$|D^k\phi_\nu(t) - D^k\phi(t)| \to 0$$

uniformly on $\mathscr{R}^n$ as $\nu \to \infty$. Take ϕ as the limit in $\mathscr{D}$ of $\{\phi_\nu\}$. It can now be shown that all the above axioms are satisfied. Thus, $\mathscr{D}$ is a sequential-convergence* linear space when it is supplied with this concept of convergence.

On the other hand, if in the definition of convergent sequences we discard the requirement that the supports of all the ϕ_ν be contained in some fixed bounded domain, then $\mathscr{D}$ will no longer be a sequential-convergence linear space. Indeed, let $\zeta(t)$ be given by Sec. 1.3, Eq. (1). With ν being a variable positive integer in $\mathscr{R}^1$, let t/ν denote the n-tuple

$\{t_1/\nu, t_2/\nu, \ldots, t_n/\nu\}$. The sequence $\{\zeta(t/\nu) \exp(1-|t|^2)\}_{\nu=1}^{\infty}$ converges uniformly on $\mathscr{R}^n$, and so does every one of its derivatives. But its limit is $\exp(-|t|^2)$, which is not in $\mathscr{D}$. Thus, the sequence is not convergent since the requirement that the limit be in $\mathscr{D}$ is violated.

PROBLEM 1.4-1. Verify that the space $\mathscr{D}$ satisfies all the axioms for a sequential-convergence* linear space when its convergent sequences are defined as in Example 1.4-1.

1.5. Seminorms and Multinorms

In this book we will not be concerned with general sequential-convergence linear spaces. Instead, our spaces of conventional functions will be restricted to two special types. In one type the rule defining the convergent sequences is derived from a topology for the space; in the second type the space is decomposed into a countable collection of subspaces each of which possesses a topology, and the convergent sequences are defined in terms of this resulting collection of topologies. In both cases the topologies considered are generated by multinorms. It is the purpose of this section to explain what is meant by a multinorm. The topologies and sequential-convergence linear spaces arising from multinorms are discussed in subsequent sections.

Let $\mathscr{V}$ be a linear space. A *seminorm on* $\mathscr{V}$ is a rule γ that assigns a real number $\gamma(\phi)$ to each $\phi \in \mathscr{V}$ and that satisfies the following axioms. Here, ϕ and ψ are arbitrary elements of $\mathscr{V}$ and α is any complex number.

1. $\gamma(\alpha\phi) = |\alpha|\gamma(\phi)$

2. $\gamma(\phi + \psi) \leq \gamma(\phi) + \gamma(\psi)$

By choosing $\alpha = 0$ in axiom 1, we see that $\gamma(\varnothing) = 0$. Also, $\gamma(\phi) \geq 0$ for every $\phi \in \mathscr{V}$ because

$$0 = \gamma(\phi - \phi) \leq \gamma(\phi) + \gamma(-\phi) = 2\gamma(\phi)$$

Still another property is the inequality

$$|\gamma(\phi) - \gamma(\psi)| \leq \gamma(\phi - \psi) \tag{1}$$

which follows from the above axioms through the inequalities

$$\gamma(\phi) \leq \gamma(\phi - \psi) + \gamma(\psi)$$

and

$$\gamma(\psi) \leq \gamma(\psi - \phi) + \gamma(\phi)$$

A seminorm is called a *norm* if it satisfies still another axiom:

3. $\gamma(\phi) = 0$ implies that $\phi = \varnothing$ (i.e., ϕ is the zero element in $\mathscr{V}$).

With a being a positive number, we define the seminorm $a\gamma$ by $(a\gamma)(\phi) \triangleq a\gamma(\phi)$. Clearly, $a\gamma$ is also a seminorm. If $\{\gamma_1, \ldots, \gamma_n\}$ is a finite collection of seminorms on $\mathscr{V}$, we define the seminorm $\gamma_1 + \cdots + \gamma_n$ by

$$(\gamma_1 + \cdots + \gamma_n)(\phi) \triangleq \gamma_1(\phi) + \cdots + \gamma_n(\phi)$$

Also, the seminorm max $(\gamma_1, \ldots, \gamma_n)$ is defined by

$$[\max(\gamma_1, \ldots, \gamma_n)](\phi) \triangleq \max[\gamma_1(\phi), \ldots, \gamma_n(\phi)]$$

That these are truly seminorms follows readily from axioms 1 and 2. Moreover, if any one of the γ_ν is a norm, then $\gamma_1 + \cdots + \gamma_n$ and max $(\gamma_1, \ldots, \gamma_n)$ are also norms.

Two seminorms γ and ρ on $\mathscr{V}$ are called *equivalent* if there exist two fixed positive numbers a and b such that for every $\phi \in \mathscr{V}$ we have $a\gamma(\phi) \leq \rho(\phi) \leq b\gamma(\phi)$. Consequently, $\gamma_1 + \cdots + \gamma_n$ and $\max(\gamma_1, \ldots, \gamma_n)$ are equivalent seminorms because

$$\max[\gamma_1(\phi), \ldots, \gamma_n(\phi)] \leq \gamma_1(\phi) + \cdots + \gamma_n(\phi)$$
$$\leq n \max[\gamma_1(\phi), \ldots, \gamma_n(\phi)]$$

Now, let $S = \{\gamma_\mu\}_{\mu \in A}$ be a collection of seminorms on $\mathscr{V}$ where the index μ traverses a finite or infinite set A. The collection S is said to be *separating* (or *S separates $\mathscr{V}$*) if for every $\phi \neq \varnothing$ in $\mathscr{V}$ there is at least one γ_μ such that $\gamma_\mu(\phi) \neq 0$. In other words, S is separating if only the zero element in $\mathscr{V}$ has the number zero assigned to it by every seminorm in S. In this case we call S a *multinorm*. Obviously, a sufficient condition for S to be separating is that at least one of the γ_μ be a norm. If S is a countable separating collection of seminorms, it is called a *countable multinorm*. We shall usually assume that a countable multinorm has been ordered into a sequence of seminorms $\{\gamma_\mu\}_{\mu=1}^{\infty}$.

PROBLEM 1.5-1. Assume that $\gamma_1, \ldots, \gamma_n$ are seminorms on the linear space $\mathscr{V}$. Show that $\gamma_1 + \cdots + \gamma_n$ and $\max(\gamma_1, \ldots, \gamma_n)$ are both seminorms on $\mathscr{V}$. Then, assuming in addition that γ_1 is a norm, show that $\gamma_1 + \cdots + \gamma_n$ and $\max(\gamma_1, \ldots, \gamma_n)$ are norms on $\mathscr{V}$.

PROBLEM 1.5-2. Let $\mathscr{V}$ be a linear space and γ a norm on $\mathscr{V}$. Let us define the sequence $\{\phi_\nu\}$ in $\mathscr{V}$ as being convergent if there exists a $\phi \in \mathscr{V}$ such that $\gamma(\phi_\nu - \phi) \to 0$ as $\nu \to \infty$. Show that $\mathscr{V}$ is then a sequential-convergence* linear space.

1.6. Multinormed Spaces

We can employ a collection of seminorms to construct the concept of a "neighborhood" in a linear space $\mathscr{V}$ as follows. Let $S = \{\gamma_\nu\}_{\nu \in A}$ be a set of seminorms on $\mathscr{V}$, which need not separate $\mathscr{V}$. Given any nonvoid finite subset $\{\gamma_{\nu_k}\}_{k=1}^{n}$ of S and the arbitrary positive numbers $\varepsilon_1, \varepsilon_2, \ldots, \varepsilon_n$, a *balloon centered at* ψ, where ψ is a fixed point in $\mathscr{V}$, is defined as the set of all $\phi \in \mathscr{V}$ such that

$$\gamma_{\nu_k}(\phi - \psi) \leq \varepsilon_k \qquad k = 1, 2, \ldots, n$$

Clearly, the intersection of two balloons centered at the same point ψ is also a balloon centered at ψ. A *neighborhood in* $\mathscr{V}$ is any set in $\mathscr{V}$ that contains a balloon, and a *neighborhood of* $\psi \in \mathscr{V}$ is any set that contains a balloon centered at ψ. A neighborhood of the origin $\varnothing$ is called a *neighborhood of zero*. We shall refer to the collection of all neighborhoods in $\mathscr{V}$ as the *topology* of $\mathscr{V}$.

(The term "topology" is more commonly used to denote the collection of "open sets" in a space. However, the collection of neighborhoods determines and is determined by a collection of open sets (Kantorovich and Akilov [1]), and so our use of the term "topology" does not lead to any inconsistency with the more common usage.)

Note that the neighborhoods of any $\psi \in \mathscr{V}$ are simply translations through the element ψ of the neighborhoods of zero. (A translation through ψ of any set Ω in $\mathscr{V}$ is obtained by adding ψ to each member of Ω.) Thus, the topology of $\mathscr{V}$ is simply the collection of all possible translations of all neighborhoods of zero. Because of this, we need merely consider the neighborhoods of zero in many of our subsequent discussions. A similar comment applies to the balloons in $\mathscr{V}$.

If two seminorms γ and ρ are equivalent, then the balloon $\{\phi\colon \gamma(\phi) \leq \varepsilon\}$ generated by γ contains a balloon $\{\phi\colon \rho(\phi) \leq \delta\}$ generated by ρ, and conversely. Thus, the neighborhoods generated by two equivalent seminorms coincide.

A *multinormed space* $\mathscr{V}$ is a linear space having a topology generated by a multinorm S (i.e., by a *separating* collection of seminorms); if S is countable, $\mathscr{V}$ is called a *countably multinormed space*. (Do not misread this. The space itself need not be countable; it is the multinorm that is countable.)

In a multinormed space $\mathscr{V}$ the intersection of all neighborhoods of a given $\phi \in \mathscr{V}$ contains no element other than ϕ. Indeed, if $\theta \in \mathscr{V}$ is also in every neighborhood of ϕ, then for every $\varepsilon > 0$ and every $\gamma \in S$ we have $\gamma(\phi - \theta) \leq \varepsilon$. Hence, $\gamma(\phi - \theta) = 0$. But, since S is separating, $\phi = \theta$.

Given a neighborhood Ω in a multinormed space $\mathscr{V}$, we say that a sequence $\{\phi_\nu\}_{\nu=1}^{\infty}$ is eventually in Ω if there exists a number N such that ϕ_ν is in Ω whenever $\nu \geq N$. A sequence $\{\phi_\nu\}$ is called *convergent in* $\mathscr{V}$ (or simply *convergent*) if all ϕ_ν are members of $\mathscr{V}$ and if there exists a $\phi \in \mathscr{V}$ such that, for every neighborhood Ω of ϕ, $\{\phi_\nu\}$ is eventually in Ω. The element ϕ is called a *limit* of the sequence. We say that $\{\phi_\nu\}$ *converges in* $\mathscr{V}$ *to* ϕ, and we write "$\phi_\nu \to \phi$ in $\mathscr{V}$ as $\nu \to \infty$" or "$\lim_{\nu\to\infty} \phi_\nu = \phi$ in $\mathscr{V}$."

As is customary, a series $\sum_{\nu=1}^{n} \phi_\nu$ is said to be convergent in $\mathscr{V}$ if the sequence of partial sums $\{\sum_{\nu=1}^{n} \phi_\nu\}_{n=1}^{\infty}$ is convergent in $\mathscr{V}$.

This concept of convergence can be extended to a directed set $\{\phi_\nu\}_{\nu\to a}$, where the numerical index ν tends to some limit a, as follows. We say that $\{\phi_\nu\}_{\nu\to a}$ converges in $\mathscr{V}$ to a limit ϕ if every sequence $\{\phi_{\nu_k}\}_{k=1}^{\infty}$ with $\nu_k \to a$ that is contained in $\{\phi_\nu\}_{\nu\to a}$ converges to ϕ.

LEMMA 1.6-1. *Let* $\mathscr{V}$ *be a multinormed space with the multinorm* S. *A sequence* $\{\phi_\nu\}_{\nu=1}^{\infty}$ *converges in* $\mathscr{V}$ *to the limit* ϕ *if and only if, for each* $\gamma \in S$, $\gamma(\phi - \phi_\nu) \to 0$ *as* $\nu \to \infty$. *The limit* ϕ *is unique.*

PROOF. By the definition of convergent sequences, $\phi_\nu \to \phi$ in $\mathscr{V}$ only if, for every balloon B centered at ϕ, $\{\phi_\nu\}$ is eventually in B. But this requires that, for every $\gamma \in S$, $\gamma(\phi - \phi_\nu) \to 0$ as $\nu \to \infty$. Conversely, assume that, for each $\gamma \in S$, $\gamma(\phi - \phi_\nu) \to 0$ as $\nu \to \infty$. Since every neighborhood of ϕ contains a balloon centered at ϕ, it follows that $\{\phi_\nu\}_{\nu=1}^{\infty}$ will eventually be in each neighborhood of ϕ. Consequently, $\phi_\nu \to \phi$ in $\mathscr{V}$.

To show that ϕ is unique, assume that both ϕ and θ are limits of $\{\phi_\nu\}_{\nu=1}^{\infty}$. Then, for every $\gamma \in S$,

$$\gamma(\phi - \theta) \leq \gamma(\phi - \phi_\nu) + \gamma(\phi_\nu - \theta) \to 0$$

as $\nu \to \infty$. Hence, $\gamma(\phi - \theta) = 0$. Since our multinorm is separating, $\phi = \theta$.

By using this lemma, it is easy to show that every multinormed space is a sequential-convergence* linear space. We call the reader's attention to Fig. 1.6-1 which illustrates the relationships between sequential-convergence spaces and multinormed spaces, as well as between all the various types of spaces that will be discussed in this chapter.

Let M be a subset of the multinormed space $\mathscr{V}$; $\phi \in \mathscr{V}$ is called a *contact point* of M if every neighborhood of ϕ intersects M. The set obtained by adding to M all of its contact points that are not already in M is called the *closure* of M and is denoted by $\overline{M}$. When $\overline{M} = \mathscr{V}$, M is said to be *dense* in $\mathscr{V}$. Obviously, M is dense in $\mathscr{V}$ if, for each $\phi \in \mathscr{V}$, there exists a sequence $\{\phi_\nu\}_{\nu=1}^{\infty}$ of elements in M which converges in $\mathscr{V}$ to ϕ. The converse of this assertion is not true in general. To get a valid statement the concept of sequences must be generalized (see Wilansky [1], Sec. 9.2). However, the converse is true when $\mathscr{V}$ is a countably multinormed space. Indeed, we have

LEMMA 1.6-2. *Assume that* $\mathscr{V}$ *is a countably multinormed space. A necessary and sufficient condition for a subset* M *of* $\mathscr{V}$ *to be dense in* $\mathscr{V}$ *is that for each* $\phi \in \mathscr{V}$ *there exist a sequence* $\{\phi_\nu\}_{\nu=1}^{\infty}$ *of elements in* M *which converges in* $\mathscr{V}$ *to* ϕ.

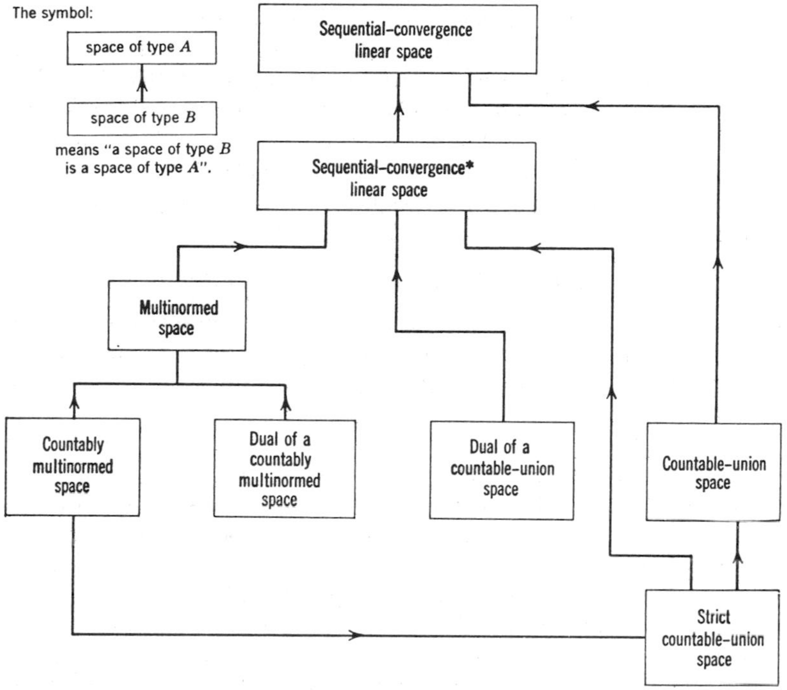

FIGURE 1.6-1. Relationships between the various types of spaces discussed in Chapter I.

PROOF. Sufficiency being obvious, we turn to necessity. We first show that the countability of the multinorm S for $\mathscr{V}$ implies that there exists a sequence $\{B_k\}_{k=1}^{\infty}$ of balloons centered at ϕ such that every neighborhood of ϕ contains at least one of the B_k. Indeed, the collection of all balloons, each of which has the form $\{\psi: \gamma(\phi - \psi) \leq \varepsilon\}$, where γ is an arbitrary seminorm in S and ε is an arbitrary positive rational number, is a countable set of balloons centered at ϕ. Moreover, all finite intersections of such balloons comprise still another countable set of balloons, which can be arranged into a sequence having the desired property.

Next, the sequence of balloons $C_1 = B_1$, $C_2 = C_1 \cap B_2$, $C_3 = C_2 \cap B_3$, ... has the same property, and in addition $C_1 \supset C_2 \supset C_3 \supset \ldots$. Now the fact that M is dense in $\mathscr{V}$ implies that we can choose from each C_k some point $\phi_k \in M$, possibly ϕ itself if $\phi \in M$. Clearly, $\{\phi_k\}_{k=1}^{\infty}$ converges to ϕ and is therefore the sequence we seek. Q.E.D.

A *Cauchy sequence* in the multinormed space $\mathscr{V}$ is a sequence $\{\phi_\nu\}_{\nu=1}^{\infty}$ of elements in $\mathscr{V}$ such that, for every neighborhood Ω of zero, there exists a positive integer N for which $\phi_\nu - \phi_\mu$ is in Ω whenever $\nu > N$ and $\mu > N$. In this case, we again say that $\phi_\nu - \phi_\mu$ is eventually in Ω as ν and μ tend to infinity independently. Through an argument that is similar to the proof of Lemma 1.6-1, we obtain the following result. $\{\phi_\nu\}$ is a Cauchy sequence in $\mathscr{V}$ if and only if all ϕ_ν are in $\mathscr{V}$, and, for each $\gamma \in S$, $\gamma(\phi_\nu - \phi_\mu) \to 0$ as ν and μ tend to infinity independently.

Observe that every convergent sequence in $\mathscr{V}$ is a Cauchy sequence because, when $\{\phi_\nu\}$ converges to ϕ in $\mathscr{V}$,

$$0 \leq \gamma(\phi_\nu - \phi_\mu) \leq \gamma(\phi_\nu - \phi) + \gamma(\phi - \phi_\mu) \to 0$$

as ν and μ tend to infinity independently. The converse need not be true. But, when it is true (i.e., when every Cauchy sequence in $\mathscr{V}$ is convergent), $\mathscr{V}$ is said to be *complete*. A complete countably multinormed space is called a *Fréchet space*.

Actually, what we have defined here is *sequential completeness*, and for arbitrary topological linear spaces the concept of *completeness* means something more. However, since only sequential completeness will be considered in this book, we shall consistently drop the adjective "sequential." Moreover, for countably multinormed spaces the two concepts happen to coincide (Robertson and Robertson [1], p. 60, Proposition 12).

In the linear space $\mathscr{V}$ let T_1 and T_2 denote two topologies generated respectively by two different multinorms $R = \{\rho_\mu\}_{\mu \in A}$ and $S = \{\gamma_\mu\}_{\mu \in B}$. The topology T_1 is said to be *weaker* than the topology T_2, and T_2 is said to be *stronger* than T_1, if every T_1 neighborhood is also a T_2 neighborhood. (This is the customary terminology, even though it would be more precise to say that T_1 is *not stronger* than T_2, and T_2 is *not weaker* than T_1.) In symbols, we write $T_1 \subset T_2$. Thus, $T_1 \subset T_2$ if and only if every balloon in T_1 centered at zero (i.e., centered at the origin) contains a balloon in T_2 centered at zero. If $T_1 \subset T_2$ and $T_2 \subset T_1$, then T_1 and T_2 are equal (i.e., the collection of neighborhoods due to R is identical to the collection of neighborhoods due to S).

If $T_1 \subset T_2$ and if a sequence $\{\phi_\nu\}$ converges in $\mathscr{V}$ to ϕ under the topology T_2, then it must also converge to ϕ under the topology T_1.

LEMMA 1.6-3. *Under the notation just defined, a necessary and sufficient condition for T_1 to be weaker than T_2 is that for each $\rho \in R$ there exist a finite set of seminorms $\gamma_1, \ldots, \gamma_n \in S$ such that for every $\phi \in \mathscr{V}$*

(1) $$\rho(\phi) \leq C \max\{\gamma_1(\phi), \ldots, \gamma_n(\phi)\}$$

or alternatively

(2) $$\rho(\phi) \leq C[\gamma_1(\phi) + \cdots + \gamma_n(\phi)]$$

where C is a positive number. Both C and n depend on the choice of ρ.

PROOF. *Sufficiency*: In the topology T_1 consider the arbitrary balloon centered at the origin:

$$A = \{\phi : \rho_k(\phi) \leq \varepsilon_k > 0,\ k = 1, \ldots, m\}$$

where the ρ_k denote arbitrary seminorms in R. In view of (1) we have for each k

$$\rho_k(\phi) \leq C_k \max [\gamma_{1,k}(\phi), \ldots, \gamma_{n,k}(\phi)]$$

where the $\gamma_{j,k}$ denote seminorms in S and n depends on ρ_k (i.e., $n = n(\rho_k)$).

Let

$$B_k = \{\phi : \gamma_{j,k}(\phi) \leq \varepsilon_k / C_k,\ j = 1, \ldots, n\}$$

The B_k are balloons in the topology T_2 centered at the origin, and $\rho_k(\phi) \leq \varepsilon_k$ for all $\phi \in B_k$. Hence, $\bigcap_{k=1}^m B_k$ is a balloon in the topology T_2 centered at the origin, and it is contained in A. Therefore, $T_1 \subset T_2$.

We can arrive at the same conclusion using (2) instead of (1) by replacing the balloons B_k by the balloons

$$B_k' = \{\phi : \gamma_{j,k}(\phi) \leq \varepsilon_k / nC_k,\ j = 1, \ldots, n\}$$

Necessity: For the arbitrary seminorm $\rho \in R$ consider the balloon in T_1

$$A = \{\psi : \psi \in \mathscr{V},\ \rho(\psi) \leq 1\}$$

Under the assumption that $T_1 \subset T_2$, there exists a finite number of seminorms $\gamma_1, \ldots, \gamma_n \in S$ and an $\varepsilon > 0$ such that the balloon

$$B = \{\theta : \theta \in \mathscr{V},\ \gamma_1(\theta) \leq \varepsilon, \ldots, \gamma_n(\theta) \leq \varepsilon\}$$

is contained in A. Now let ϕ be any member of $\mathscr{V}$ such that $\max \{\gamma_1(\phi), \ldots, \gamma_n(\phi)\} \neq 0$, and set

$$\theta' = \frac{\varepsilon \phi}{\max \{\gamma_1(\phi), \ldots, \gamma_n(\phi)\}} \tag{3}$$

Then, θ' will be in B. Hence $\theta' \in A$, so that $\rho(\theta') \leq 1$. Substituting (3) into $\rho(\theta') \leq 1$ and setting $C = \varepsilon^{-1}$, we get (1).

On the other hand, if ϕ is such that $\max\{\gamma_1(\phi), \ldots, \gamma_n(\phi)\} = 0$, then, for every positive number a, $a\phi$ is in B, so that $\rho(\phi) \leq 1/\mathrm{a}$. This implies that $\rho(\phi) = 0$, and consequently (1) is still satisfied. If (1) is fulfilled, (2) will certainly be fulfilled. Q.E.D.

EXAMPLE 1.6 1. For each nonnegative integer $k \in \mathscr{R}^n$ we define a seminorm γ_k on the linear space $\mathscr{D}_K$, which was defined in Example 1.3-1, as follows:

$$\gamma_k(\phi) \triangleq \sup_{t \in R^n} |D^k \phi(t)| \qquad \phi \in \mathscr{D}_K \tag{4}$$

Here, γ_0 is a norm, and therefore the set $\{\gamma_k\}$ of all γ_k is a countable multinorm on $\mathscr{D}_K$. We assign the topology generated by $\{\gamma_k\}$ to $\mathscr{D}_K$ and obtain thereby a countably multinormed space. $\mathscr{D}_K$ is complete. (Prove this.)

PROBLEM 1.6-1. Show that every multinormed space is a sequential-convergence* linear space.

PROBLEM 1.6-2. Prove that $\mathscr{D}_K$ is complete.

PROBLEM 1.6-3. Show that every γ_k defined by (4) is a norm on $\mathscr{D}_K$.

PROBLEM 1.6-4. Let $\mathscr{S}$ denote the so-called *space of smooth functions of rapid descent* defined as follows: ϕ is a member of $\mathscr{S}$ if and only if it is a complex-valued smooth function on $\mathscr{R}^n$ and for every choice of the nonnegative integers m and k it satisfies

$$\gamma_{m,k}(\phi) \triangleq \sup_{t \in R^n} |(1 + |t|^{2m}) D^k \phi(t)| < \infty$$

The topology of $\mathscr{S}$ is that generated by the set of seminorms $\{\gamma_{m,k}\}$ where m and k range through the nonnegative integers independently. Show that $\mathscr{S}$ is a complete countably multinormed space. Also, show that the space $\mathscr{D}$, which was defined in Example 1.3-2, is a dense subspace of $\mathscr{S}$.

1.7. Countable-Union Spaces

Let $\mathscr{V}$ be a multinormed space and $\mathscr{U}$ a linear subspace of $\mathscr{V}$. Also, let S denote the multinorm for $\mathscr{V}$. Clearly, S is also a multinorm on $\mathscr{U}$. The topology generated in $\mathscr{U}$ by S is called the *topology induced on* $\mathscr{U}$ *by* $\mathscr{V}$ or simply the *induced topology in* $\mathscr{U}$. It follows that the neighborhoods in $\mathscr{U}$ under the induced topology are simply the intersections of $\mathscr{U}$ with the neighborhoods in $\mathscr{V}$.

LEMMA 1.7-1. *Let $\mathscr{U}$ and $\mathscr{V}$ be multinormed spaces and assume that $\mathscr{U}$ is a linear subspace of $\mathscr{V}$. Also, assume that the topology of $\mathscr{U}$ is stronger than the topology induced on $\mathscr{U}$ by $\mathscr{V}$. If $\{\phi_\nu\}_{\nu=1}^\infty$ converges in $\mathscr{U}$ to ϕ, then $\{\phi_\nu\}$ converges in $\mathscr{V}$ and has the same limit in $\mathscr{V}$ as it does in $\mathscr{U}$. Also, if $\{\psi_\nu\}$ is a Cauchy sequence in $\mathscr{U}$,* then *$\{\psi_\nu\}$ is a Cauchy sequence in $\mathscr{V}$ as well.*

PROOF. We prove only the first conclusion, the proof of the second one being almost the same. Let Ω be an arbitrary neighborhood of ϕ in $\mathscr{V}$. $\mathscr{U} \cap \Omega$ is a neighborhood of ϕ in the induced topology of $\mathscr{U}$, and consequently a neighborhood of ϕ in $\mathscr{U}$. By hypothesis, $\{\phi_\nu\}$ is eventually in $\mathscr{U} \cap \Omega$ as $\nu \to \infty$. Since $\mathscr{U} \cap \Omega \subset \Omega$, $\{\phi_\nu\}$ is eventually in Ω. Thus, $\{\phi_\nu\}$ also converges in $\mathscr{V}$ to ϕ. Q.E.D.

Let $\{\mathscr{V}_m\}_{m=1}^{\infty}$ be a sequence of countably multinormed spaces such that $\mathscr{V}_1 \subset \mathscr{V}_2 \subset \mathscr{V}_3 \subset \ldots$. Furthermore, assume that the topology of each $\mathscr{V}_m$ is stronger than the topology induced on it by $\mathscr{V}_{m+1}$. Let $\mathscr{V}$ denote the union of these spaces: $\mathscr{V} = \bigcup_{m=1}^{\infty} \mathscr{V}_m$. $\mathscr{V}$ is a linear space. We assign a concept of convergence to $\mathscr{V}$ as follows: A sequence $\{\phi_\nu\}_{\nu=1}^{\infty}$ is said to *converge in* $\mathscr{V}$ *to* ϕ (or simply to be *convergent*) if all the ϕ_ν and ϕ belong to some particular $\mathscr{V}_m$ and $\{\phi_\nu\}$ converges to ϕ in $\mathscr{V}_m$ (and therefore in $\mathscr{V}_{m+1}, \mathscr{V}_{m+2}, \ldots$, as well, according to Lemma 1.7-1). Under these circumstances $\mathscr{V}$ is called a *countable-union space*. Spaces of this type were introduced by Gelfand and Shilov [2].

A sequence $\{\phi_\nu\}_{\nu=1}^{\infty}$ is called a *Cauchy sequence* in the countable-union space $\mathscr{V}$ if it is a Cauchy sequence in one of the spaces $\mathscr{V}_m$. Moreover, when all the Cauchy sequences in $\mathscr{V}$ are convergent, $\mathscr{V}$ is called *complete*. In view of Lemma 1.7-1, if every $\mathscr{V}_m$ is complete, then every Cauchy sequence in $\mathscr{V}$ converges to a unique limit in $\mathscr{V}$. Thus, the countable-union space $\mathscr{V} = \bigcup_{m=1}^{\infty} \mathscr{V}_m$ is complete whenever all the $\mathscr{V}_m$ are complete countably multinormed spaces.

Our definition of convergence in a countable-union space and Lemma 1.6-1 immediately yield

LEMMA 1.7-2. *Let* $\mathscr{V} = \bigcup_{m=1}^{\infty} \mathscr{V}_m$ *be a countable-union space, and for each* m *let* S_m *be the multinorm for* $\mathscr{V}_m$. *Then, a sequence* $\{\phi_\nu\}$ *converges in* $\mathscr{V}$ *to* ϕ *if and only if all the* ϕ_ν *and* ϕ *are contained in one of the* $\mathscr{V}_m$ *and, for each* $\gamma \in S_m$, $\gamma(\phi - \phi_\nu) \to 0$ *as* $\nu \to \infty$.

A countable-union space $\mathscr{V} = \bigcup_{m=1}^{\infty} \mathscr{V}_m$ will be called a *strict countable-union space* if for each m the topology of $\mathscr{V}_m$ is identical to the topology induced on $\mathscr{V}_m$ by $\mathscr{V}_{m+1}$. It follows that the topology of $\mathscr{V}_m$ is identical to the topology induced on $\mathscr{V}_m$ by $\mathscr{V}_q$ for every $q > m$. $\mathscr{V}$ will certainly be a strict countable-union space if the following condition is satisfied: If, for each m, $\{\gamma_{m,k}\}_{k=1}^{\infty}$ denotes the multinorm for $\mathscr{V}_m$, then $\gamma_{m,k}(\phi) = \gamma_{m+1,k}(\phi)$ for all $\phi \in \mathscr{V}_m$, all m, and all k.

An essential distinction should be made here. A countably multinormed space $\mathscr{U}$ is a special case of a strict countable-union space $\mathscr{V} = \bigcup_{m=1}^{\infty} \mathscr{V}_m$ because we may choose $\mathscr{V}_m = \mathscr{U}$ for every m. On the other hand, a strict countable-union space $\mathscr{V} = \bigcup_{m=1}^{\infty} \mathscr{V}_m$ need not be a countably multinormed space even when a single multinorm S generates the topology of every one of the spaces $\mathscr{V}_m$. Indeed, if the set consisting of all the elements of $\mathscr{V}$ were assigned the topology generated by S, the convergence rule in the resulting countably multinormed space would be weaker than that in the original countable-union space because the requirement that a convergent sequence be contained in one of the $\mathscr{V}_m$ would be lost.

Every countable-union space is a sequential-convergence linear space. (See Fig. 1.6-1.) Indeed, it is easy to show that the first five axioms of Sec. 1.4 are satisfied. Furthermore, a strict countable-union space is a sequential-convergence* linear space. To verify the fulfillment of axiom 6 in Sec. 1.4, assume it is false that $\phi_\nu \to \phi$ in $\mathscr{V}$ as $\nu \to \infty$. There are only two ways in which this can occur: (i) Either there is no $\mathscr{V}_m$ that contains all the ϕ_ν, or (ii), for each m such that all $\phi_\nu \in \mathscr{V}_m$ and $\phi \in \mathscr{V}_m$, the sequence $\{\phi_\nu\}$ does not converge in $\mathscr{V}_m$ to ϕ.

In the first case (i), choose an element ϕ_{ν_1} from $\{\phi_\nu\}$ which is not a member of $\mathscr{V}_1$. The sequence obtained after the deletion of ϕ_{ν_1} also cannot be contained in any $\mathscr{V}_m$. Choose an element ϕ_{ν_2} from it which is not in $\mathscr{V}_2$, and therefore not in $\mathscr{V}_1$ either. Continuing in this fashion and setting $\psi_\mu = \phi_{\nu_\mu}$, we find for each m a ψ_m that is not contained in any $\mathscr{V}_p$ where $1 \leq p \leq m$; in other words, $\mathscr{V}_m$ does not contain any ψ_μ where $\mu \geq m$. Thus, we obtain a sequence $\{\psi_\mu\}$ all of whose subsequences are not contained in any $\mathscr{V}_m$ and hence cannot converge in $\mathscr{V}$ to ϕ. Thus, axiom 6 is fulfilled.

In the second case (ii), let p be the smallest integer for which all $\phi_\nu \in \mathscr{V}_p$ and $\phi \in \mathscr{V}_p$. Then, all ϕ_ν and ϕ are members of $\mathscr{V}_m$ for every $m \geq p$. Moreover, there exists in $\mathscr{V}_p$ a neighborhood Ω_p of ϕ and a subsequence $\{\psi_\mu\}$ of $\{\phi_\nu\}$ such that $\psi_\mu \notin \Omega_p$ for every μ. But since the topology of $\mathscr{V}_p$ is identical to the topology induced on $\mathscr{V}_p$ by $\mathscr{V}_m$ $(m > p)$, there must be in $\mathscr{V}_m$ at least one neighborhood Ω_m of ϕ such that $\psi_\mu \notin \Omega_m$ for every μ. (Indeed, choose Ω_m such that its intersection with $\mathscr{V}_p$ is Ω_p. There is at least one such Ω_m.) Thus, every subsequence of $\{\psi_\mu\}$ also does not converge in $\mathscr{V}_m$ to ϕ for any $m > p$ and therefore not in $\mathscr{V}$ as well. Hence, axiom 6 is again fulfilled.

Let $\mathscr{V}$ be a countable-union space, which is not necessarily strict. We shall say that a subset M of $\mathscr{V}$ is *dense in* $\mathscr{V}$ if for each $\phi \in \mathscr{V}$ there exists a sequence $\{\phi_\nu\}$ whose members are all in M and which converges in $\mathscr{V}$ to ϕ. In view of Lemma 1.6-2, this definition is consistent with the one given previously for countably multinormed spaces.

EXAMPLE 1.7-1. Consider the countably multinormed spaces $\mathscr{D}_K$ and $\mathscr{D}_J$, where K and J are compact subsets of $\mathscr{R}^n$ and $K \subset J$. Such spaces were last discussed in Example 1.6-1. For each nonnegative integer k in $\mathscr{R}^n$ we have the same seminorm for $\mathscr{D}_K$ and $\mathscr{D}_J$, namely,

$$\gamma_k(\phi) = \sup_{t \in R^n} |D^k \phi(t)| \tag{1}$$

The collection $\{\gamma_k\}_{k=0}^{\infty}$ is the countable multinorm for both $\mathscr{D}_K$ and $\mathscr{D}_J$. Thus, $\mathscr{D}_K \subset \mathscr{D}_J$, and the topology of $\mathscr{D}_K$ is the same as the topology induced on $\mathscr{D}_K$ by $\mathscr{D}_J$.

Now let $\{K_m\}_{m=1}^{\infty}$ be a sequence of compact spheres in $\mathscr{R}^n$ such that $K_1 \subset K_2 \subset K_3 \subset \cdots$ and $\mathscr{R}^n = \bigcup_{m=1}^{\infty} K_m$. For example, we can choose $K_m = \{t\colon |t| \leq m\}$. $\mathscr{D}$ is the strict countable-union space generated by the $\mathscr{D}_{K_m}$. That is, $\mathscr{D} = \bigcup_{m=1}^{\infty} \mathscr{D}_{K_m}$, and a sequence $\{\phi_\nu\}$ is defined as being convergent in $\mathscr{D}$ if all ϕ_ν are in a particular $\mathscr{D}_{K_m}$ and $\{\phi_\nu\}$ converges in $\mathscr{D}_{K_m}$. Since each $\mathscr{D}_{K_m}$ is complete, $\mathscr{D}$ is too. Note that $\mathscr{D}$, as a sequential-convergence* linear space, is independent of the choice of the sequence $\{K_m\}_{m=1}^{\infty}$ of compact spheres. Had we chosen some other sequence $\{J_m\}_{m=1}^{\infty}$ of compact spheres with $J_1 \subset J_2 \subset J_3 \subset \cdots$ and $\mathscr{R}^n = \bigcup_{m=1}^{\infty} J_m$, $\mathscr{D}$ would still consist of the same elements and its convergent sequences would be precisely the same.

In view of Lemma 1.7-2, a sequence $\{\phi_\nu\}$ converges in $\mathscr{D}$ if and only if all the ϕ_ν are in $\mathscr{D}_K$ where K is some fixed compact subset of $\mathscr{R}^n$ and, for each nonnegative integer $k \in \mathscr{R}^n$, $\{D^k \phi_\nu(t)\}$ converges uniformly on $\mathscr{R}^n$. This convergence concept is the same as that assigned to $\mathscr{D}$ in Example 1.4-1.

PROBLEM 1.7-1. Verify that a countable-union space is a sequential-convergence linear space,

PROBLEM 1.7-2. Instead of treating $\mathscr{D}$ as a countable-union space, let us convert it into a countably multinormed space by assigning to it the topology generated by $\{\gamma_k\}_{k=0}^{\infty}$ where each γ_k is defined by (1). Show that the resulting space is not complete.

PROBLEM 1.7-3. It was indicated in Problem 1.6-4 that $\mathscr{D}$ is a dense subspace of $\mathscr{S}$. Now, show that convergence in $\mathscr{D}$ implies convergence in $\mathscr{S}$.

1.8. Duals of Countably Multinormed Spaces

Let $\mathscr{V}$ be a countably multinormed space. A rule that assigns a unique complex number to each $\phi \in \mathscr{V}$ is called a *functional* on $\mathscr{V}$. We shall denote this complex number by $\langle f, \phi \rangle$. The functional f is said to be *linear* if for any $\phi, \psi \in \mathscr{V}$ and any $\alpha, \beta \in \mathscr{C}^1$ we always have

$$\langle f, \alpha\phi + \beta\psi \rangle = \alpha \langle f, \phi \rangle + \beta \langle f, \psi \rangle.$$

This implies that $\langle f, \varnothing \rangle = 0$. The functional f is called *continuous at the point* $\phi \in \mathscr{V}$ if for each $\varepsilon > 0$ there is a neighborhood Ω of ϕ in $\mathscr{V}$ such that $|\langle f, \psi \rangle - \langle f, \phi \rangle| < \varepsilon$ whenever $\psi \in \Omega$. Also, f is called simply *continuous* if it is continuous at every point of $\mathscr{V}$. A useful and rather apparent fact (which is also an immediate consequence of Lemma 1.8-2) is that a linear functional is continuous if and only if it is continuous at the origin.

LEMMA 1.8-1. *Let $\mathscr{V}$ be a countably multinormed space. A necessary and sufficient condition for a functional f on $\mathscr{V}$ to be continuous is that, for every $\phi \in \mathscr{V}$ and for every sequence $\{\phi_\nu\}_{\nu=1}^{\infty}$ that converges in $\mathscr{V}$ to ϕ, we have*

$$\lim_{\nu\to\infty} \langle f, \phi_\nu\rangle = \langle f, \phi\rangle$$

PROOF. Necessity is obvious. To prove sufficiency we shall use the fact established in the proof of Lemma 1.6-2 that, given any $\phi \in \mathscr{V}$, there exists a sequence $\{C_\nu\}_{\nu=1}^{\infty}$ of balloons centered at ϕ such that $C_1 \supset C_2 \supset C_3 \supset \cdots$ and every neighborhood of ϕ contains at least one of the C_ν and therefore $C_{\nu+1}, C_{\nu+2}, \ldots$ as well. Assume that f is not continuous. Consequently, f is not continuous at some $\phi \in \mathscr{V}$. This means that there exists an $\varepsilon > 0$ such that there is no neighborhood Ω of ϕ with $|\langle f, \psi\rangle - \langle f, \phi\rangle| < \varepsilon$ for every $\psi \in \Omega$. In other words, there exists an $\varepsilon > 0$ such that for each C_ν (centered at ϕ) there is at least one $\phi_\nu \in C_\nu$ with $|\langle f, \phi_\nu\rangle - \langle f, \phi\rangle| \geq \varepsilon$. Clearly, $\phi_\nu \to \phi$ in $\mathscr{V}$ as $\nu \to \infty$, but $\lim_{\nu\to\infty}\langle f, \phi_\nu\rangle \neq \langle f, \phi\rangle$. Q.E.D.

LEMMA 1.8-2. *Let $\mathscr{V}$ be a countably multinormed space. A necessary and sufficient condition for a linear functional f on $\mathscr{V}$ to be continuous is that, for every sequence $\{\phi_\nu\}_{\nu=1}^{\infty}$ that converges in $\mathscr{V}$ to zero, we have* $\lim_{\nu\to\infty}\langle f, \phi_\nu\rangle = 0$.

PROOF. Necessity is again obvious. Next, choose an arbitrary $\psi \in \mathscr{V}$ and any sequence $\{\psi_\nu\}$ that converges in $\mathscr{V}$ to ψ. Then, $\phi_\nu \triangleq \psi_\nu - \psi \to \varnothing$ in $\mathscr{V}$. By the linearity of f and the hypothesis for the sufficiency condition,

$$|\langle f, \psi_\nu\rangle - \langle f, \psi\rangle| = |\langle f, \psi_\nu - \psi\rangle| \to 0.$$

as $\nu \to \infty$. The preceding lemma completes the proof.

Now assume that the topology T_1 of the linear space $\mathscr{V}$ is generated by the countable multinorm $\{\gamma_\mu\}_{\mu=1}^{\infty}$, where γ_1 is a norm. Define the countable multinorm $\{\rho_\mu\}_{\mu=1}^{\infty}$ by $\rho_\mu \triangleq \max\{\gamma_1, \ldots, \gamma_\mu\}$, and let T_2 be the topology that $\{\rho_\mu\}$ generates in $\mathscr{V}$. T_1 and T_2 are equal. Indeed, for every $\phi \in \mathscr{V}$, $\rho_\mu(\phi) = \max\{\gamma_1(\phi), \ldots, \gamma_\mu(\phi)\}$ so that, by Lemma 1.6-3, $T_2 \subset T_1$; also, $\gamma_\mu(\phi) \leq \rho_\mu(\phi)$, and therefore by the same lemma $T_1 \subset T_2$.

Note that each ρ_μ is a norm, and for every $\phi \in \mathscr{V}$ other than the zero element

$$0 < \rho_1(\phi) \leq \rho_2(\phi) \leq \rho_3(\phi) \leq \cdots$$

a result that we shall exploit in the next proof.

THEOREM 1.8-1. *Let $\mathscr{V}$ be a linear space with a topology generated by the countable multinorm $\{\gamma_\mu\}_{\mu=1}^{\infty}$ where γ_1 is a norm. Let the countable multinorm $\{\rho_\mu\}_{\mu=1}^{\infty}$ be defined by $\rho_\mu \triangleq$ max $\{\gamma_1, \ldots, \gamma_\mu\}$. For each continuous* *linear functional f defined on $\mathscr{V}$ there exist a positive constant C and a nonnegative integer r such that for every $\phi \in \mathscr{V}$*

$$(1) \qquad |\langle f, \phi\rangle| \leq C\rho_r(\phi)$$

Here, C and r depend on f but not on ϕ.

PROOF. Assume that there are no values of C and r for which the inequality (1) holds for all $\phi \in \mathscr{V}$. This means that for each positive integer ν there exists a $\phi_\nu \in \mathscr{V}$ such that

$$(2) \qquad |\langle f, \phi_\nu\rangle| > \nu\rho_\nu(\phi_\nu)$$

We have already pointed out that ρ_ν is a norm. Therefore, $\rho_\nu(\phi_\nu) > 0$ because ϕ_ν cannot be the zero element in $\mathscr{V}$. (Otherwise, we would have equality in (2) since both sides there would be zero.) Set

$$\theta_\nu = \frac{\phi_\nu}{\nu\rho_\nu(\phi_\nu)} \in \mathscr{V}$$

With k being an arbitrary but fixed nonnegative integer, we have for $\nu > k$

$$\rho_k(\theta_\nu) \leq \rho_\nu(\theta_\nu) = \frac{\rho_\nu(\phi_\nu)}{\nu\rho_\nu(\phi_\nu)} = \frac{1}{\nu} \to 0$$

as $\nu \to \infty$. Since the topology generated by $\{\rho_\mu\}$ is equal to that generated by $\{\gamma_\mu\}$, it follows from Lemma 1.6-1 that $\{\theta_\nu\}$ converges to zero in $\mathscr{V}$. Consequently, $\langle f, \theta_\nu\rangle \to 0$ because f is a continuous functional on $\mathscr{V}$. But, from (2) we have that

$$|\langle f, \theta_\nu\rangle| = \frac{|\langle f, \phi_\nu\rangle|}{\nu\rho_\nu(\phi_\nu)} > 1$$

This contradiction proves the theorem.

We have just shown that condition (1) is necessary for a linear functional f on $\mathscr{V}$ to be continuous. By Lemma 1.8-2, it is also sufficient since, if $\phi_\nu \to \varnothing$ in $\mathscr{V}$ as $\nu \to \infty$, then $|\langle f, \phi_\nu\rangle| \leq C\rho(\phi_\nu) \to 0$.

The collection of all continuous linear functionals on the countably multinormed space $\mathscr{V}$ is called the *dual of* $\mathscr{V}$ (also the *conjugate of* $\mathscr{V}$) and is denoted by $\mathscr{V}'$. Two members f and g of $\mathscr{V}'$ are said to be *equal in* $\mathscr{V}'$ (or simply *equal*) if for every $\phi \in \mathscr{V}$ we have $\langle f, \phi\rangle = \langle g, \phi\rangle$. Also, addition and multiplication by a complex number α is defined in $\mathscr{V}'$ by the following two equations respectively.

$$\langle f + g, \phi\rangle \triangleq \langle f, \phi\rangle + \langle g, \phi\rangle$$

$$\langle \alpha f, \phi\rangle \triangleq \alpha\langle f, \phi\rangle$$

With these definitions $\mathscr{V}'$ becomes a linear space, the zero element in $\mathscr{V}'$ being the functional that assigns the number zero to every $\phi \in \mathscr{V}$.

Assume that $\mathscr{U}$ and $\mathscr{V}$ are countably multinormed spaces with $\mathscr{U}$ being a subspace of $\mathscr{V}$. By the *restriction of* $f \in \mathscr{V}'$ *to* $\mathscr{U}$ we mean that unique functional g on $\mathscr{U}$ defined by $\langle g, \phi \rangle \triangleq \langle f, \phi \rangle$ for every $\phi \in \mathscr{U}$. Clearly, g is linear. If the topology of $\mathscr{U}$ is stronger than that induced on it by $\mathscr{V}$, then g is also continuous and therefore in $\mathscr{U}'$. Indeed, the assumption that $\phi_\nu \to \varnothing$ in $\mathscr{U}$ implies that $\phi_\nu \to \varnothing$ in $\mathscr{V}$, and consequently $\langle g, \phi_\nu \rangle = \langle f, \phi_\nu \rangle \to 0$. Thus, by Lemma 1.8-2, g is a continuous functional on $\mathscr{U}$. Hence, the restriction of $f \in \mathscr{V}'$ to $\mathscr{U}$ is a member of $\mathscr{U}'$.

However, we cannot conclude that $\mathscr{V}'$ is a subspace of $\mathscr{U}'$ because two different members of $\mathscr{V}'$ may have the same restriction to $\mathscr{U}$ (i.e., there need not exist a one-to-one correspondence between $\mathscr{V}'$ and a subset of $\mathscr{U}'$). On the other hand, if we add the hypothesis that $\mathscr{U}$ is dense in $\mathscr{V}$, then we can say that $\mathscr{V}'$ is a subspace of $\mathscr{U}'$, as we shall show presently.

EXAMPLE 1.8-1. $\mathscr{D}_K'$ is the dual of $\mathscr{D}_K$. An example of a member of $\mathscr{D}_K'$ is the (Dirac) delta functional at the point $\tau \in \mathscr{R}^n$. It is denoted by $\delta(t-\tau)$ and defined on $\mathscr{D}_K$ by

$$\langle \delta(t-\tau), \phi(t) \rangle \triangleq \phi(\tau) \qquad \phi \in \mathscr{D}_K$$

Now, let K be the set $\{t: |t| \leq 1\}$ and J the set $\{t: |t| \leq 2\}$. $\mathscr{D}_K$ is a subspace of $\mathscr{D}_J$, and the topology of $\mathscr{D}_K$ is the same as the topology induced on $\mathscr{D}_K$ by $\mathscr{D}_J$. Therefore, the restriction of any $f \in \mathscr{D}_J'$ to $\mathscr{D}_K$ is in $\mathscr{D}_K'$. However, as was just pointed out, we cannot say that $\mathscr{D}_J'$ is a subspace of $\mathscr{D}_K'$ because there does not exist a one-to-one correspondence between $\mathscr{D}_J'$ and some subspace of $\mathscr{D}_K'$. For instance, let $\tau = \{3/2, 0, 0, \ldots, 0\} \in \mathscr{R}^n$; then, $\delta(t-\tau)$ is in both $\mathscr{D}_K'$ and $\mathscr{D}_J'$. (More precisely, the restrictions of $\delta(t-\tau)$ to $\mathscr{D}_K$ and $\mathscr{D}_J$ are in $\mathscr{D}_K'$ and $\mathscr{D}_J'$, respectively.) $\delta(t-\tau)$ is the zero element in $\mathscr{D}_K'$ since $\phi(\tau) = 0$ for every $\phi \in \mathscr{D}_K$, but it is not the zero element in $\mathscr{D}_J'$. Thus, we have an example of two members of $\mathscr{D}_J'$, namely, $\delta(t-\tau)$ and the zero element in $\mathscr{D}_J'$, whose restrictions to $\mathscr{D}_K$ are the same element in $\mathscr{D}_K'$, the zero element.

THEOREM 1.8-2. *If $\mathscr{U}$ and $\mathscr{V}$ are countably multinormed spaces with $\mathscr{U}$ being a dense subspace of $\mathscr{V}$ and if the topology of $\mathscr{U}$ is stronger than that induced on it by $\mathscr{V}$, then $\mathscr{V}'$ is a subspace of $\mathscr{U}'$ (more precisely, there is a one to one correspondence between $\mathscr{V}'$ and a subspace of $\mathscr{U}'$, the correspondence being the one between the members of $\mathscr{V}'$ and their restrictions to $\mathscr{U}$).*

PROOF. Since $\mathscr{U} \subset \mathscr{V}$, it is obviously impossible for two different members of $\mathscr{U}'$ to correspond to the same member of $\mathscr{V}'$. Thus, we need merely prove the following: If $f \in \mathscr{V}'$, then the values of $\langle f, \phi \rangle$ as ϕ traverses $\mathscr{U}$ uniquely determine the values of $\langle f, \phi \rangle$ as ϕ traverses $\mathscr{V}$. This will prohibit two different members of $\mathscr{V}'$ from having the same restriction on $\mathscr{U}$.

Because $\mathscr{U}$ is dense in $\mathscr{V}$, we can find for each $\phi \in \mathscr{V}$ a sequence $\{\phi_\nu\}$ of elements of $\mathscr{U}$ which converges in $\mathscr{V}$ to ϕ (see Lemma 1.6-2). Thus, for any $f \in \mathscr{V}'$, $\langle f, \phi_\nu \rangle \to \langle f, \phi \rangle$ as $\nu \to \infty$; hence, $\langle f, \phi \rangle$ is uniquely determined by the $\langle f, \phi_\nu \rangle$. Q.E.D.

Our arguments have also established the following alternative to Theorem 1.8-2.

COROLLARY 1.8-2a. *Let $\mathscr{U}$ and $\mathscr{V}$ be countably multinormed spaces with $\mathscr{U}$ being a dense subspace of $\mathscr{V}$. Assume that the convergence of any sequence to zero in $\mathscr{U}$ implies its convergence to zero in $\mathscr{V}$. Then, $\mathscr{V}'$ is a subspace of $\mathscr{U}'$.*

We shall assign what is known as the *weak topology* to the dual $\mathscr{V}'$ of a countably multinormed space $\mathscr{V}$. This is the topology generated by the multinorm $\{\xi_\phi\}_{\phi \in \mathscr{V}}$ where for each $\phi \in \mathscr{V}$ we have a seminorm ξ_ϕ on $\mathscr{V}'$ defined by $\xi_\phi(f) \triangleq |\langle f, \phi \rangle|$. That $\{\xi_\phi\}$ is separating follows from the fact that $\langle f, \phi \rangle$ cannot be zero for all $\phi \in \mathscr{V}$ without f being the zero element in $\mathscr{V}'$.

Thus, in the space $\mathscr{V}'$ a *balloon centered at* $g \in \mathscr{V}'$ is any set of the form

$$\{f : f \in \mathscr{V}', |\langle f - g, \phi_k \rangle| \le \varepsilon_k, k = 1, \ldots, n\}$$

where n is any positive integer, the ϕ_k are any elements in $\mathscr{V}$, and the ε_k are any positive numbers. A *neighborhood in* $\mathscr{V}'$ (*of a point* $g \in \mathscr{V}'$) is any set containing a balloon (centered at g). The collection of all neighborhoods in $\mathscr{V}'$ is called the weak topology of $\mathscr{V}'$; we shall discard the adjective "weak" because no other topology in $\mathscr{V}'$ will be considered. In this way $\mathscr{V}'$ is a multinormed space, but not necessarily a countably multinormed space, as is illustrated in Figure 1.6-1. Therefore, $\mathscr{V}'$ is also a sequential-convergence* linear space (see Problem 1.6-1).

If a sequence in $\mathscr{V}'$ converges with respect to this topology, it is customarily called *weakly convergent*, but we shall simply call it *convergent*. According to Lemma 1.6-1, a sequence $\{f_\nu\}_{\nu=1}^\infty$ of elements of $\mathscr{V}'$ is convergent if and only if there exists an $f \in \mathscr{V}'$ such that, for each $\phi \in \mathscr{V}$, $\langle f_\nu - f, \phi \rangle \to 0$ as $\nu \to \infty$. In this case the limit f is unique. Analogous assertions can be made for directed sets $\{f_\nu\}_{\nu \to a}$ and series $\sum_{\nu=1}^\infty f_\nu$. Similarly, a sequence $\{f_\nu\}_{\nu=1}^\infty$ of elements of $\mathscr{V}'$ is a (weak) *Cauchy sequence* in $\mathscr{V}'$ if and only if, for each $\phi \in \mathscr{V}$, $\langle f_\nu - f_\mu, \phi \rangle \to 0$ as ν and μ tend to infinity independently. As before, $\mathscr{V}'$ is called *complete* (we should really say "sequentially complete") if every Cauchy sequence in it is convergent.

THEOREM 1.8-3. *If $\mathscr{V}$ is a complete countably multinormed space, then its dual $\mathscr{V}'$ is also complete.*

PROOF. Let $\{f_\nu\}$ be a Cauchy sequence in $\mathscr{V}'$. By the completeness of the complex plane $\mathscr{C}^1$, for each $\phi \in \mathscr{V}$ there exists a unique number, which we denote by $\langle f, \phi \rangle$, such that $\lim_{\nu\to\infty}\langle f_\nu, \phi\rangle = \langle f, \phi\rangle$. As ϕ traverses $\mathscr{V}$, these limits $\langle f, \phi\rangle$ define a functional f on $\mathscr{V}$. It is easily seen that the linearity of each f_ν implies the linearity of f. To complete the proof we shall show that f is continuous at the origin of $\mathscr{V}$.

Assume the opposite; namely, that there exists a sequence that converges in $\mathscr{V}$ to zero such that the corresponding sequence of numbers assigned to it by f does not converge. We can choose a subsequence $\{\phi_\nu\}_{\nu=1}^\infty$ from the original sequence in $\mathscr{V}$ such that

$$(3) \qquad |\langle f, \phi_\nu\rangle| \geq c > 0 \qquad \nu = 1, 2, 3, \ldots$$

where c is some fixed positive number, and

$$(4) \qquad \gamma_k(\phi_\nu) \leq 4^{-\nu} \qquad k = 1, \ldots, \nu$$

where $\{\gamma_k\}_{k=1}^\infty$ is the multinorm for $\mathscr{V}$. Set $\psi_\nu = 2^\nu\phi_\nu$. By (4) and Lemma 1.6-1, $\{\psi_\nu\}_{\nu=1}^\infty$ converges in $\mathscr{V}$ to zero, and by (3) the numerical sequence $\{|\langle f, \psi_\nu\rangle|\}_{\nu=1}^\infty$ diverges to infinity. Next, we choose a subsequence $\{\psi_\nu'\}_{\nu=1}^\infty$ from $\{\psi_\nu\}_{\nu=1}^\infty$ and a subsequence $\{f_\nu'\}_{\nu=1}^\infty$ from the Cauchy sequence $\{f_\nu\}_{\nu=1}^\infty$ as follows.

First, choose ψ_1' such that $|\langle f, \psi_1'\rangle| > 1$. Since, for every $\psi \in \mathscr{V}$, $\langle f_\nu, \psi\rangle \to \langle f, \psi\rangle$ as $\nu \to \infty$, f_1' can be so chosen that $|\langle f_1', \psi_1',\rangle| > 1$. Then, assuming that the first $\nu - 1$ elements of these subsequences have been chosen, select ψ_ν' as an element from $\{\psi_\nu\}$ with index higher than those of $\psi_1', \ldots, \psi_{\nu-1}'$ such that

$$(5) \qquad |\langle f_j', \psi_\nu'\rangle| < 2^{j-\nu} \qquad j = 1, \ldots, \nu - 1$$

and

$$(6) \qquad |\langle f, \psi_\nu'\rangle| > \sum_{\mu=1}^{\nu-1} |\langle f, \psi_\mu'\rangle| + \nu$$

Condition (5) can be satisfied because, for each fixed f_j', $\langle f_j', \psi_\nu'\rangle \to 0$ as $\nu \to \infty$. Condition (6) can also be satisfied because $|\langle f, \psi_\nu'\rangle| \to \infty$ as $\nu \to \infty$.

For any $\phi \in \mathscr{V}$, $\langle f_\nu, \phi\rangle \to \langle f, \phi\rangle$ as $\nu \to \infty$. In view of (6), f_ν' can now be chosen as an element from $\{f_\nu\}$ with index higher than those of $f_1', \ldots, f_{\nu-1}'$ such that

$$(7) \qquad |\langle f_\nu', \psi_\nu'\rangle| > \sum_{\mu=1}^{\nu-1} |\langle f_\nu', \psi_\mu'\rangle| + \nu$$

Now that $\{\psi_\nu'\}$ and $\{f_\nu'\}$ have been specified, consider the series:

$$(8) \qquad \psi = \sum_{\nu=1}^\infty \psi_\nu'$$

With $m > n > k$ we have

$$\gamma_k\left(\sum_{\nu=n}^m \psi_\nu'\right) \leq \sum_{\nu=n}^m \gamma_k(\psi_\nu') \leq \sum_{\nu=n}^\infty \gamma_k(\psi_\nu) \leq \sum_{\nu=n}^\infty 2^{-\nu}$$

The right-hand side tends to zero as $n \to \infty$, which proves that the partial sums of (8) form a Cauchy sequence in $\mathscr{V}$. By the completeness of $\mathscr{V}$, ψ is a member of $\mathscr{V}$.

Finally,

$$\langle f_\nu', \psi \rangle = \sum_{\mu=1}^{\nu-1} \langle f_\nu', \psi_\mu' \rangle + \langle f_\nu', \psi_\nu' \rangle + \sum_{\mu=\nu+1}^{\infty} \langle f_\nu', \psi_\mu' \rangle \tag{9}$$

By (5)

$$\left| \sum_{\mu=\nu+1}^{\infty} \langle f_\nu', \psi_\mu' \rangle \right| < \sum_{\mu=\nu+1}^{\infty} 2^{\nu-\mu} = 1 \tag{10}$$

For any three complex numbers α, β, and η, we have $|\alpha + \beta + \eta| \geq |\alpha| - |\beta| - |\eta|$. Therefore, by (7), (9), and (10),

$$|\langle f_\nu', \psi \rangle| \geq |\langle f_\nu', \psi_\nu' \rangle| - \left| \sum_{\mu=1}^{\nu-1} \langle f_\nu', \psi_\mu' \rangle \right| - \left| \sum_{\mu=\nu+1}^{\infty} \langle f_\nu', \psi_\mu' \rangle \right|$$
$$> \nu - 1$$

This proves that, as $\nu \to \infty$, $|\langle f_\nu', \psi \rangle| \to \infty$, which contradicts the assumption that $\{f_\nu\}$ is a Cauchy sequence in $\mathscr{V}'$.

Altogether then, f is a continuous linear functional on $\mathscr{V}$. Q.E.D.

PROBLEM 1.8-1. Verify that the dual $\mathscr{V}'$ of a countably multinormed space is a linear space.

PROBLEM 1.8-2. Spell out the proof of Corollary 1.8-2a.

PROBLEM 1.8-3. Let K be a fixed compact subset of $\mathscr{R}^n$, and let $h(t)$ be a locally integrable function on $\mathscr{R}^n$. Define a functional f on $\mathscr{D}_K$ by

$$\langle f, \phi \rangle \triangleq \int_{\mathscr{R}^n} h(t)\phi(t)\, dt \tag{11}$$

where $\phi \in \mathscr{D}_K$. Show that f is a member of $\mathscr{D}_K'$, and determine for f the smallest possible values that C and r can have in (1).

PROBLEM 1.8-4. $\mathscr{S}'$ denotes the dual of $\mathscr{S}$, the space defined in Problem 1.6-4. The members of $\mathscr{S}'$ are called *distributions of slow growth* or *tempered distributions*. Again let $h(t)$ be a locally integrable function on $\mathscr{R}^n$. State a condition on the growth of $h(t)$ as $|t| \to \infty$ which insures that $h(t)$ generates a member f of $\mathscr{S}'$ in accordance with (11) where now $\phi \in \mathscr{S}$. Is your condition also necessary in order for (11) to hold?

PROBLEM 1.8-5. Let $t \in \mathscr{R}^1$, and assume that the members of $\mathscr{S}$ are defined on $\mathscr{R}^1$. Define the expression $f(t) = \sum_{n=1}^{\infty} a_n \, \delta(t - n)$, where the a_n are complex numbers, as a functional on $\mathscr{S}$ by

$$\langle f(t), \phi(t) \rangle \triangleq \sum_{n=1}^{\infty} a_n \phi(n) \qquad \phi \in \mathscr{S}$$

if the right-hand side converges. State a sufficient condition on the coefficients a_n in order for f to be a member of $\mathscr{S}'$.

1.9. Duals of Countable-Union Spaces

Let $\mathscr{V} = \bigcup_{m=1}^{\infty} \mathscr{V}_m$ denote the countable-union space generated by the sequence $\{\mathscr{V}_m\}_{m=1}^{\infty}$ of countably multinormed spaces. As before, a *functional f on $\mathscr{V}$* is a rule that assigns a complex number $\langle f, \phi \rangle$ to each $\phi \in \mathscr{V}$. Also, f is called *linear* if for any $\phi, \psi \in \mathscr{V}$ and any $\alpha, \beta \in \mathscr{C}^1$ we have $\langle f, \alpha\phi + \beta\psi \rangle = \alpha\langle f, \phi \rangle + \beta\langle f, \psi \rangle$. Furthermore, f is called *continuous on* $\mathscr{V}$ if it is a continuous functional on every $\mathscr{V}_m$. By definition a sequence converges in $\mathscr{V}$ if and only if it converges in one of the $\mathscr{V}_m$. Consequently, by Lemma 1.8-2, a linear functional f on $\mathscr{V}$ is continuous if and only if $\langle f, \phi_\nu \rangle \to 0$ whenever $\phi_\nu \to \varnothing$ in $\mathscr{V}$.

Once again, the collection of all continuous linear functionals on $\mathscr{V}$ is called the *dual of* $\mathscr{V}$ and is denoted by $\mathscr{V}'$. Equality, addition, and multiplication by a complex number are defined in $\mathscr{V}'$ just as they were in Sec. 1.8. This in turn makes $\mathscr{V}'$ a linear space.

Let $\mathscr{U}$ and $\mathscr{V}$ be countable-union spaces with $\mathscr{U}$ being a linear subspace of $\mathscr{V}$. (As was pointed out in Sec. 1.7, this assumption does not prohibit either $\mathscr{U}$ or $\mathscr{V}$ from being a countably multinormed space.) Assume that the convergence concept for $\mathscr{U}$ is *stronger* than the one for $\mathscr{V}$ in the following sense: The convergence in $\mathscr{U}$ of a sequence to some limit ϕ implies its convergence in $\mathscr{V}$ to the same limit ϕ. (In this case, we also say that the convergence concept for $\mathscr{V}$ is *weaker* than the one for $\mathscr{U}$.) When both $\mathscr{U}$ and $\mathscr{V}$ are countably multinormed spaces, this condition is certainly fulfilled when the topology of $\mathscr{U}$ is stronger than that induced on it by $\mathscr{V}$.

If f is a member of $\mathscr{V}'$, its restriction to $\mathscr{U}$ is that unique functional g on $\mathscr{U}$ for which $\langle g, \phi \rangle = \langle f, \phi \rangle$ whenever $\phi \in \mathscr{U}$. It follows as in Sec. 1.8 that g is in $\mathscr{U}'$. To conclude that $\mathscr{V}'$ is a subspace of $\mathscr{U}'$, we need to assume more. It suffices to have $\mathscr{U}$ as a dense subset of $\mathscr{V}$, the proof of this being the same as the proof of Theorem 1.8-2. Thus, we can state

THEOREM 1.9-1. *Let $\mathscr{U}$ and $\mathscr{V}$ be countable-union spaces. Assume that $\mathscr{U}$ is a dense subspace of $\mathscr{V}$ and that the convergence concept for $\mathscr{U}$ is stronger than the one for $\mathscr{V}$ in the aforementioned sense. Then, $\mathscr{V}'$ is a subspace of $\mathscr{U}'$.*

The convergence concept for the dual $\mathscr{V}'$ of a countable-union space $\mathscr{V}$ is the following. A sequence $\{f_\nu\}$ is said to *converge in* $\mathscr{V}'$, and the sequence is called *convergent*, if all $f_\nu \in \mathscr{V}'$ and there exists an $f \in \mathscr{V}'$ such that for every $\phi \in \mathscr{V}$ we have that $\langle f_\nu, \phi \rangle \to \langle f, \phi \rangle$ as $\nu \to \infty$. (This type of convergence is also called *weak convergence in* $\mathscr{V}'$.) Clearly, there can be no more than one limit f in $\mathscr{V}'$ for any given convergent sequence. Once again, this definition of convergence is extended in the customary way to directed sets and series. We have not specified a topology for $\mathscr{V}'$, as was also the case for countable-union spaces. However, it is readily shown that $\mathscr{V}'$ is a sequential-convergence* linear space. This is illustrated in Fig. 1.6-1.

A sequence $\{f_\nu\}$ is called a *Cauchy sequence in* $\mathscr{V}'$ (more precisely, a *weak Cauchy sequence in* $\mathscr{V}'$) if all $f_\nu \in \mathscr{V}'$ and for each $\phi \in \mathscr{V}$ we have $\langle f_\nu - f_\mu, \phi\rangle \to 0$ as ν and μ tend to infinity independently. $\mathscr{V}'$ is called *complete* when all its Cauchy sequences are convergent.

THEOREM 1.9-2. *Let $\mathscr{V} = \bigcup_{m=1}^{\infty} \mathscr{V}_m$ be a countable-union space obtained from the sequence $\{\mathscr{V}_m\}_{m=1}^{\infty}$ of complete countably multinormed spaces. Then, the dual space $\mathscr{V}'$ is also complete.*

PROOF. Let $\{f_\nu\}_{\nu=1}^{\infty}$ be a Cauchy sequence in $\mathscr{V}'$. As in the proof of Theorem 1.8-3, $\{f_\nu\}$ defines a unique linear functional f on $\mathscr{V}$ through the relation $\langle f, \phi\rangle \triangleq \lim_{\nu\to\infty}\langle f_\nu, \phi\rangle (\phi \in \mathscr{V})$. We have to show that f is also continuous on $\mathscr{V}$. By definition each f_ν is a continuous linear functional on every $\mathscr{V}_m$, and $\{f_\nu\}$ is a Cauchy sequence in $\mathscr{V}_m'$. Hence, by Theorem 1.8-3, f is continuous on every $\mathscr{V}_m$, which means that f is continuous on $\mathscr{V}$. Q.E.D.

EXAMPLE 1.9-1. $\mathscr{D}'$ is the dual of the strict countable-union space $\mathscr{D}$, which was discussed in Example 1.7-1. Thus, f is in $\mathscr{D}'$ if and only if, for every compact subset K of $\mathscr{R}^n$, f is in $\mathscr{D}_K'$ (more precisely, the restriction of f to each $\mathscr{D}_k$ is in $\mathscr{D}_K'$). By Theorem 1.9-2, $\mathscr{D}'$ is complete. The members of $\mathscr{D}'$ are called *distributions on* $\mathscr{R}^n$ or simply *distributions*.

PROBLEM 1.9-1. Verify that the dual of a countable-union space is a sequential-convergence* linear space.

PROBLEM 1.9-2. Define the expression $f(t) = \sum_{n=1}^{\infty}\delta(t - \tau_n)$, where t is a variable in $\mathscr{R}^n$ and the τ_n are fixed points of $\mathscr{R}^n$, as a functional on $\mathscr{D}$ by

$$\langle f(t), \phi(t)\rangle \triangleq \sum_{n=1}^{\infty} \phi(\tau_n) \qquad \phi \in \mathscr{D}$$

if the right-hand side converges. Find a necessary and sufficient condition on the set of points $\{\tau_n\}$ in order for f to be a distribution.

PROBLEM 1.9-3. $\mathscr{S}'$ is a subspace of $\mathscr{D}'$. Why?

1.10. Operators and Adjoint Operators

Let $\mathscr{U}$ and $\mathscr{V}$ be linear spaces. An *operator* (or *operation* or *mapping*) *from* $\mathscr{U}$ *into* $\mathscr{V}$ is a rule $\mathfrak{N}$ that assigns precisely one element in $\mathscr{V}$ to each element in some subset of $\mathscr{U}$. The set of elements in $\mathscr{U}$ on which $\mathfrak{N}$ is defined is called the *domain of* $\mathfrak{N}$ and is denoted by $d(\mathfrak{N})$. The set of all elements in $\mathscr{V}$ that result from the application of $\mathfrak{N}$ to the members of $d(\mathfrak{N})$ is called the *range of* $\mathfrak{N}$ and is denoted by $r(\mathfrak{N})$. If $d(\mathfrak{N}) = \mathscr{U}$, we say that $\mathfrak{N}$ is a mapping *of* (or *on*) $\mathscr{U}$ into $\mathscr{V}$, or that $\mathfrak{N}$ *maps* $\mathscr{U}$ into $\mathscr{V}$. If $r(\mathfrak{N}) = \mathscr{V}$, then $\mathfrak{N}$ is said to be a mapping from $\mathscr{U}$ *onto* $\mathscr{V}$. The notations for this correspondence are $\psi = \mathfrak{N}\phi$, or $\mathfrak{N}\colon \phi \mapsto \psi$, or simply $\phi \mapsto \psi$, where $\phi \in d(\mathfrak{N})$ and $\psi \in r(\mathfrak{N})$.

It can happen that more than one element in $\mathscr{U}$ is mapped into the same element in $\mathscr{V}$, but, when this is prohibited (i.e., when the condition $\mathfrak{N}\phi = \mathfrak{N}\theta$ implies that $\phi = \theta$), the mapping is called *one-to-one*. If $\mathfrak{N}$ is a one-to-one mapping from $\mathscr{U}$ into $\mathscr{V}$, the *inverse mapping* $\mathfrak{N}^{-1}$ from $\mathscr{V}$ into $\mathscr{U}$ is simply the rule that assigns to each $\psi \in r(\mathfrak{N})$ the unique element $\phi \in d(\mathfrak{N})$ for which $\mathfrak{N}\phi = \psi$; now we write $\phi = \mathfrak{N}^{-1}\psi$.

An operator $\mathfrak{N}$ from the linear space $\mathscr{U}$ into the linear space $\mathscr{V}$ is called *linear* if $d(\mathfrak{N})$ is a linear space (i.e., a linear subspace of $\mathscr{U}$ or possibly $\mathscr{U}$ itself) and for every ϕ, $\theta \in d(\mathfrak{N})$ and every α, $\beta \in \mathscr{C}^1$ we always have

$$\mathfrak{N}(\alpha\phi + \beta\theta) = \alpha\mathfrak{N}\phi + \beta\mathfrak{N}\theta$$

Consequently, a linear operator $\mathfrak{N}$ from $\mathscr{U}$ into $\mathscr{V}$ maps the origin of $\mathscr{U}$ into the origin of $\mathscr{V}$. Moreover, the range of $\mathfrak{N}$ is a linear subspace of $\mathscr{V}$ or possibly $\mathscr{V}$ itself. If in addition $\mathfrak{N}$ is one-to-one, then its inverse operator $\mathfrak{N}^{-1}$ is linear too. (Prove the last two statements.)

Now, let $\mathscr{U}$ and $\mathscr{V}$ be multinormed spaces, and let $\mathfrak{N}$ be a mapping of $\mathscr{U}$ into $\mathscr{V}$. $\mathfrak{N}$ is said to be *continuous at some point* $\phi \in \mathscr{U}$ if, for every neighborhood Λ of $\mathfrak{N}\phi$ in $\mathscr{V}$, there exists a neighborhood Ω of ϕ in $\mathscr{U}$ such that $\mathfrak{N}\psi \in \Lambda$ whenever $\psi \in \Omega$. $\mathfrak{N}$ is called *continuous* if it is continuous at every point of $\mathscr{U}$. It is readily seen that a linear mapping $\mathfrak{N}$ of $\mathscr{U}$ into $\mathscr{V}$ is continuous if and only if it is continuous at the origin of $\mathscr{U}$.

LEMMA 1.10-1. *Let $\mathfrak{N}$ be a linear mapping of the multinormed space $\mathscr{U}$ into the multinormed space $\mathscr{V}$, and let S and R be the multinorms for $\mathscr{U}$ and $\mathscr{V}$ respectively. A necessary and sufficient condition for $\mathfrak{N}$ to be continuous is that to every $\rho \in R$ there correspond a finite number of seminorms $\gamma_1, \dots, \gamma_n \in S$ and a positive number C such that for all $\phi \in \mathscr{U}$ we have*

$$\rho(\mathfrak{N}\phi) \le C \max\,\{\gamma_1(\phi), \dots, \gamma_n(\phi)\}$$

or alternatively

$$\rho(\mathfrak{N}\phi) \le C[\gamma_1(\phi) + \cdots + \gamma_n(\phi)]$$

The proof of this lemma is almost identical to that of Lemma 1.6-3. In the present case the balloons A are taken in $\mathscr{V}$ and the balloons B and B_k are taken in $\mathscr{U}$.

A slight modification of the proofs of Lemmas 1.8-1 and 1.8-2 establishes the following result. (Replace the neighborhoods in the complex plane $\mathscr{C}^1$ in the previous case by the neighborhoods in $\mathscr{V}$ in the present case.)

LEMMA 1.10-2. *A mapping* $\mathfrak{N}$ *of the countably multinormed space* $\mathscr{U}$ *into the multinormed space* $\mathscr{V}$ *is continuous if and only if* $\mathfrak{N}\phi_\nu \to \mathfrak{N}\phi$ *in* $\mathscr{V}$ *whenever* $\phi_\nu \to \phi$ *in* $\mathscr{U}$. *Furthermore, under the additional assumption that* $\mathfrak{N}$ *is linear,* $\mathfrak{N}$ *is continuous if and only if* $\mathfrak{N}\phi_\nu \to \varnothing$ *in* $\mathscr{V}$ *whenever* $\phi_\nu \to \varnothing$ *in* $\mathscr{U}$.

The multinormed spaces $\mathscr{U}$ and $\mathscr{V}$ are called *isomorphic*, if there exists a one-to-one continuous linear mapping $\mathfrak{N}$ of $\mathscr{U}$ onto $\mathscr{V}$ such that its inverse $\mathfrak{N}^{-1}$ is a continuous linear mapping of $\mathscr{V}$ onto $\mathscr{U}$. In this case $\mathfrak{N}$ is called an *isomorphism from* $\mathscr{U}$ *onto* $\mathscr{V}$; if in addition $\mathscr{U} = \mathscr{V}$, then $\mathfrak{N}$ is called an *automorphism on* $\mathscr{U}$.

So far as the linear operations and topologies are concerned, we can replace a multinormed space $\mathscr{U}$ by an isomorphic space $\mathscr{V}$. For example, if $\mathfrak{N}$ is an isomorphism from $\mathscr{U}$ onto $\mathscr{V}$ and if $\mathscr{U}$ is complete, then $\mathscr{V}$ must also be complete. Also, a continuous linear operator A mapping $\mathscr{U}$ into another multinormed space $\mathscr{W}$ defines an operator B mapping $\mathscr{V}$ into $\mathscr{W}$ as follows: $B\psi \triangleq A\phi \in \mathscr{W}$ whenever $\phi \in \mathscr{U}$ and $\psi = \mathfrak{N}\phi \in \mathscr{V}$. It then follows that B must also be linear and continuous. The proofs of these assertions are left to the reader.

Next, let $\mathscr{U} = \bigcup_{m=1}^{\infty} \mathscr{U}_m$ be a countable-union space generated by the sequence $\{\mathscr{U}_m\}_{m=1}^{\infty}$ of countably multinormed spaces. Also, let $\mathscr{V}$ be a multinormed space. An operator $\mathfrak{N}$ on $\mathscr{U}$ into $\mathscr{V}$ is called *continuous* if it is continuous on each $\mathscr{U}_m$. We have from Lemma 1.10-2 that $\mathfrak{N}$ is continuous if and only if $\mathfrak{N}\phi_\nu \to \mathfrak{N}\phi$ in $\mathscr{V}$ whenever the sequence $\{\phi_\nu\}_{\nu=1}^{\infty}$ converges in $\mathscr{U}$ to ϕ; moreover, when $\mathfrak{N}$ is linear, it is again continuous if and only if $\mathfrak{N}\phi_\nu \to \varnothing$ in $\mathscr{V}$ whenever the sequence $\{\phi_\nu\}_{\nu=1}^{\infty}$ converges in $\mathscr{U}$ to zero.

Finally, if both $\mathscr{U}$ and $\mathscr{V}$ are countable-union spaces, the operator $\mathfrak{N}$ on $\mathscr{U}$ into $\mathscr{V}$ is called *continuous* if $\mathfrak{N}\phi_\nu \to \mathfrak{N}\phi$ in $\mathscr{V}$ whenever the sequence $\{\phi_\nu\}$ converges in $\mathscr{U}$ to ϕ. Here again, we see that, when $\mathfrak{N}$ is linear, it is continuous if and only if $\mathfrak{N}\phi_\nu \to \varnothing$ in $\mathscr{V}$ whenever the sequence $\{\phi_\nu\}$ converges in $\mathscr{U}$ to zero. All these definitions of continuity are consistent with one another. In the present case, the definitions of isomorphic, isomorphism, and automorphism are exactly the same as before. Namely. the countable-union spaces $\mathscr{U}$ and $\mathscr{V}$ are called *isomorphic* if there exists a one-to-one continuous linear mapping $\mathfrak{N}$ of $\mathscr{U}$ onto $\mathscr{V}$ such that its inverse $\mathfrak{N}^{-1}$ is a continuous linear mapping of $\mathscr{V}$ onto $\mathscr{U}$; in this case, $\mathfrak{N}$ is called an *isomorphism from* $\mathscr{U}$ *onto* $\mathscr{V}$, and, if $\mathscr{U} = \mathscr{V}$, an *automorphism on* $\mathscr{U}$. Now, however, this is an abuse of their customary meanings since topologies have not been defined in our countable-union spaces.

Since the dual of a countably multinormed space is a multinormed space, our previous definitions of *continuity*, *isomorphic*, *isomorphism*, and *automorphism* also apply to operators mapping the duals of countably multinormed spaces into other such duals. On the other hand, when $\mathscr{U}$ and $\mathscr{V}$ are countable-union spaces, an operator $\mathfrak{N}$ on $\mathscr{V}'$ into $\mathscr{U}'$ is called *continuous* if $\mathfrak{N}f_\nu \to \mathfrak{N}f$ in $\mathscr{U}'$ whenever $f_\nu \to f$ in $\mathscr{V}'$. It is readily seen once again that a linear operator $\mathfrak{N}$ on $\mathscr{V}'$ into $\mathscr{U}'$ is continuous if and only if $\mathfrak{N}f_\nu \to \varnothing$ in $\mathscr{U}'$ whenever $f_\nu \to \varnothing$ in $\mathscr{V}'$. An *isomorphism* is defined as before.

We now turn to the concept of the adjoint operator. Assume that $\mathscr{U}$ and $\mathscr{V}$ are both countably multinormed spaces or both countable-union spaces, and let $\mathfrak{N}$ be a continuous linear mapping of $\mathscr{U}$ into $\mathscr{V}$. We define the *adjoint operator* $\mathfrak{N}'$ on the dual space $\mathscr{V}'$ by

$$\langle \mathfrak{N}'f, \phi \rangle \triangleq \langle f, \mathfrak{N}\phi \rangle \tag{1}$$

where $f \in \mathscr{V}'$ and ϕ traverses all of $\mathscr{U}$. Here, $\mathfrak{N}\phi$ is in $\mathscr{V}$ so that the right-hand side has a sense. Equation (1) defines $\mathfrak{N}'f$ as a functional on $\mathscr{U}$; Namely, $\mathfrak{N}'f$ is that functional on $\mathscr{U}$ which assigns to each $\phi \in \mathscr{U}$ the same number that $f \in \mathscr{V}'$ assigns to $\mathfrak{N}\phi \in \mathscr{V}$.

Actually, $\mathfrak{N}'f$ is a member of $\mathscr{U}'$. Indeed, for any $\phi, \psi \in \mathscr{U}$ and $\alpha, \beta \in \mathscr{C}^1$,

$$\begin{aligned}\langle \mathfrak{N}'f, \alpha\phi + \beta\psi \rangle &= \langle f, \mathfrak{N}(\alpha\phi + \beta\psi) \rangle = \langle f, \alpha\mathfrak{N}\phi + \beta\mathfrak{N}\psi \rangle \\ &= \alpha\langle f, \mathfrak{N}\phi \rangle + \beta\langle f, \mathfrak{N}\psi \rangle = \alpha\langle \mathfrak{N}'f, \phi \rangle + \beta\langle \mathfrak{N}'f, \psi \rangle\end{aligned}$$

which shows that $\mathfrak{N}'f$ is a linear functional on $\mathscr{U}$. Furthermore, let $\{\phi_\nu\}_{\nu=1}^\infty$ converge in $\mathscr{U}$ to zero. Then, as $\nu \to \infty$, $\mathfrak{N}\phi_\nu \to \varnothing$ in $\mathscr{V}$, and

$$\langle \mathfrak{N}'f, \phi_\nu \rangle = \langle f, \mathfrak{N}\phi_\nu \rangle \to 0$$

Hence, $\mathfrak{N}'f$ is a continuous linear functional on $\mathscr{U}$.

Thus, $\mathfrak{N}'$ is a mapping on $\mathscr{V}'$ into $\mathscr{U}'$. We now prove that $\mathfrak{N}'$ is linear and continuous. Let $\phi \in \mathscr{U}, f, g \in \mathscr{V}'$, and $\alpha, \beta \in \mathscr{C}^1$. Then,

$$\begin{aligned}\langle \mathfrak{N}'(\alpha f + \beta g), \phi \rangle &= \langle \alpha f + \beta g, \mathfrak{N}\phi \rangle = \alpha\langle f, \mathfrak{N}\phi \rangle + \beta\langle g, \mathfrak{N}\phi \rangle \\ &= \alpha\langle \mathfrak{N}'f, \phi \rangle + \beta\langle \mathfrak{N}'g, \phi \rangle = \langle \alpha\mathfrak{N}'f + \beta\mathfrak{N}'g, \phi \rangle\end{aligned}$$

and thus $\mathfrak{N}'$ is linear.

To show that $\mathfrak{N}'$ is continuous, assume first that both $\mathscr{U}$ and $\mathscr{V}$ are countably multinormed spaces. For fixed $\phi \in \mathscr{U}$ the magnitude of the left-hand side of (1) is the value of a seminorm on $\mathscr{U}'$, whereas the magnitude of the right-hand side of (1) is the value of a seminorm on $\mathscr{V}'$. Thus, the continuity of $\mathfrak{N}'$ follows directly from Lemma 1.10-1. Second, assume that $\mathscr{U}$ and $\mathscr{V}$ are countable-union spaces, and let $f_\nu \to \varnothing$ in $\mathscr{V}'$. Then, for every $\phi \in \mathscr{U}$, $\langle \mathfrak{N}'f_\nu, \phi \rangle = \langle f_\nu, \mathfrak{N}\phi \rangle \to 0$, so that $\mathfrak{N}'f_\nu \to \varnothing$ in $\mathscr{U}'$. Consequently, $\mathfrak{N}'$ is again continuous.

Altogether then, we have proven

THEOREM 1.10-1. *If $\mathscr{U}$ and $\mathscr{V}$ are both countably multinormed spaces or both countable-union spaces and if $\mathfrak{N}$ is a continuous linear mapping of $\mathscr{U}$ into $\mathscr{V}$, then the adjoint operator $\mathfrak{N}'$ is a continuous linear mapping of $\mathscr{V}'$ into $\mathscr{U}'$.*

A closely related result is

THEOREM 1.10-2. *If $\mathscr{U}$ and $\mathscr{V}$ are both countably multinormed spaces or both countable-union spaces and if $\mathfrak{N}$ is an isomorphism from $\mathscr{U}$ onto $\mathscr{V}$, then $\mathfrak{N}'$ is an isomorphism from $\mathscr{V}'$ onto $\mathscr{U}'$. Moreover, $(\mathfrak{N}')^{-1} = (\mathfrak{N}^{-1})'$.*

PROOF. By definition of an isomorphism, $\mathfrak{N}$ is a one-to-one continuous linear mapping of $\mathscr{U}$ onto $\mathscr{V}$ and its inverse $\mathfrak{N}^{-1}$ is a continuous linear mapping of $\mathscr{V}$ onto $\mathscr{U}$. Theorem 1.10-1 asserts that $\mathfrak{N}'$ is a continuous linear mapping of $\mathscr{V}'$ into $\mathscr{U}'$ and that $(\mathfrak{N}^{-1})'$ is a continuous linear mapping of $\mathscr{U}'$ into $\mathscr{V}'$.

Now, let $\psi = \mathfrak{N}\phi \in \mathscr{V}$. For any $f \in \mathscr{V}'$ we have $(\mathfrak{N}^{-1})'\mathfrak{N}'f = f$ because

$$\langle f, \psi \rangle = \langle f, \mathfrak{N}\mathfrak{N}^{-1}\psi \rangle = \langle (\mathfrak{N}^{-1})'\mathfrak{N}'f, \psi \rangle.$$

Similarly, for any $g \in \mathscr{U}'$ we have $\mathfrak{N}'(\mathfrak{N}^{-1})'g = g$ because for $\phi \in \mathscr{U}$

$$\langle g, \phi \rangle = \langle g, \mathfrak{N}^{-1}\mathfrak{N}\phi \rangle = \langle \mathfrak{N}'(\mathfrak{N}^{-1})'g, \phi \rangle$$

It follows that $\mathfrak{N}'$ is a one-to-one mapping of $\mathscr{V}'$ onto $\mathscr{U}'$ and that $(\mathfrak{N}^{-1})' = (\mathfrak{N}')^{-1}$. Q.E.D.

EXAMPLE 1.10-1. As before, let K be a compact subset of $\mathscr{R}^n$. The partial differentiation operator $\partial/\partial t_\nu$, where t_ν is a component of $t \in \mathscr{R}^n$, is a continuous linear mapping of $\mathscr{D}_K$ into $\mathscr{D}_K$, as well as of $\mathscr{D}$ into $\mathscr{D}$. Indeed, for any $\phi \in \mathscr{D}_K$ and any nonnegative integer $k \in \mathscr{R}^n$, we have

$$\gamma_k\left(\frac{\partial \phi}{\partial t_\nu}\right) = \sup_{t \in R^n} \left| D^k \frac{\partial \phi}{\partial t_\nu} \right| = \gamma_p(\phi)$$

where $D^p = D^k \partial/\partial t_\nu$. Thus, our assertion follows from Lemma 1.10-1 and the definition of continuity for an operator on a countable-union space into itself.

The adjoint of $\partial/\partial t_\nu$ is denoted by $-\partial/\partial t_\nu$ and defined by

$$(2) \qquad \left\langle -\frac{\partial f}{\partial t_\nu}, \phi \right\rangle \triangleq \left\langle f, \frac{\partial \phi}{\partial t_\nu} \right\rangle \qquad f \in \mathscr{D}', \phi \in \mathscr{D}$$

This is consistent with an integration by parts when for instance f and $\partial f/\partial t_\nu$ are continuous functions on $\mathscr{R}^n$; in this case both sides of (2) happen to be integrals on $\mathscr{R}^n$, and the limit terms arising from the integration by parts are equal to zero. In the general case the adjoint operator $-\partial/\partial t_\nu$ is interpreted as a *generalized* differential operator acting on $\mathscr{D}_K'$ or $\mathscr{D}'$. For example, for any fixed $\tau \in \mathscr{R}^n$ and with $\delta(t-\tau)$ denoting the delta functional at $t = \tau$ (see Example 1.8-1), the functional $-(\partial/\partial t_\nu)\delta(t-\tau)$ is a member of $\mathscr{D}'$ and is determined on $\mathscr{D}$ by the expression

$$\left\langle -\frac{\partial}{\partial t_\nu}\,\delta(t-\tau),\,\phi(t)\right\rangle \triangleq \left\langle \delta(t-\tau),\,\frac{\partial}{\partial t_\nu}\,\phi(t)\right\rangle = \left.\frac{\partial\phi}{\partial t_\nu}\right|_{t=\tau}$$

EXAMPLE 1.10-2. Let τ be a fixed real number. The shifting operator S_τ: $\phi(t) \mapsto \phi(t-\tau)$ is clearly a continuous linear mapping of $\mathscr{D}$ into $\mathscr{D}$. Its inverse mapping is $S_{-\tau}$ and has the same properties. Moreover, S_τ is a one-to-one mapping of $\mathscr{D}$ onto $\mathscr{D}$. Thus, S_τ is an automorphism on $\mathscr{D}$.

By definition, the adjoint S_τ' of S_τ is given by

$$\langle S_\tau' f, \phi\rangle \triangleq \langle f, S_\tau \phi\rangle \qquad f \in \mathscr{D}',\ \phi \in \mathscr{D} \tag{3}$$

It is customary to denote $S_\tau' f(t)$ by $S_{-\tau} f(t) \triangleq f(t+\tau)$ since this is what we would have if f were a locally integrable function and (3) were an equation between two integrals:

$$\int_{R^n} f(t+\tau)\phi(t)\,dt = \int_{R^n} f(t)\phi(t-\tau)\,dt$$

Thus, S_τ' is interpreted as the *generalized* shifting operator $S_{-\tau}$: $f(t) \mapsto f(t+\tau)$. It is an automorphism on $\mathscr{D}'$ according to Theorem 1.10-2.

PROBLEM 1.10-1. Show that the range of a linear operator $\mathfrak{N}$ is a linear space. Also show that, if $\mathfrak{N}$ is linear and one-to-one, then $\mathfrak{N}^{-1}$ is also linear.

PROBLEM 1.10-2. Prove Lemma 1.10-1.

PROBLEM 1.10-3. Prove Lemma 1.10-2.

PROBLEM 1.10-4. (*a*) Let $\mathscr{U}$ and $\mathscr{V}$ be multinormed spaces, and let $\mathfrak{N}$ be an isomorphism from $\mathscr{U}$ onto $\mathscr{V}$. Show that, if $\mathscr{U}$ is complete, then $\mathscr{V}$ is complete.

(*b*) In addition, let A be a continuous linear mapping of $\mathscr{U}$ into another multinormed space $\mathscr{W}$. Define the operator B on $\mathscr{V}$ into $\mathscr{W}$ by $B\psi \triangleq A\phi$ where $\psi = \mathfrak{N}\phi$, $\phi \in \mathscr{U}$. Show that B is linear and continuous.

Problem 1.10-5. (*a*) Let $\theta(t)$ be a smooth function on $\mathscr{R}^n$. Show that $\phi \mapsto \theta\phi$ is a continuous linear mapping of $\mathscr{D}$ into $\mathscr{D}$. Upon defining the adjoint mapping $f \mapsto \theta f$ by

$$\langle \theta f, \phi \rangle \triangleq \langle f, \theta\phi \rangle \qquad \phi \in \mathscr{D} \tag{4}$$

we can conclude that $f \mapsto \theta f$ is a continuous linear mapping of $\mathscr{D}'$ into $\mathscr{D}'$.

(*b*) State a condition on the growth of $\theta(t)$ as $|t| \to \infty$ which insures that the operator $f \mapsto \theta f$, defined by (4) with $\phi \in \mathscr{S}$, is a continuous linear mapping of $\mathscr{S}'$ into $\mathscr{S}'$.

CHAPTER II

Distributions and Generalized Functions

2.1. Introduction

In this chapter we present some elements of the theory of distributions and generalized functions. As was the case in the preceding chapter, we discuss only those results that we shall have need of in the subsequent chapters. A more thorough discussion of these ideas appears in many other books; see for example Friedman [1], Horvath [3], Schwartz [1], and Zemanian [1]. At the end of Sec. 2.2 we merely quote some deeper properties of distributions as local phenomena and refer the reader to other sources for their proofs.

2.2. The Spaces $\mathscr{D}_K(I)$, $\mathscr{D}(I)$, and Their Duals: Distributions

Let $t = \{t_1, t_2, \ldots, t_n\} \in \mathscr{R}^n$, and let I be a nonvoid open set in $\mathscr{R}^n$ (possibly $I = \mathscr{R}^n$). Furthermore, let K be a compact subset of I. $\mathscr{D}_K(I)$ is the set of all complex-valued smooth functions defined on I which vanish on those points of I that are not in K. When $I = \mathscr{R}^n$, this set is denoted simply by $\mathscr{D}_K$. We have already discussed $\mathscr{D}_K$ to some extent in the examples of the preceding chapter.

$\mathscr{D}_K(I)$ is a linear space under the usual definitions of addition of functions and their multiplication by complex numbers. The zero element in $\mathscr{D}_K(I)$ is the identically zero function on I. We denote it by 0 rather than by $\varnothing$. When $K = \{t: |t| \leq 1\}$ and I is any open set containing K, an example of a member of $\mathscr{D}_K(I)$ is the restriction to I of the function defined by Sec. 1.3, Eq. (1). For each nonnegative integer $k \in \mathscr{R}^n$, we define the seminorm

$$(1) \qquad \gamma_k(\phi) \triangleq \sup_{t \in I} |D^k \phi(t)| \qquad \phi \in \mathscr{D}_K(I)$$

Note that γ_0 is a norm. Thus, as k traverses the nonnegative integers in

$\mathscr{R}^n$, we obtain a countable multinorm $\{\gamma_k\}$ defined on $\mathscr{D}_K(I)$. We assign to $\mathscr{D}_K(I)$ the topology generated by $\{\gamma_k\}$, and thus $\mathscr{D}_K(I)$ is a countably multi-normed space.

$\mathscr{D}_K(I)$ is complete and hence a Fréchet space. For, if $\{\phi_\nu\}$ is a Cauchy sequence in $\mathscr{D}_K(I)$, it converges uniformly on I because the complex plane is complete and

$$\gamma_0(\phi_\nu - \phi_\mu) \triangleq \sup_{t \in I} |\phi_\nu(t) - \phi_\mu(t)| \to 0$$

as ν and μ tend to infinity independently. Hence, by a standard theorem (Apostol [1], p. 394) the limit ϕ of $\{\phi_\nu\}$ is a continuous function on I; ϕ also vanishes on those points of I that are not in K. Moreover, in view of the definition of γ_k, the sequence $\{D^k\phi_\nu\}$ converges uniformly on I. By repeatedly using another standard theorem (Apostol [1], p. 402), we see that $\{D^k\phi_\nu\}$ converges on I to $D^k\phi$, which is therefore also continuous on I. Thus, ϕ is in $\mathscr{D}_K(I)$ and, for each k, $\gamma_k(\phi_\nu - \phi) \to 0$ as $\nu \to \infty$. This verifies our assertion.

We now turn to the definition of the function space $\mathscr{D}(I)$ where I again denotes some nonvoid open set in $\mathscr{R}^n$. If $I = \mathscr{R}^n$, we denote $\mathscr{D}(I)$ by $\mathscr{D}$. Let $\{K_m\}_{m-1}^{\infty}$ be a sequence of compact subsets of I with the following two properties: (i) $K_1 \subset K_2 \subset K_3 \subset \cdots$. (ii) Each compact subset of I is contained in one of the K_m. Consequently, $I = \bigcup_{m=1}^{\infty} K_m$. For any nonvoid open set I such a sequence always exists. (See Problem 2.2-2.)

If $m < p$, then $\mathscr{D}_{K_m}(I) \subset \mathscr{D}_{K_p}(I)$. Moreover, all the $\mathscr{D}_{K_m}(I)$ have topologies generated by the same multinorm, namely, $\{\gamma_k\}$ where the γ_k are defined by (1) and k traverses the nonnegative integers in $\mathscr{R}^n$. Thus, we can construct a strict countable-union space $\mathscr{D}(I) = \bigcup_{m=1}^{\infty} \mathscr{D}_{K_m}(I)$. By definition a sequence $\{\phi_\nu\}_{\nu=1}^{\infty}$ converges in $\mathscr{D}(I)$ if all the ϕ_ν are members of a particular $\mathscr{D}_{K_m}(I)$ and $\{\phi_\nu\}$ converges with respect to the multinorm $\{\gamma_k\}$. The space $\mathscr{D}(I)$ is independent of the choice of $\{K_m\}_{m=1}^{\infty}$ since any other such choice of compact sets would lead to the same elements in $\mathscr{D}(I)$ and the same convergent sequences. Since every $\mathscr{D}_{K_m}(I)$ is complete, $\mathscr{D}(I)$ is also.

This is not the usual way of defining $\mathscr{D}(I)$. Schwartz [1], Vol. I, pp. 64–71, assigns a topology to $\mathscr{D}(I)$ in such a way that the convergent sequences in $\mathscr{D}(I)$ under his topology coincide with the convergent sequences described here. However the definition of this topology is complicated. For our purposes it is quite adequate and considerably simpler to treat $\mathscr{D}(I)$ as a countable-union space.

We turn now to the dual spaces. $\mathscr{D}_K'(I)$ (or $\mathscr{D}_K'$ if $I = \mathscr{R}^n$) denotes the dual of $\mathscr{D}_K(I)$ (or respectively $\mathscr{D}_K$). By Theorem 1.8-3, $\mathscr{D}_K'(I)$ is complete. Moreover, since γ_0 is a norm on $\mathscr{D}_K(I)$, Theorem 1.8-1 yields

THEOREM 2.2-1. If $f \in \mathscr{D}_K'(I)$, *then there exist a positive constant* C *and a nonnegative integer* r, *which depend on* f, *such that for every* $\phi \in \mathscr{D}_K(I)$

$$|\langle f, \phi \rangle| \leq C \max_{0 \leq |k| \leq r} \sup_{t \in I} |D^k \phi(t)| \tag{2}$$

If K and J are compact sets such that I contains J and J contains a neighborhood of K, then the restriction of $f \in \mathscr{D}_J'(I)$ to $\mathscr{D}_K(I)$ is in $\mathscr{D}_K'(I)$. But, it is not true that $\mathscr{D}_J'(I)$ can be identified in a one-to-one fashion with a subset of $\mathscr{D}_K'(I)$; an illustration of this was given in Example 1.8-1.

$\mathscr{D}'(I)$(*or* $\mathscr{D}'$) is the dual of the strict countable-union space $\mathscr{D}(I)$ (*or* respectively $\mathscr{D}$), and by Theorem 1.9-2 it too is complete. The elements of $\mathscr{D}'(I)$ are called *distributions on* I or simply *distributions*. (We shall not refer to the elements of $\mathscr{D}_K'(I)$ as distributions, but instead as generalized functions. See Sec. 2.4.) By definition, f is a member of $\mathscr{D}'(I)$ if and only if, for every compact set K contained in I, the restriction of f to $\mathscr{D}_K(I)$ is a member of $\mathscr{D}_K'(I)$. Therefore, from Theorem 2.2-1 we get

THEOREM 2.2-2. *If* K *is a compact set contained in* I *and if* $f \in \mathscr{D}'(I)$, *then there exist a positive constant* C *and a nonnegative integer* r, *which depend upon* f *and* K, *such that for every* $\phi \in \mathscr{D}_K(I)$ *the inequality* (2) *is satisfied.*

If H and I are both open sets with $H \subset I$, then the restriction of $f \in \mathscr{D}'(I)$ to $\mathscr{D}(H)$ is in $\mathscr{D}'(H)$.

Let $f(t)$ be a locally integrable function on I. This generates a distribution in $\mathscr{D}'(I)$ (which we also denote by f or $f(t)$) through the definition

$$\langle f, \phi \rangle \triangleq \int_I f(t)\phi(t)\, dt \qquad \phi \in \mathscr{D}(I)$$

Indeed, this clearly defines a linear functional on $\mathscr{D}(I)$. That it is also continuous follows from the inequality

$$|\langle f, \phi \rangle| \leq \int_J |f(t)|\, dt \sup_{t \in J} |\phi(t)|$$

$$= \int_J |f(t)|\, dt\, \gamma_0(\phi)$$

where J denotes the support of ϕ. Actually, there is a one-to-one correspondence between all such distributions f and the equivalence classes of locally integrable functions on I, where any two functions in a given class differ on no more than a set of measure zero (Zemanian [1], pp. 8–9). Such distributions are called *regular*.

If $f \in \mathscr{D}'(I)$ cannot be related to a locally integrable function on I in this way, it is called *singular*. Examples of singular distributions are the delta functional (see Example 1.8-1) and its derivatives (see Example 1.10-1). Still other examples of singular distributions are discussed in Zemanian [1], Secs. 1.4 and 2.5.

As is customary, we shall also use the notation $f(t)$ for both regular and singular distributions. This does not mean that f is a function of t. It is merely a convenient way of indicating that the testing functions on which f is defined have t as their independent variable.

The support of a distribution on I is defined as follows. First of all, $f \in \mathscr{D}'(I)$ is said to be zero on an open subset J of I if $\langle f, \phi \rangle = 0$ for every $\phi \in \mathscr{D}(J)$. The union N of all open subsets of I on which f is zero is called the *null set of* f. N is also an open subset of I since it is the union of open sets (Williamson [1], p. 9). Moreover, f is zero on N as well; this fact requires a fairly substantial proof (see Schwartz [1], Vol. I, pp. 26–28, or Zemanian [1], Sec. 1.8). The complement $I \backslash N$ of N with respect to I is called the *support of* f and denoted by *supp* f. Thus, the support of $f \in \mathscr{D}'(I)$ is the smallest closed (with respect to I) subset S of I for which f is zero on $I \backslash S$. If a subset T of I contains the support of $f \in \mathscr{D}'(I)$, f is said to be *concentrated on* T.

Note that the support of $f \in \mathscr{D}'(I)$ is a closed set with respect to I but need not be closed with respect to $\mathscr{R}^n$. For example, if $I = \{t: |t| < 1\}$, then the function, $f(t) \equiv 1$ for $|t| < 1$, defines a regular distribution in $\mathscr{D}'(I)$. The support of f is all of I. It is closed with respect to I since I is now the universal set. On the other hand, it is not closed with respect to $\mathscr{R}^n$.

Another property of distributions is the following: *Let* $\{J_m\}$ *be a collection of open sets, and let* I *be the union of all the* J_m. *On each* J_m *let there be defined an* $f_m \in \mathscr{D}'(J_m)$. *Finally, whenever two sets* J_m *and* J_p *have a nonvoid intersection, let the restrictions of* f_m *and* f_p *to* $\mathscr{D}(J_m \cap J_p)$ *be equal. Then, there is one and only one* $f \in \mathscr{D}'(I)$ *whose restriction to each* $\mathscr{D}(J_m)$ *is equal to* f_m. A proof of this is given in Zemanian [1], Sec. 1.8 in the case where $I = \mathscr{R}^n$. Essentially the same proof can be used when $I \neq \mathscr{R}^n$. See Schwartz [1], Vol I, pp. 26–28, or Friedman [1], p. 59.

A consquence of this property is that, for $f \in \mathscr{D}'(I)$ and $\phi \in \mathscr{D}(I)$, $\langle f, \phi \rangle$ depends only on the values that ϕ assumes on any neighborhood of *supp* f (Zemanian [1], Sec. 1.8). Thus, if θ is any smooth function on I that is identical to ϕ on some neighborhood of *supp* f, then we are free to define the number $\langle f, \theta \rangle$ by $\langle f, \theta \rangle \triangleq \langle f, \phi \rangle$.

PROBLEM 2.2-1. For the case where $t \in \mathscr{R}^1$, $I = \mathscr{R}^1$, and $K = \{t: |t| \leq a\}$, show that every $\gamma_k(k = 0, 1, 2, \ldots)$ defined by (1) is a norm on $\mathscr{D}_K$ and that for every $\phi(t) \not\equiv 0$ on I

$$0 < \gamma_0(\phi) \leq 2a\gamma_1(\phi) \leq (2a)^2\gamma_2(\phi) \leq \cdots$$

Then, simplify the conclusion of Theorem 2.2-1.

PROBLEM 2.2-2. Let I be an arbitrary nonvoid open set in $\mathscr{R}^n$. Show that there exists a sequence $\{K_m\}_{m=1}^{\infty}$ of compact subsets of I such that $K_1 \subset K_2 \subset K_3 \subset \cdots$ and each compact subset is contained in one of the K_m. *Hint*: A closed rational interval in $\mathscr{R}^n$ is a set of the form $\{t\colon a_\nu \leq t_\nu \leq b_\nu, \nu = 1, \ldots, n\}$ where the t_ν are the components of t, and a_ν and b_ν are finite rational numbers in $\mathscr{R}^1$. I is the union of the interiors of all closed rational intervals contained in I. Also, if a compact set is contained in the union of a collection of open sets, it is also contained in the union of a finite subset of the collection. Use these facts.

PROBLEM 2.2-3. Let ϕ be a smooth function, and let

$$\xi_{\theta,k}(\phi) \triangleq \sup_{t \in R^n} |\theta(t) D^k \phi(t)|$$

where θ is any smooth function on $\mathscr{R}^n$ and k is any nonnegative integer in $\mathscr{R}^n$. Let $\hat{\mathscr{D}}$ be the linear space of all smooth functions ϕ on $\mathscr{R}^n$ for which $\xi_{\theta,k}(\phi) < \infty$ for every permissible θ and k. Show that the collection S of all $\xi_{\theta,k}$ is a multinorm on $\hat{\mathscr{D}}$. Assign to $\hat{\mathscr{D}}$ the topology generated by S. Then, show that $\mathscr{D}$ and $\hat{\mathscr{D}}$ have precisely the same members and that a sequence $\{\phi_\nu\}$ converges in $\mathscr{D}$ to ϕ if and only if it converges in $\hat{\mathscr{D}}$ to the same limit ϕ.

2.3. The Space $\mathscr{E}(I)$ and Its Dual; Distributions of Compact Support

As before, t is a variable point in $\mathscr{R}^n$, and I is an open subset of $\mathscr{R}^n$. $\mathscr{E}(I)$ is the space of all complex-valued smooth functions on I. When $I = \mathscr{R}^n$, we also denote $\mathscr{E}(I)$ by simply $\mathscr{E}$. There is no restriction imposed on the rate of growth of any function $\phi(t) \in \mathscr{E}(I)$ as $t \in I$ approaches the boundary of I. $\mathscr{E}(I)$ is a linear space under the usual definitions. Moreover, $\mathscr{D}(I)$ is contained in $\mathscr{E}(I)$.

$\mathscr{E}(I)$ is always taken to be a multinormed space by assigning to it the following topology: For each compact subset K of I and each nonnegative integer $k \in \mathscr{R}^n$, let $\gamma_{K,k}$ be the seminorm on $\mathscr{E}(I)$ defined by

$$\gamma_{K,k}(\phi) \triangleq \sup_{t \in K} |D^k \phi(t)| \qquad \phi \in \mathscr{E}(I)$$

For every $\phi \in \mathscr{E}(I)$ other than the identically zero function on I there is at least one $\gamma_{K,0}$ such that $\gamma_{K,0}(\phi) \neq 0$. Consequently, the collection R of all $\gamma_{K,k}$ is separating and therefore is a multinorm on $\mathscr{E}(I)$, even though no single $\gamma_{K,k}$ is a norm. The topology of $\mathscr{E}(I)$ is that generated by R. $\mathscr{E}(I)$ is also complete, the proof of this being quite similar to the corresponding proof for $\mathscr{D}_K(I)$.

Actually, the same topology is also generated in $\mathscr{E}(I)$ by a countable subset of the multinorm R, as we shall now show. Let $\{K_m\}_{m=1}^{\infty}$ be a sequence of compact subsets of I such that $K_1 \subset K_2 \subset K_3 \subset \cdots$ and each compact subset of I is contained in one of the K_m. The topology that is generated by the countable multinorm $S = \{\gamma_{K_m, k}\}_{m, k}$ is the same as that generated by R. Indeed, the former is obviously weaker than the latter. Conversely, for any compact subset K of I, there exists a K_m such that $K \subset K_m$. Then, for each $\phi \in \mathscr{E}(I)$ we have that $\gamma_{K,k}(\phi) \leq \gamma_{K_m,k}(\phi)$. Hence, by Lemma 1.6-3 the topology due to R is weaker than the topology due to S. Thus, our assertion is established. This proves that $\mathscr{E}(I)$ is a countably multinormed space, and indeed a Fréchet space in view of its completeness.

A sequence $\{\phi_\nu\}_{\nu=1}^{\infty}$ converges in $\mathscr{E}(I)$ to ϕ if and only if all ϕ_ν and ϕ are in $\mathscr{E}(I)$ and, for each nonnegative integer $k \in \mathscr{R}^n$, $\{D^k\phi_\nu\}_{\nu=1}^{\infty}$ converges to $D^k\phi$ uniformly on every compact subset K of I. This concept of convergence is clearly weaker than that for $\mathscr{D}(I)$; that is, if a sequence converges in $\mathscr{D}(I)$, it certainly converges in $\mathscr{E}(I)$ to the same limit.

Moreover, $\mathscr{D}(I)$ is dense in $\mathscr{E}(I)$. To show this, let K be any compact subset of I. Choose an open set J such that $K \subset J \subset \bar{J} \subset I$, where $\bar{J}$ denotes the closure of J. (Prove that this can be done.) According to Zemanian [1], Sec. 1.8, Lemma 1, there exists a smooth function $\lambda(t)$ defined on I such that $\lambda(t) \equiv 1$ on K and $\lambda(t) \equiv 0$ outside J. Now, for any $\phi \in \mathscr{E}(I)$ and any k, we have $\gamma_{K,k}(\lambda\phi) = \lambda_{K,k}(\phi)$. Since $\lambda\phi \in \mathscr{D}(I)$, it readily follows that every neighborhood of ϕ in the $\mathscr{E}(I)$ topology contains a member of $\mathscr{D}(I)$. Indeed, any such neighborhood contains a balloon of the form

$$\{\psi: \psi \in \mathscr{E}(I), \gamma_{K_\nu, k_\nu}(\psi - \phi) \leq \varepsilon_\nu > 0, \nu = 1, \ldots, m\}$$

where the k_ν are nonnegative integers in $\mathscr{R}^n$ and the K_ν are not necessarily distinct compact subsets of I. Setting $K = \bigcup_{\nu=1}^{m} K_\nu$ and constructing λ as above, we find that $\lambda\phi$ is a member of the balloon. This verifies our assertion.

$\mathscr{E}'(I)$ is the dual of $\mathscr{E}(I)$. When $I = \mathscr{R}^n$, we usually denote $\mathscr{E}'(I)$ by $\mathscr{E}'$. According to Theorem 1.8-3, $\mathscr{E}'(I)$ is complete. Also, by Theorem 1.9-1, $\mathscr{E}'(I)$ is a subspace of $\mathscr{D}'(I)$.

An informative characterization of the distributions in $\mathscr{E}'(I)$ is the following.

THEOREM 2.3-1. *Let $f \in \mathscr{D}'(I)$. A necessary and sufficient condition for f to be in $\mathscr{E}'(I)$ is that the support of f be a compact subset of I.*

PROOF. *Sufficiency*: Let the support of $f \in \mathscr{D}'(I)$ be a compact subset of I. Choose $\lambda(t) \in \mathscr{D}(I)$ such that $\lambda(t) \equiv 1$ on a neighborhood J of *supp* f where $\bar{J} \subset I$. Then, by the last paragraph of the preceding section, f can be defined on $\mathscr{E}(I)$ by $\langle f, \phi \rangle \triangleq \langle f, \lambda\phi \rangle$ for any $\phi \in \mathscr{E}(I)$. Clearly, f is linear on $\mathscr{E}(I)$. That it is continuous on $\mathscr{E}(I)$ follows from the fact that $\lambda\phi_\nu \to 0$ in $\mathscr{D}(I)$ whenever $\phi_\nu \to 0$ in $\mathscr{E}(I)$. Hence, $f \in \mathscr{E}'(I)$.

Necessity: Let $\{K_m\}_{m=1}^{\infty}$ be a sequence of compact subsets of I such that $K_1 \subset K_2 \subset K_3 \subset \cdots$ and every compact subset of I is contained in one of the K_m. Now, suppose that f is not of compact support. Then, for every m, there exists a $\phi_m \in \mathscr{D}(I)$ that vanishes on a neighborhood of K_m for which $\langle f, \phi_m \rangle \neq 0$. Set $\theta_m = \phi_m / \langle f, \phi_m \rangle$. Hence, $\langle f, \theta_m \rangle = 1$ for every m. Now, the sequence $\{\theta_m\}$ converges in $\mathscr{E}(I)$ to zero because every compact subset of I intersects only a finite number of the sets $I \backslash K_m$; here, $I \backslash K_m$ denotes the complement of K_m with respect to I. Since $f \in \mathscr{E}'(I)$, $\langle f, \theta_m \rangle \to 0$ as $m \to \infty$. This contradiction completes the proof.

PROBLEM 2.3-1. Show that $\mathscr{E}(I)$ is complete.

PROBLEM 2.3-2. Given a compact subset K of an open set I in $\mathscr{R}^n$, show that there exists an open set J such that $K \subset J \subset \bar{J} \subset I$.

PROBLEM 2.3-3. (*a*) Let $\mathscr{Q}_{R,n}$ denote the space of smooth complex-valued functions on $\mathscr{R}^1$ whose supports are bounded on the right at the fixed positive integer n (i.e., *supp* ϕ(t) is contained in the interval $-\infty < t \leq n$). Assign to $\mathscr{Q}_{R,n}$ the topology generated by the collection of seminorms:

$$\mu_{m,k}(\phi) \triangleq \sup_{-m<t<\infty} |D^k\phi(t)| \qquad m = 0, 1, 2, \ldots; k = 0, 1, 2, \ldots$$

Show that $\mathscr{Q}_{R,n}$ is a complete countably multinormed space.

(*b*) Show that $\mathscr{Q}_R \triangleq \bigcup_{n=1}^{\infty} \mathscr{Q}_{R,n}$ is a complete strict countable-union space when the natural convergence rule is assigned to it.

(*c*) Also, show that $\mathscr{D}$ is dense in $\mathscr{Q}_R$.

(*d*) Next, let $\mathscr{Q}_R'$ be the dual of $\mathscr{Q}_R$. Prove that $\mathscr{Q}_R'$ consists precisely of those distributions whose supports are bounded on the left. (In Zemanian [1] we denoted this space by $\mathscr{D}_R'$.)

2.4. Generalized Functions

The purpose of this section is to point out the distinction we shall make in this book between the concepts of a distribution and a generalized function. We start by defining a generalized function. I is now allowed to be an open set in either $\mathscr{R}^n$ or $\mathscr{C}^n$, where $\mathscr{C}^n$ is the complex n-dimensional euclidean space. A set $\mathscr{V}(I)$ is said to be a *testing-function space* (*on* I) if the following three conditions are satisfied:

1. $\mathscr{V}(I)$ consists entirely of smooth complex-valued functions defined on I.

2. $\mathscr{V}(I)$ is either a complete countably multinormed space or a complete countable-union space.

3. If $\{\phi_\nu\}_{\nu=1}^{\infty}$ converges in $\mathscr{V}(I)$ to zero, then, for every nonnegative integer $k \in \mathscr{R}^n$, $\{D^k\phi_\nu\}_{\nu=1}^{\infty}$ converges to the zero function uniformly on every compact subset of I.

A *generalized function on I* (or simply *generalized function*) is any continuous linear functional on any testing-function space on I. In other words, f is called a generalized function if it is a member of the dual $\mathscr{V}'(I)$ of some testing-function space $\mathscr{V}(I)$. We shall also use the term "generalized function" to designate a functional whose domain contains (but is larger than) a testing-function space $\mathscr{V}(I)$ and whose restriction to $\mathscr{V}(I)$ is in $\mathscr{V}'(I)$.

Observe that condition *2* and Theorems 1.8-3 and 1.9-2 imply that $\mathscr{V}'(I)$ is complete too.

Henceforth, we shall always denote the zero element in either a testing-function space or its dual by 0 rather than $\varnothing$.

Note that the spaces $\mathscr{D}_K(I)$, $\mathscr{D}(I)$, and $\mathscr{E}(I)$ satisfy conditions *1*, *2*, and *3*. Thus, the elements of $\mathscr{D}_K'(I)$, $\mathscr{D}'(I)$, and $\mathscr{E}'(I)$ are all generalized functions.

In contrast to the phrase "generalized function," the name "distribution" will be reserved for the members of $\mathscr{D}'(I)$ (where I is now any open set in $\mathscr{R}^n$) or for the members of any space of generalized functions, such as $\mathscr{E}'(I)$, which can be identified in a one-to-one fashion with a subspace of $\mathscr{D}'(I)$, the identification being that between any generalized function and its restriction to $\mathscr{D}(I)$. Thus, every distribution is a generalized function but not conversely.

Because of this convention, the members of $\mathscr{D}_K'(I)$ will be called generalized functions but not distributions because $\mathscr{D}_K(I)$ does not contain $\mathscr{D}(I)$. Moreover, there is no one-to-one correspondence between $\mathscr{D}'(I)$ and a subset of $\mathscr{D}_K'(I)$; for example, if $I = \mathscr{R}^1$ and K is the interval $1 \le t \le 2$, then both $\delta(t)$ and the zero member of $\mathscr{D}'$ have the same restriction to $\mathscr{D}_K$.

In some of the papers (Zemanian [2]–[6], [9], [12]) on which this book is based, we used the word "distribution" in a somewhat more general sense. In particular, any generalized function which possessed a restriction to $\mathscr{D}(I)$ that was in $\mathscr{D}'(I)$ was also called a distribution, even though there did not exist the aforementioned one-to-one correspondence. Apparently, we will conform more to current practice if we restrict the use of the name "distribution" as explained above.

Note that, when I is $\mathscr{R}^n$ or an open subset of $\mathscr{R}^n$, any testing-function space $\mathscr{V}(I)$ is contained in $\mathscr{E}(I)$, and, because of condition *3* above, the convergence to zero in $\mathscr{V}(I)$ of any sequence implies its convergence to zero in $\mathscr{E}(I)$. It follows that the restriction of any $f \in \mathscr{E}'(I)$ to $\mathscr{V}(I)$ is a member of $\mathscr{V}'(I)$.

Let $f(t)$ be a conventional function on I $(I \subset \mathscr{R}^n)$ such that, for every $\phi \in \mathscr{V}(I)$,

$$\int_I f(t)\phi(t)\, dt \tag{1}$$

converges in the Lebesgue sense and, for every sequence $\{\phi_\nu\}$ which converges in $\mathscr{V}(I)$ to zero,

$$\int_I f(t)\phi_\nu(t)\, dt \to 0 \qquad \nu \to \infty$$

Then, by setting $\langle f, \phi\rangle$ equal to (1), we define a member f of $\mathscr{V}'(I)$, which we shall refer to either as a *regular generalized function in* $\mathscr{V}'(I)$ or as a *function in* $\mathscr{V}'(I)$. $f \in \mathscr{V}'(I)$ is called *singular* if it is not regular.

EXAMPLE 2.4-1. As an illustration of these ideas we shall construct a space of generalized functions on I whose restrictions to $\mathscr{D}(I)$ are in $\mathscr{D}'(I)$, but which cannot be identified with a subspace of $\mathscr{D}'(I)$ because there will be two different generalized functions having the same restriction to $\mathscr{D}(I)$.

Let I be the finite open interval $0 < t < 1$ in $\mathscr{R}^1$. $\mathscr{B}(I)$ is the space of all complex-valued smooth functions $\phi(t)$ on I such that

$$\beta_k(\phi) \triangleq \sup_{t \in I} |D^k\phi(t)| < \infty \qquad k = 0, 1, 2, \ldots$$

We assign to $\mathscr{B}(I)$ the topology generated by the multinorm $\{\beta_k\}_{k=0}^{\infty}$ and obtain thereby a testing-function space which contains $\mathscr{D}(I)$. However, $\mathscr{D}(I)$ is not dense in $\mathscr{B}(I)$. For example, set $\phi(t) \equiv 1$ on I. Then, ϕ is in $\mathscr{B}(I)$, but it cannot be the limit in $\mathscr{B}(I)$ of a sequence of functions in $\mathscr{D}(I)$ because $\lim_{t\to 0+}\psi(t) = 0$ for every $\psi \in \mathscr{D}(I)$.

For every $\phi \in \mathscr{B}(I)$, $\lim_{t\to 0+}\phi(t)$ and $\lim_{t\to 1-}\phi(t)$ both exist. Indeed,

$$\phi(t) = \int_{\frac{1}{2}}^{t} D\phi(x)\, dx + \phi(\tfrac{1}{2}) \qquad 0 < t < 1$$

and the right-hand side tends to a limit as either $t \to 0+$ or $t \to 1-$ because $D\phi(t)$ is bounded and continuous on $0 < t < 1$.

Define f as a functional on $\mathscr{B}(I)$ by $\langle f, \phi\rangle \triangleq \lim_{t\to 0+}\phi(t)$. Clearly, f is a nonzero member of the dual space $\mathscr{B}'(I)$ of $\mathscr{B}(I)$. Also, the restriction of f to $\mathscr{D}(I)$ is the zero element in $\mathscr{D}'(I)$. Thus, there exist two distinct members of $\mathscr{B}'(I)$, namely, f and the zero element in $\mathscr{B}'(I)$, whose restrictions to $\mathscr{D}(I)$ are the same, the zero distribution on I.

EXAMPLE 2.4-2. Let I now denote the entire real line $\mathscr{R}^1$, and let $\mathscr{B}$ be the testing-function space $\mathscr{B}(I)$, which is defined as in the preceding example. Here again, $\mathscr{D}$ is contained in $\mathscr{B}$ but is not dense in $\mathscr{B}$. The dual $\mathscr{B}'$ of $\mathscr{B}$ is a space of generalized functions on $\mathscr{R}^1$. We cannot now use the argument of the preceding example to construct a nonzero member of $\mathscr{B}'$ whose restriction to $\mathscr{D}$ is the zero distribution on $\mathscr{R}^1$. This is because neither $\lim_{t\to-\infty}\phi(t)$ nor $\lim_{t\to\infty}\phi(t)$ will exist for every member of $\mathscr{B}$. (For example, set $\phi(t) = \sin t$.) However, we can prove the existence of such a member of $\mathscr{B}'$ through an argument suggested by Vaclav Dolezal.

Let $\mathscr{B}_L$ denote that subset of $\mathscr{B}$ for which $\lim_{t\to\infty}\phi(t)$ exists. $\mathscr{B}_L$ is a linear space. Define the linear functional f on $\mathscr{B}_L$ by

$$\langle f, \phi\rangle \triangleq \lim_{t\to\infty} \phi(t) \qquad \phi \in \mathscr{B}_L$$

Then, for every $\phi \in \mathscr{B}_L$

$$|\langle f, \phi\rangle| \leq \beta_0(\phi) \triangleq \sup_{-\infty<t<\infty} |\phi(t)|.$$

Note that β_0 is a norm on $\mathscr{B}$. By the Hahn-Banach theorem (Robertson and Robertson [1], p. 29) there exists a linear functional f_1 on $\mathscr{B}$ such that $\langle f_1, \phi\rangle = \langle f, \phi\rangle$ for each $\phi \in \mathscr{B}_L$ and $|\langle f_1, \phi\rangle| \leq \beta_0(\phi)$ for every $\phi \in \mathscr{B}$. Hence, f_1 is a continuous linear functional on $\mathscr{B}$.

Since $\mathscr{D} \subset \mathscr{B}_L$, the restriction of f_1 to $\mathscr{D}$ is the zero distribution because $\langle f_1, \phi\rangle = \lim_{t\to\infty}\phi(t) = 0$ for every $\phi \in \mathscr{D}$. Thus, there exist two distinct members of $\mathscr{B}'$, namely, f_1 and the zero element in $\mathscr{B}'$, whose restrictions to $\mathscr{D}$ are the same, the zero distribution. This shows that the generalized-function space $\mathscr{B}'$ cannot be identified as a subspace of $\mathscr{D}'$.

PROBLEM 2.4-1. Let I be an open set in $\mathscr{C}^1$, and let $\mathscr{H}(I)$ be the space of all analytic functions on I. For each compact subset K of I define the seminorm η_K on $\mathscr{H}(I)$ by

$$\eta_K(\phi) \triangleq \sup_{s\in K} |\phi(s)| \qquad \phi \in \mathscr{H}(I)$$

Assign to $\mathscr{H}(I)$ the topology generated by $\{\eta_K\}$ where K varies through all compact subsets of I. Show that $\mathscr{H}(I)$ is a testing-function space.

The delta functional $\delta(s-a)$ concentrated on $a \in I$ is defined by

$$\langle \delta(s-a), \phi(s)\rangle \triangleq \phi(a)$$

and is a member of the dual space $\mathscr{H}'(I)$. Find an integral respresentation for $\delta(s-a)$.

2.5. Linear Partial Differential Operators Acting on Generalized Functions

In this section we discuss a type of operator that may be applied to generalized functions under certain conditions. The operators in question are defined as the adjoints of certain linear partial differential operators acting on testing-function spaces.

Let I be some open set in either $\mathscr{R}^n$ or $\mathscr{C}^n$, let $t = \{t_1, t_2, \cdots, t_n\} \in I$, and let $\theta_\nu(t)(\nu = 0, 1, \ldots, m)$ denote complex-valued smooth functions on I. Consider the linear partial differential operator

$$\mathfrak{N} = (-1)^{|k|}\theta_0 D^{k_1}\theta_1 D^{k_2} \cdots \theta_{m-1} D^{k_m}\theta_m \tag{1}$$

where the k_ν now denote nonnegative integers in $\mathscr{R}^n$ and $|k|$ is the one-dimensional integer $|k_1| + |k_2| + \cdots + |k_m|$. The *order of* $\mathfrak{N}$ is $|k|$. (The symbol (1) denotes the sequence of operations: multiply by θ_m, differentiate according to D^{k_m}, multiply by θ_{m-1}, etc. Moreover, when I is an open set in $\mathscr{C}^n$, D has the customary definition wherein the limit is required to be independent of the direction in which the complex increment goes to zero.) Finally, let $\mathscr{U}(I)$ and $\mathscr{V}(I)$ be testing-function spaces on I.

If $\mathfrak{N}$ is a continuous linear mapping of $\mathscr{U}(I)$ into $\mathscr{V}(I)$, we define the adjoint $\mathfrak{N}'$ of $\mathfrak{N}$ as an operator on $\mathscr{V}'(I)$ by

$$\langle \mathfrak{N}'f, \phi\rangle \triangleq \langle f, \mathfrak{N}\phi\rangle \qquad \phi \in \mathscr{U}(I), f \in \mathscr{V}'(I) \tag{2}$$

According to Theorem 1.10-1, $\mathfrak{N}'$ is a continuous linear mapping of $\mathscr{V}'(I)$ into $\mathscr{U}'(I)$. Note that the order of differentiations in each D^{k_ν} ($|k_\nu| > 1$) can be changed in any fashion without altering $\mathfrak{N}\phi$ because ϕ and the θ_ν are smooth functions. Therefore, $\mathfrak{N}'f$ also does not depend on this order. For this reason we never specify what the order of differentiations in D^{k_ν} is. On the other hand, $\mathfrak{N}\phi$ and therefore $\mathfrak{N}'f$ do depend on the order in which the multiplications by θ_ν and the differential operators D^{k_ν} are applied.

When I is an open set in $\mathscr{R}^n$ and when f is a smooth function whose support is a compact subset of I (i.e., $f \in \mathscr{D}(I)$), f and each of its derivatives define regular distributions in $\mathscr{E}'(I)$. As was noted in the preceding section, the restrictions of the members of $\mathscr{E}'(I)$ to $\mathscr{V}(I)$ are in $\mathscr{V}'(I)$. Thus, for $f \in \mathscr{D}(I)$ and $\phi \in \mathscr{U}(I)$, (2) can be written as

$$\langle \mathfrak{N}'f, \phi\rangle = \langle f, \mathfrak{N}\phi\rangle = \int_I f(t)\mathfrak{N}\phi(t)\, dt$$

By successive integrations by parts, this becomes

$$\int_I \phi[\theta_m D^{k_m} \cdots \theta_1 D^{k_1}\theta_0 f]\, dt$$

Thus, in this case we can identify $\mathfrak{N}'$ as the operator

$$\theta_m D^{k_m} \cdots \theta_1 D^{k_1}\theta_0 \tag{3}$$

where D denotes conventional differentiation. The symbolism (3) for $\mathfrak{N}'$ will still be used when f is an arbitrary member of $\mathscr{V}'(I)$ and even when I

is an open set in $\mathscr{C}^n$, but in this case it is understood that D denotes *generalized differentiation* specified implicitly by (2). It should be emphasized that a conventional derivative and a generalized derivative need not be the same thing, as is illustrated by Example 2.5-1 below. (See also Zemanian [1], Example 2.4-1.) In this book, whenever linear differential operators, such as (3), act on generalized functions, it is always understood that they are generalized operators defined by (2). On the other hand, when they act on testing functions, the operations are taken in the conventional sense.

EXAMPLE 2.5-1. Let I be the open interval $0 < t < 1$ ($t \in \mathscr{R}^1$), and let $\mathscr{B}(I)$ be the testing-function space defined in Example 2.4-1. In this case differentiation is a continuous linear mapping of $\mathscr{B}(I)$ into $\mathscr{B}(I)$. Therefore, the (generalized) derivative Df of any $f \in \mathscr{B}'(I)$ is defined by

$$\langle Df, \phi\rangle \triangleq \langle f, -D\phi\rangle \qquad \phi \in \mathscr{B}(I)$$

Now, let f be defined by

$$\langle f, \phi\rangle \triangleq \int_0^1 \phi(t)\, dt$$

Thus, f is the regular generalized function in $\mathscr{B}'(I)$ corresponding to the conventional function $f(t) \equiv 1 (0 < t < 1)$. The conventional derivative of this function is equal to zero everywhere on I and therefore generates the zero element in $\mathscr{B}'(I)$. On the other hand, the generalized derivative is not the zero element in $\mathscr{B}'(I)$. Instead, we have for every $\phi \in \mathscr{B}(I)$

$$\langle Df, \phi\rangle = \langle f, -D\phi\rangle = -\int_0^1 D\phi(t)\, dt$$

$$= \lim_{t\to 0+} \phi(t) - \lim_{t \to 1-} \phi(t)$$

As a special case of our definition (2), we have the operation of multiplication by certain smooth functions. In particular, if θ is a smooth function on I and if $\phi \mapsto \theta\phi$ is a continuous linear mapping of $\mathscr{U}(I)$ into $\mathscr{V}(I)$, then $f \mapsto \theta f$ is a continuous linear mapping of $\mathscr{V}'(I)$ into $\mathscr{U}'(I)$ defined by

$$\text{(4)} \qquad \langle \theta f, \phi\rangle \triangleq \langle f, \theta\phi\rangle \qquad \phi \in \mathscr{U}(I), f \in \mathscr{V}'(I)$$

When $\mathscr{U}(I)$ and $\mathscr{V}(I)$ happen to be the same testing-function space, θ is called a *multiplier for* $\mathscr{U}(I)$. For example, every smooth function on I is a multiplier for $\mathscr{D}(I)$. (Show this.)

PROBLEM 2.5-1. Assume that the first-order differentiation $\partial/\partial t_\nu$ and multiplication by the smooth function θ are both continuous linear mappings of the testing-function space $\mathscr{V}(I)$ into itself. Show that for $f \in \mathscr{V}'(I)$

Liebniz's rule for the differentiation of a product:

$$\frac{\partial}{\partial t_\nu}(\theta f) = \theta \frac{\partial f}{\partial t_\nu} + f \frac{\partial \theta}{\partial t_\nu}$$

is valid.

PROBLEM 2.5-2. Prove the following two assertions. Every smooth function $\theta(\tau)$ on the open set I is a multiplier for $\mathscr{D}(I)$. If, in addition, $\theta(\tau)$ is never equal to zero on I, then the mapping $\phi \leftrightarrow \theta\phi$ is an automorphism on $\mathscr{D}(I)$.

PROBLEM 2.5-3. Let $\delta(s-a)$ be the delta functional defined in Problem 2.4-1. Show that its kth-order generalized derivative $D^k\delta(s-a)$ is a member of $\mathscr{H}'(I)$. Also, find an integral representation for $D^k\delta(s-a)$.

2.6. Generalized Functions That Depend upon a Parameter and Parametric Differentiation

Let $\tau \in \mathscr{R}^1$ be a parameter restricted to some open set J in $\mathscr{R}^1$, let $t \in \mathscr{R}^n$ be restricted to the open set I in $\mathscr{R}^n$, and, as before, let $\mathscr{V}(I)$ be a testing-function space. We say that $f_\tau(t)$ is a *generalized function in* $\mathscr{V}'(I)$ *depending on the parameter* τ when, for each fixed value of τ, $f_\tau(t)$ is a member of $\mathscr{V}'(I)$. Thus, for each $\phi(t) \in \mathscr{V}(I)$, $\langle f_\tau(t), \phi(t)\rangle$ is a conventional function of τ. We also say that $f_\tau(t)$ is *differentiable with respect to* τ *at the fixed point* τ_0 if, for every $\phi(t) \in \mathscr{V}(I)$, $\langle f_\tau(t), \phi(t)\rangle$ possesses a conventional derivative at τ_0. This requires that

$$(1) \qquad \lim_{\Delta\tau \to 0} \frac{1}{\Delta\tau}[\langle f_{\tau_0+\Delta\tau}(t), \phi(t)\rangle - \langle f_{\tau_0}(t), \phi(t)\rangle]$$

exist. Moreover, we say that $f_\tau(t)$ is *differentiable on the open set* J if (1) exists for every $\tau_0 \in J$. After replacing τ_0 by τ, we denote (1) by $\langle D_\tau f_\tau(t), \phi(t)\rangle$, and this defines the *parametric* derivative $D_\tau f_\tau(t)$ as a functional on $\mathscr{V}(I)$. In symbols,

$$(2) \qquad \langle D_\tau f_\tau(t), \phi(t)\rangle \triangleq D_\tau\langle f_\tau(t), \phi(t)\rangle$$

Furthermore, it is a fact that $D_\tau f_\tau \in \mathscr{V}'(I)$. Indeed, for $\Delta\tau \neq 0$, $(\Delta\tau)^{-1}(f_{\tau+\Delta\tau} - f_\tau)$ is a member of $\mathscr{V}'(I)$ and, according to our definition, is a Cauchy directed set as $\Delta\tau \to 0$. Our assertion is therefore a consequence of the completeness of $\mathscr{V}'(I)$.

The higher order parametric derivatives $D_\tau^k f_\tau$ of f_τ, as well as the parametric partial derivatives of f_τ in the case where $\tau \in \mathscr{R}^n$, are defined in the same way by repeating the above definition for each successive (first order) differentiation. Thus, all such derivatives, if they exist, are also members of $\mathscr{V}'(I)$.

A sense can also be assigned to the integral of f_τ with respect to τ. This situation is discussed, for example, in Zemanian [1], Sec. 2.8, in the case where $\mathscr{V}(I) = \mathscr{D}(I)$. However, we won't need this concept in this book.

PROBLEM 2.6-1. Let $f_\tau(t)$ be a regular generalized function in $\mathscr{V}'(I)$ $(I \subset \mathscr{R}^1)$ depending on the parameter $\tau \in J$, where J is a fixed open interval in $\mathscr{R}^1$. Assume that $f_\tau(t)$ is a regular generalized function generated by the conventional function $h(\tau, t)$ which satisfies the following conditions. Both $h(\tau, t)$ and $D_\tau h(\tau, t)$ are continuous functions on the domain $\{(\tau, t)\colon \tau \in J, t \in I\}$. $\int_I h(\tau, t)\phi(t)\, dt$ converges pointwise on J and $\int_I D_\tau h(\tau, t)\phi(t)\, dt$ converges uniformly on J for every $\phi \in \mathscr{V}(I)$. Finally, the conventional derivative $D_\tau h(\tau, t)$ also generates a regular generalized function $g_\tau(t)$ in $\mathscr{V}'(I)$. Show that, in this case, $D_\tau f_\tau(t) = g_\tau(t)$ in the sense of equality in $\mathscr{V}'(I)$. In other words, parametric differentiation and conventional differentiation agree under these circumstances.

2.7. Generalized Functions That Are Concentrated on Compact Sets

We noted in Sec. 2.4 that, when I is a subset of $\mathscr{R}^n$, the restriction of any $f \in \mathscr{E}'(I)$ to the testing-function space $\mathscr{V}(I)$ is in $\mathscr{V}'(I)$. Under certain circumstances, $\mathscr{E}'(I)$ can be viewed as a subspace of $\mathscr{V}'(I)$, and a characterization of these elements in $\mathscr{V}'(I)$ similar to Theorem 2.3-1 can be obtained.

We shall say that $f \in \mathscr{V}'(I)$ is *concentrated on a subset* Ω *of* I whenever $\langle f, \phi\rangle = 0$ for every $\phi \in \mathscr{V}(I)$ that vanishes on a neighborhood of Ω. For example, the delta functional $\delta(t)$ is concentrated on every set that contains the origin. If f is a distribution, its support is the smallest closed set on which it is concentrated; see Sec. 2.2.

THEOREM 2.7-1. *Let I be an open set in $\mathscr{R}^n$, and let the testing-function space $\mathscr{V}(I)$ possess the following property: $\mathscr{D}(I)$ is contained in $\mathscr{V}(I)$, and convergence in $\mathscr{D}(I)$ to some limit implies convergence in $\mathscr{V}(I)$ to the same limit. Finally, let $f \in \mathscr{V}'(I)$.*

Then, a necessary and sufficient condition for f to be in $\mathscr{E}'(I)$ is that f be concentrated on a compact subset Ω of I.

PROOF. By our hypothesis on $\mathscr{V}(I)$ and the first requirement in the definition of a testing-function space (see Sec. 2.4), $\mathscr{D}(I) \subset \mathscr{V}(I) \subset \mathscr{E}(I)$. Since $\mathscr{D}(I)$ is dense in $\mathscr{E}(I)$, so too is $\mathscr{V}(I)$. Furthermore, by the third requirement in the definition of a testing-function space, convergence in $\mathscr{V}(I)$ to some limit implies convergence in $\mathscr{E}(I)$ to the same limit. Theorem 1.9-1 now shows that $\mathscr{E}'(I)$ is a subspace of $\mathscr{V}'(I)$ (more precisely, $\mathscr{E}'(I)$ can be identified with a subspace of $\mathscr{V}'(I)$ in a one-to-one fashion). Having established this preliminary result, we now turn to the proof of our necessary and sufficient condition.

Sufficiency: First note that, by the hypothesis on $\mathscr{V}(I)$, the restriction of f to $\mathscr{D}(I)$ is in $\mathscr{D}'(I)$.

Next, let $\lambda(t) \in \mathscr{D}(I)$ be such that $\lambda(t) \equiv 1$ on a neighborhood of Ω. λ is a multiplier for $\mathscr{V}(I)$. Indeed, $\phi \mapsto \lambda\phi$ is a linear mapping of $\mathscr{V}(I)$ into $\mathscr{D}(I) \subset \mathscr{V}(I)$. It is also continuous. For, by property *3* of a testing-function space (see Sec. 2.4), if $\phi_\nu \to 0$ in $\mathscr{V}(I)$ as $\nu \to \infty$, then $\lambda\phi_\nu \to 0$ in $\mathscr{D}(I)$ and therefore in $\mathscr{V}(I)$.

Next, define $\xi(t)$ on I by the condition $\xi(t) = 1 - \lambda(t)$ on I. Since 1 is certainly a multiplier for $\mathscr{V}(I)$, so too is ξ. For any $\phi \in \mathscr{V}(I)$, $\xi\phi$ vanishes on a neighborhood of Ω, and so $\langle f, \xi\phi \rangle = 0$ by the assumption that f is concentrated on Ω. Also,

$$\langle f, \phi \rangle = \langle f, (\lambda + \xi)\phi \rangle = \langle f, \lambda\phi \rangle + \langle f, \xi\phi \rangle = \langle f, \lambda\phi \rangle$$

Since $\lambda\phi \in \mathscr{D}(I)$, this shows that a knowledge of f on $\mathscr{D}(I)$ determines f on $\mathscr{V}(I)$. Thus, f is a distribution on I (i.e., $f \in \mathscr{D}'(I)$).

The support of $f \in \mathscr{D}'(I)$ is a bounded set since it is contained in the compact set Ω. Moreover, this support is a closed set (see Sec. 2.2), and therefore it too is a compact set. Theorem 2.3-1 now implies that f is in $\mathscr{E}'(I)$.

Necessity: This is proved exactly as is the necessity part of Theorem 2.3-1. The only change is that now the ϕ_m are in $\mathscr{V}(I)$ instead of in $\mathscr{D}(I)$.

CHAPTER III

The Two–Sided Laplace Transformation

3.1. Introduction

The *conventional two-sided Laplace transformation* is defined by

$$F(s) = \int_{-\infty}^{\infty} f(t) e^{-st} \, dt \tag{1}$$

where $f(t)$ is a suitably restricted conventional function on the real line $-\infty < t < \infty$. Thus, this transformation maps $f(t)$ into a function $F(s)$ of the complex variable s. The adjective *two-sided* designates that the integration in (1) is over the entire real line $-\infty < t < \infty$ and that the support of f is unrestricted (i.e., $f(t)$ need not be zero anywhere). On the other hand, f may be such that the lower (upper) limit in (1) can be replaced by a finite number, in which case this transformation is called *one-sided*, and also *right-sided* (*left-sided*).

There are a variety of methods for extending (1) to generalized functions, some of which are restricted to the one-sided Laplace transformation. (See Benedetto [1], Cooper [1], Dolezal [1], Garnir and Munster [1], Ishihara [1], Jones [1], Korevaar [1], Lavoine [1], Liverman [1], Miller [1], Myers [1], Rehberg [1], Schwartz [2], Weston [1], and Zemanian [1], [2].) The original method is due to Schwartz [2] and is based upon his definition of the generalized Fourier transformation. In particular, Schwartz defines the Laplace transform of a distribution f as the Fourier transform of $e^{-\sigma t} f(t)$, where the real number σ is restricted to the set Γ_f for which $e^{-\sigma t} f(t)$ is a tempered distribution.

In contrast to this, the method presented here defines the Laplace transform $F(s)$ of a generalized function f directly as the application of $f(t)$ to e^{-st}:

$$F(s) = \langle f(t), e^{-st} \rangle \tag{2}$$

This requires the construction of certain spaces of testing functions on $-\infty < t < \infty$ which contain e^{-st} for various values of the complex parameter s. If Γ_f^o is the interior of Γ_f and if Γ_f^o is not empty, then this definition gives the same results as Schwartz's definition for all s with $\sigma = \operatorname{Re} s \in \Gamma_f^o$. The present theory is independent of Schwartz's theory, although the two can be related (Zemanian [2]). The construction of our testing-function spaces requires a somewhat longer development as compared to Schwartz's approach, but subsequently a number of proofs and derivations of the properties of (2) become simplified.

In addition, our approach allows one to define the Mellin transform $G(s)$ of a certain type of generalized function $g(x)$ on $0 < x < \infty$ as

$$G(s) = \langle g(x), x^{s-1} \rangle$$

by simply applying to (2) the change of variables $t = -\log x$, $g(x) = f(t)$; this is discussed in the next chapter. Furthermore, the Weierstrass transformation can also be related to the Laplace transformation, as we shall see in Chapter VII.

In Secs. 3.2 through 3.10 we discuss the one-dimensional Laplace transformation wherein t and s are variables in $\mathscr{R}^1$ and $\mathscr{C}^1$ respectively. Sec. 3.9 is devoted to an application to the Cauchy problem for the wave equation in one-dimensional space. The one-sided Laplace transformation of generalized functions is taken up in Sec. 3.10; our discussion there is somewhat different but entirely equivalent to that given in Zemanian [1]. The n-dimensional Laplace transformation, in which t and s are now variables in $\mathscr{R}^n$ and $\mathscr{C}^n$ respectively, is defined and its most important properties summarized in Sec. 3.11. An application of this latter transformation to the inhomogeneous wave equation appears in Sec. 3.12.

3.2. The Testing-Function Spaces $\mathscr{L}_{a,b}$ and $\mathscr{L}(w, z)$ and Their Duals

Throughout this chapter we assume that $a, b, c, d, t \in \mathscr{R}^1$ and $s \in \mathscr{C}^1$, except in Sec. 3.11 where $\mathscr{R}^1$ and $\mathscr{C}^1$ are replaced by $\mathscr{R}^n$ and $\mathscr{C}^n$, respectively. Let $\kappa_{a,b}(t)$ be the function

$$\kappa_{a,b}(t) = \begin{cases} e^{at} & 0 \leq t < \infty \\ e^{bt} & -\infty < t < 0 \end{cases}$$

$\mathscr{L}_{a,b}$ denotes the space of all complex-valued smooth functions $\phi(t)$ on $-\infty < t < \infty$ on which the functionals γ_k defined by

$$(1) \qquad \gamma_k(\phi) \triangleq \gamma_{a,b,k}(\phi) \triangleq \sup_{-\infty < t < \infty} |\kappa_{a,b}(t) D^k \phi(t)| \qquad k = 0, 1, 2, \ldots$$

assume finite values. We will usually use the more concise notation γ_k for these functionals instead of $\gamma_{a,b,k}$. $\mathscr{L}_{a,b}$ is a linear space under the pointwise addition of functions and their multiplication by complex numbers. Each γ_k is clearly a seminorm on $\mathscr{L}_{a,b}$, and γ_0 is a norm. Consequently, $\{\gamma_k\}_{k=0}^{\infty}$ is a multinorm on $\mathscr{L}_{a,b}$. We assign to $\mathscr{L}_{a,b}$ the topology generated by $\{\gamma_k\}_{k=0}^{\infty}$, thereby making it a countably multinormed space.

Note that, for each fixed s, e^{-st} is a member of $\mathscr{L}_{a,b}$ if and only if s satisfies $a \leq \operatorname{Re} s \leq b$. On the other hand, for every positive integer k, $t^k e^{-st}$ is in $\mathscr{L}_{a,b}$ if and only if s satisfies $a < \operatorname{Re} s < b$.

LEMMA 3.2-1. *$\mathscr{L}_{a,b}$ is complete and therefore a Fréchet space.*

PROOF. We see from (1) that $\{\phi_\nu\}_{\nu=1}^{\infty}$ is a Cauchy sequence in $\mathscr{L}_{a,b}$ if and only if each ϕ_ν is in $\mathscr{L}_{a,b}$ and, for each fixed k, the functions $\kappa_{a,b}(t)D^k\phi_\nu(t)$ comprise a uniform Cauchy sequence on $-\infty < t < \infty$ as $\nu \to \infty$. In this case, it follows from a standard theorem (Apostol [1], p. 402) that there exists a smooth function $\phi(t)$ such that, for each k and t, $D^k\phi_\nu(t) \to D^k\phi(t)$ as $\nu \to \infty$. Moreover, given any $\varepsilon > 0$ there exists an N_k such that, for every $\nu, \mu > N_k$,

$$|\kappa_{a,b}(t)D^k[\phi_\nu(t) - \phi_\mu(t)]| < \varepsilon$$

for all t. Taking the limit as $\mu \to \infty$, we obtain

$$(2) \qquad |\kappa_{a,b}(t)D^k[\phi_\nu(t) - \phi(t)]| \leq \varepsilon \qquad -\infty < t < \infty, \ \nu > N_k$$

Thus, as $\nu \to \infty$, $\gamma_k(\phi_\nu - \phi) \to 0$ for each k.

Finally, because of the uniformity of the convergence and the fact that each $\kappa_{a,b}(t)D^k\phi_\nu(t)$ is bounded on $-\infty < t < \infty$, there exists a constant C_k not depending on ν such that $|\kappa_{a,b}(t)D^k\phi_\nu(t)| < C_k$ for all t. Therefore, (2) implies that $|\kappa_{a,b}(t)D^k\phi(t)| \leq C_k + \varepsilon$, which shows that the limit function ϕ is a member of $\mathscr{L}_{a,b}$. By Lemma 1.6-1, our proof is complete.

$\mathscr{L}'_{a,b}$ denotes the dual of $\mathscr{L}_{a,b}$. Thus, f is a member of $\mathscr{L}'_{a,b}$ if and only if it is a continuous linear functional on $\mathscr{L}_{a,b}$. By what we have already shown, $\mathscr{L}_{a,b}$ is a testing function space, and therefore $\mathscr{L}'_{a,b}$ is a space of generalized functions. (Note that $\kappa_{a,b}(t)$ has a positive infimum on every compact set, so that condition *3* of Sec. 2.4 is satisfied.) Under the usual definitions of equality, addition, and multiplication by a complex number (see Sec. 1.8), $\mathscr{L}'_{a,b}$ is a linear space. We assign to $\mathscr{L}'_{a,b}$ its customary (weak) topology. It follows from Theorem 1.8-3 that $\mathscr{L}'_{a,b}$ is also complete.

If $a \leq c$ and $d \leq b$, then $\mathscr{L}_{c,d} \subset \mathscr{L}_{a,b}$, and the topology of $\mathscr{L}_{c,d}$ is stronger than the topology induced on $\mathscr{L}_{c,d}$ by $\mathscr{L}_{a,b}$. To see this, first note that $0 < \kappa_{a,b}(t) \leq \kappa_{c,d}(t)$ on $-\infty < t < \infty$. Therefore,

$$|\kappa_{a,b}(t)D^k\phi(t)| \leq |\kappa_{c,d}(t)D^k\phi(t)|$$

so that $\gamma_{a,b,k}(\phi) \leq \gamma_{c,d,k}(\phi)$. Our assertion is implied by the last inequality and Lemma 1.6-3.

We can now conclude (see Sec. 1.8) that the restriction of any $f \in \mathscr{L}'_{a,b}$ to $\mathscr{L}_{c,d}$ is in $\mathscr{L}'_{c,d}$ (speaking more loosely, we simply say that $f \in \mathscr{L}'_{c,d}$). However, if $a < c$ or $d < b$, $\mathscr{L}'_{a,b}$ cannot be identified in a one-to-one fashion with a subspace of $\mathscr{L}'_{c,d}$. For instance, consider

EXAMPLE 3.2-1. Assume that $a < c$ and $d < b$. The argument of Example 2.4-2 can be modified to show the existence of two different members of $\mathscr{L}'_{a,b}$ having the same restriction to $\mathscr{L}_{c,d}$. For certain purposes that will arise subsequently, we shall construct here a somewhat more complicated example than we presently need.

Let $\mathscr{B}_{a,b}$ be the set of all functions $\phi(t)$ in $\mathscr{L}_{a,b}$ for which both $\lim_{t\to\infty} e^{at}\phi(t)$ and $\lim_{t\to-\infty} e^{bt}\phi(t)$ exist. $\mathscr{B}_{a,b}$ is a subspace of $\mathscr{L}_{a,b}$. Define the functional f on $\mathscr{B}_{a,b}$ by

$$\langle f, \phi\rangle \triangleq \lim_{t\to\infty} e^{at}\phi(t) + \lim_{t\to-\infty} e^{bt}\phi(t) \qquad \phi \in \mathscr{B}_{a,b}$$

Then, for every $\phi \in \mathscr{B}_{a,b}$, $|\langle f, \phi\rangle| \le 2\gamma_{a,b,0}(\phi)$. Consequently, by the Hahn-Banach theorem (Robertson and Robertson [1], p. 29) there exists a linear functional f_1 on $\mathscr{L}_{a,b}$ such that $\langle f_1, \phi\rangle = \langle f, \phi\rangle$ whenever $\phi \in \mathscr{B}_{a,b}$ and $|\langle f_1, \phi\rangle| \le 2\gamma_{a,b,0}(\phi)$ for every $\phi \in \mathscr{L}_{a,b}$. The last inequality shows that $f_1 \in \mathscr{L}'_{a,b}$. It is not the zero member of $\mathscr{L}'_{a,b}$ since $\langle f_1, \phi\rangle = 2$ if $\phi \in \mathscr{L}_{a,b}$ is such that $\phi(t) = e^{-bt}$ for $t < -1$ and $\phi(t) = e^{-at}$ for $t > 1$.

$\mathscr{L}_{c,d}$ is a subspace of $\mathscr{B}_{a,b}$. Indeed, for $\psi \in \mathscr{L}_{c,d}$,

$$e^{at}\psi(t) = e^{(a-c)t}e^{ct}\psi(t)$$

As $t \to \infty$, $e^{(a-c)t} \to 0$ and $e^{ct}\psi(t) = O(1)$. Thus, $\lim_{t\to\infty} e^{at}\psi(t) = 0$. Similarly, $\lim_{t\to-\infty} e^{bt}\psi(t) = 0$. Consequently, $\langle f_1, \psi\rangle = \langle f, \psi\rangle = 0$. Therefore, both f_1 and the zero member of $\mathscr{L}'_{a,b}$ have the same restriction to $\mathscr{L}_{c,d}$.

We turn now to certain countable-union spaces generated from the $\mathscr{L}_{a,b}$ spaces. Let w denote either a finite real number or $-\infty$, and let z denote either a finite real number or $+\infty$. This is in contrast to the symbols a, b, c, and d which always denote finite real numbers. Choose two monotonic sequences of real numbers $\{a_\nu\}_{\nu=1}^{\infty}$ and $\{b_\nu\}_{\nu=1}^{\infty}$ such that $a_\nu \to w+$ and $b_\nu \to z-$. Define $\mathscr{L}(w, z)$ as the countable-union space of all $\mathscr{L}_{a_\nu, b_\nu}$ spaces; thus, $\mathscr{L}(w, z) = \bigcup_{\nu=1}^{\infty} \mathscr{L}_{a_\nu, b_\nu}$, and a sequence converges in $\mathscr{L}(w, z)$ if and only if it converges in one of the $\mathscr{L}_{a_\nu, b_\nu}$ spaces. This definition does not depend on the choices of $\{a_\nu\}$ and $\{b_\nu\}$. (Show this.) For each $k = 0, 1, 2, \ldots$, $t^k e^{-st}$ is a member of $\mathscr{L}(w, z)$ if and only if $w < \operatorname{Re} s < z$. As usual, $\mathscr{L}'(w, z)$ denotes the dual of $\mathscr{L}(w, z)$. Both $\mathscr{L}(w, z)$ and $\mathscr{L}'(w, z)$ are complete spaces (see Sec 1.7 and Theorem 1.9-2).

Now, let $w \leq u$ and $v \leq z$. Clearly, $\mathscr{L}(u, v) \subset \mathscr{L}(w, z)$, and convergence in $\mathscr{L}(u, v)$ implies convergence in $\mathscr{L}(w, z)$. Therefore, the restriction of any $f \in \mathscr{L}'(w, z)$ to $\mathscr{L}(u, v)$ is a member of $\mathscr{L}'(u, v)$. We shall show in a moment that $\mathscr{L}'(w, z)$ is actually a subspace of $\mathscr{L}'(u, v)$. This is in contrast to the fact that $\mathscr{L}'_{a,b}$ cannot be identified with a subspace of $\mathscr{L}'_{c,d}$ where $a < c$ and $d < b$, as we have already indicated.

The results obtained so far show that $\mathscr{L}(a, b)$ is contained in $\mathscr{L}_{a,b}$. Actually, $\mathscr{L}(a, b)$ is smaller than $\mathscr{L}_{a,b}$. For example, the function that is identically equal to 1 for all t is a member of $\mathscr{L}_{0,0}$. But, it is not a member of $\mathscr{L}(0, 0)$ since it is not in $\mathscr{L}_{a_\nu, b_\nu}$ for any $a_\nu > 0$ and any $b_\nu < 0$.

The restriction of any $f \in \mathscr{L}'_{a,b}$ to $\mathscr{L}(a, b)$ is a member of $\mathscr{L}'(a, b)$. On the other hand, there exist members of $\mathscr{L}'(a, b)$ that are not defined on all of $\mathscr{L}_{a,b}$. Such a functional f is the one defined by

$$\langle f, \phi \rangle \triangleq \lim_{t \to \infty} t e^{at} \phi(t)$$

(It is the zero member of $\mathscr{L}'(a, b)$.) The reader should carefully distinguish between the notations $\mathscr{L}(a, b)$ and $\mathscr{L}_{a,b}$ because both these spaces as well as their duals will be used.

We now number several results, which we will refer to later on.

I. Clearly, $\mathscr{D}$ is a subspace of $\mathscr{L}_{a,b}$ as well as of $\mathscr{L}(w, z)$ whatever be the values of a, b, w, or z; moreover, convergence in $\mathscr{D}$ implies convergence in $\mathscr{L}_{a,b}$ and also convergence in $\mathscr{L}(w, z)$. Consequently, the restriction of any member of $\mathscr{L}'_{a,b}$ or $\mathscr{L}'(w, z)$ to $\mathscr{D}$ is a member of $\mathscr{D}'$; see Sec. 1.9.

However, $\mathscr{D}$ is not dense in $\mathscr{L}_{a,b}$. Nor can we identify $\mathscr{L}'_{a,b}$ with a subspace of $\mathscr{D}'$; indeed, different members of $\mathscr{L}'_{a,b}$ can be found whose restrictions to $\mathscr{D}$ are identical. This is illustrated by Example 3.2-1 and the fact that $\mathscr{D}$ is contained in the space $\mathscr{L}_{c,d}$ of that example.

On the other hand, $\mathscr{D}$ is dense in $\mathscr{L}(w, z)$ for every w and z. Indeed, let ϕ be an arbitrary member of $\mathscr{L}(w, z)$. Then, $\phi \in \mathscr{L}_{c,d}$ for some $c > w$ and some $d < z$. Choose a and b such that $w < a < c$ and $d < b < z$. We prove our assertion by constructing a sequence whose members are in $\mathscr{D}$ and which converges to ϕ in $\mathscr{L}_{a,b}$ and therefore in $\mathscr{L}(w, z)$. Let $\lambda(t) \in \mathscr{D}$ be such that $\lambda(t) = 1$ for $|t| < 1$ and $\lambda(t) = 0$ for $|t| > 2$. Also, let $\nu > 1$. We may write

$$(3) \quad \kappa_{a,b}(t) D^k \left[\lambda\left(\frac{t}{\nu}\right) \phi(t) - \phi(t) \right] = \sum_{\mu=0}^{k} \binom{k}{\mu} \left\{ D^{k-\mu} \left[\lambda\left(\frac{t}{\nu}\right) - 1 \right] \right\} \kappa_{a,b}(t) D^\mu \phi(t)$$

Now, $D^{k-\mu}[\lambda(t/\nu) - 1]$ is identically equal to zero for $|t| \leq \nu$, and for $|t| > \nu$ it is bounded by a constant that is independent of ν. Moreover, for each μ

$$\sup_{|t|>\nu} |\kappa_{a,b}(t) D^\mu \phi(t)| \leq \gamma_{c,d,\mu}(\phi) \sup_{|t|>\nu} \frac{\kappa_{a,b}(t)}{\kappa_{c,d}(t)} \to 0 \qquad \nu \to \infty$$

Consequently, as $\nu \to \infty$ the right-hand side of (3) converges uniformly to zero on $-\infty < t < \infty$. Thus, $\{\lambda(t/\nu)\phi(t)\}_{\nu=1}^{\infty}$ is the sequence we seek. This proves our assertion.

Theorem 1.9-1 now implies that $\mathscr{L}'(w, z)$ is a subspace of $\mathscr{D}'$ for every w and z. Thus, the members of $\mathscr{L}'(w, z)$ are distributions, but this is not true for the members of $\mathscr{L}'_{a,b}$. Another useful consequence is that the values that $f \in \mathscr{L}'(w, z)$ assigns to $\mathscr{D}$ uniquely determine the values that f assigns to $\mathscr{L}(w, z)$. This can be proven by choosing (as above) for each $\phi \in \mathscr{L}(w, z)$ a sequence $\{\phi_\nu\}_{\nu=1}^{\infty}$ such that all $\phi_\nu \in \mathscr{D}$ and $\phi_\nu \to \phi$ in $\mathscr{L}(w, z)$ as $\nu \to \infty$. Then, $\langle f, \phi \rangle = \lim_{\nu\to\infty}\langle f, \phi_\nu \rangle$.

Once again, let $w \le u$ and $v \le z$. We have already pointed out that $\mathscr{L}(u, v) \subset \mathscr{L}(w, z)$ and that convergence in $\mathscr{L}(u, v)$ implies convergence in $\mathscr{L}(w, z)$. Since $\mathscr{D} \subset \mathscr{L}(u, v)$ and $\mathscr{D}$ is dense in $\mathscr{L}(w, z)$, $\mathscr{L}(u, v)$ is also dense in $\mathscr{L}(w, z)$. Therefore, by Theorem 1.9-1, $\mathscr{L}'(w, z)$ is subspace of $\mathscr{L}'(u, v)$. Here again, the $\mathscr{L}'(w, z)$ spaces have a nicer structure than do the $\mathscr{L}'_{a,b}$ spaces; this is why we use them, even though all of the subsequent theory could be developed exclusively in terms of the $\mathscr{L}'_{a,b}$ spaces.

II. For any choices of a and b, $\mathscr{L}_{a,b}$ is a dense subspace of $\mathscr{E}$, and the topology of $\mathscr{L}_{a,b}$ is stronger than the topology induced on $\mathscr{L}_{a,b}$ by $\mathscr{E}$. Indeed, $\mathscr{L}_{a,b}$ is clearly a subspace of $\mathscr{E}$. Moreover, since $\mathscr{D}$ is dense in $\mathscr{E}$ (see Sec. 2.3) and since $\mathscr{D} \subset \mathscr{L}_{a,b} \subset \mathscr{E}$, it follows that $\mathscr{L}_{a,b}$ is also dense in $\mathscr{E}$. Next, we recall that the seminorms for $\mathscr{E}$ are defined by

$$\gamma_{K,k}(\psi) \triangleq \sup_{t \in K} |D^k\psi(t)| \qquad \psi \in \mathscr{E}$$

where $k = 0, 1, 2, \ldots$ and K varies through the compact sets in $\mathscr{R}^1$. Let C be the infimum of $\kappa_{a,b}(t)$ on a given K. By the definition of $\kappa_{a,b}(t)$, C is a positive number. Hence, for any $\phi \in \mathscr{L}_{a,b}$

$$\gamma_{a,b,k}(\phi) \ge \sup_{t \in K} |\kappa_{a,b}(t) D^k\phi(t)| \ge C \sup_{t \in K} |D^k\phi(t)| = C\gamma_{K,k}(\phi).$$

Our assertion concerning the topologies of $\mathscr{L}_{a,b}$ and $\mathscr{E}$ now follows from Lemma 1.6-3.

We can immediately conclude that, for every choice of w and z, $\mathscr{L}(w, z)$ is a dense subspace of $\mathscr{E}$ and that convergence in $\mathscr{L}(w, z)$ implies convergence in $\mathscr{E}$. Moreover, Theorems 1.8-2 and 1.9-1 now imply that $\mathscr{E}'$ is a subspace of $\mathscr{L}'_{a,b}$ as well as of $\mathscr{L}'(w, z)$ no matter what the choices of a, b, w, and z are.

III. For each $f \in \mathscr{L}'_{a,b}$ there exists a positive constant C and a nonnegative integer r such that, for all $\phi \in \mathscr{L}_{a,b}$,

$$|\langle f, \phi \rangle| \le C \max_{0 \le k \le r} \gamma_{a,b,k}(\phi).$$

This is simply a special case of Theorem 1.8-1.

IV. As was shown before, if $a \le \sigma \le b$, then the restriction of $f \in \mathscr{L}'_{a,b}$ to $\mathscr{L}_{\sigma,\sigma}$ is in $\mathscr{L}'_{\sigma,\sigma}$. There is a converse to this statement. Assume that the functional f is a member of $\mathscr{L}'_{a,a}$ and also of $\mathscr{L}'_{b,b}$ where $a < b$. This means of course that f is uniquely specified on $\mathscr{L}_{a,a} \cup \mathscr{L}_{b,b}$. Let $\lambda(t)$ be a fixed smooth function on $\mathscr{R}^1$ such that $\lambda(t) = 0$ for $t < -1$ and $\lambda(t) = 1$ for $t > 1$. Let us extend the definition of f onto the space $\mathscr{L}_{a,b}$ by means of the equation

$$(4) \qquad \langle f, \psi \rangle \triangleq \langle f, \lambda\psi \rangle + \langle f, (1 - \lambda)\psi \rangle \qquad \psi \in \mathscr{L}_{a,b}$$

The right-hand side has a sense since $\lambda\psi \in \mathscr{L}_{a,a}$ and $(1 - \lambda)\psi \in \mathscr{L}_{b,b}$ whenever $\psi \in \mathscr{L}_{a,b}$.

This extension of f onto $\mathscr{L}_{a,b}$ is clearly linear. To show that it is continuous, we invoke note *III*: There exists a constant C and a nonnegative integer r such that

$$|\langle f, \lambda\psi \rangle| \le C \max_{0 \le k \le r} \sup_t |\kappa_{a,a}(t) D^k[\lambda(t)\psi(t)]|$$

$$\le C \max_{0 \le k \le r} \sum_{p=0}^{k} \binom{k}{p} \sup_t \left| \left(\frac{\kappa_{a,a} D^{k-p}\lambda}{\kappa_{a,b}} \right) (\kappa_{a,b} D^p \psi) \right|$$

Note that $\kappa_{a,a} \kappa_{a,b}^{-1} D^{k-p}\lambda$ is a bounded function on $-\infty < t < \infty$. Consequently, $\langle f, \lambda\psi_\nu \rangle \to 0$ as $\nu \to \infty$, whenever $\psi_\nu \to 0$ in $\mathscr{L}_{a,b}$. By similar considerations, $\langle f, (1 - \lambda)\psi_\nu \rangle \to 0$ as $\nu \to \infty$, and therefore $\langle f, \psi_\nu \rangle \to 0$, whenever $\psi_\nu \to 0$ in $\mathscr{L}_{a,b}$. So truly, the definition (4) extends f into a member of $\mathscr{L}'_{a,b}$.

Finally, f can be extended into a member of $\mathscr{L}'_{a,b}$ in only one way. Indeed, since $\mathscr{L}_{a,a} \subset \mathscr{L}_{a,b}$ and $\mathscr{L}_{b,b} \subset \mathscr{L}_{a,b}$, it follows from the decomposition indicated in (4) that there cannot be two different members of $\mathscr{L}'_{a,b}$ having identical restrictions to $\mathscr{L}_{a,a} \cup \mathscr{L}_{b,b}$.

We can restate these results as follows:

LEMMA 3.2-2. *Let f be a member of $\mathscr{L}'_{a,a}$ and also of $\mathscr{L}'_{b,b}$, where $a < b$. Then, there exists a unique member of $\mathscr{L}'_{a,b}$ whose restriction to $\mathscr{L}_{a,a} \cup \mathscr{L}_{b,b}$ coincides with the given f. If f is also used to denote this unique member of $\mathscr{L}'_{a,b}$, then its values on $\mathscr{L}_{a,b}$ are given by* (4), *where λ is chosen as stated above.*

V. If $f(t)$ is a locally integrable function such that $f(t)/\kappa_{a,b}(t)$ is absolutely integrable on $-\infty < t < \infty$, then $f(t)$ generates a member of $\mathscr{L}'_{a,b}$ through the definition:

$$(5) \qquad \langle f, \phi \rangle \triangleq \int_{-\infty}^{\infty} f(t)\phi(t)\, dt \qquad \phi \in \mathscr{L}_{a,b}.$$

Indeed,

$$|\langle f, \phi\rangle| = \left| \int_{-\infty}^{\infty} \frac{f(t)}{\kappa_{a,b}(t)} \kappa_{a,b}(t)\phi(t)\, dt \right|$$
$$\leq \gamma_0(\phi) \int_{-\infty}^{\infty} \left| \frac{f(t)}{\kappa_{a,b}(t)} \right| dt$$

which shows that (5) truly defines a functional f on $\mathscr{L}_{a,b}$. This functional is clearly a linear one. Moreover, if $\{\phi_\nu\}_{\nu=1}^{\infty}$ converges in $\mathscr{L}_{a,b}$ to zero, then $\gamma_0(\phi_\nu) \to 0$ so that $|\langle f, \phi_\nu\rangle| \to 0$. Thus, by Lemma 1.8-2, f is also continuous on $\mathscr{L}_{a,b}$. Members of $\mathscr{L}'_{a,b}$ that are generated in this way from conventional functions of the stated type are called *regular*, as was indicated in Sec. 2.4.

Similarly, if $f(t)$ is a locally integrable function such that $f(t)/\kappa_{a,b}(t)$ is absolutely integrable on $-\infty < t < \infty$ for every choice of a and b satisfying $w < a$ and $b < z$, then f generates a regular member of $\mathscr{L}'(w, z)$ through the definition (5) where now $\phi \in \mathscr{L}(w, z)$. This condition on $f(t)/\kappa_{a,b}(t)$ will be satisfied if $e^{-\sigma t}f(t)$ is bounded on $-\infty < t < \infty$ for every choice of σ in the interval $w < \sigma < z$. Indeed, for any $a > w$ and $b < z$, choose a' and b' such that $w < a' < a$ and $b < b' < z$. We can conclude that $f(t)/\kappa_{a,b}(t)$ is absolutely integrable on $-\infty < t < \infty$ from the inequalities

$$\int_0^{\infty} |e^{-at}f(t)|\, dt \leq \int_0^{\infty} e^{(a'-a)t}\, dt \sup_{0<t<\infty} |e^{-a't}f(t)| < \infty$$

and

$$\int_{-\infty}^{0} |e^{-bt}f(t)|\, dt \leq \int_{-\infty}^{0} e^{(b'-b)t}\, dt \sup_{-\infty<t<0} |e^{-b't}f(t)| < \infty$$

PROBLEM 3.2-1. Assume that f is a member of $\mathscr{L}'(w, z)$ and is concentrated on some interval $T \leq t < \infty$ where T is finite. Let $\lambda(t)$ be a smooth-function such that $\lambda(t) = 0$ for $-\infty < t < T - 2$ and $\lambda(t) = 1$ for $T - 1 < t < \infty$. Define f_e as a functional on $\mathscr{L}(w, \infty)$ by $\langle f_e, \phi\rangle \triangleq \langle f, \lambda\phi\rangle$. (This is reasonable since $(1 - \lambda)\phi$ is zero on $T - 1 < t < \infty$ so that $\langle f, (1 - \lambda)\phi\rangle = 0$ when $\phi \in \mathscr{L}(w, \infty)$.) Prove that f_e is a member of $\mathscr{L}'(w, \infty)$.

PROBLEM 3.2-2. Let $f(t)$ be the conventional function $f(t) = \kappa_{-1,1}(t)$. Show that the corresponding regular generalized function is in $\mathscr{L}'(-1, 1)$ but is not in $\mathscr{L}'_{-1,1}$. Then, construct another regular generalized function that is in $\mathscr{L}'_{-1,1}$ but is not in $\mathscr{L}'_{a,b}$ if either $a < -1$ or $b > 1$.

PROBLEM 3.2-3. Prove that the definition of $\mathscr{L}(w, z)$ does not depend on the choices of the monotonic sequences $\{a_\nu\}_{\nu=1}^{\infty}$ and $\{b_\nu\}_{\nu=1}^{\infty}$ where $a_\nu \to w+$ and $b_\nu \to z-$.

PROBLEM 3.2-4. Prove that $\mathscr{D}$ is not dense in $\mathscr{L}_{a,b}$.

PROBLEM 3.2-5. Let $\mathscr{S}$ be the space of testing functions of rapid descent described in Problem 1.6-4, where now $t \in \mathscr{R}^1$. Show that $\mathscr{L}_{a,b} \subset \mathscr{S}$ and that the topology of $\mathscr{L}_{a,b}$ is stronger than the topology induced on it by $\mathscr{S}$ if and only if $a > 0$ and $b < 0$. Also, show that $\mathscr{L}(w, z) \subset \mathscr{S}$ and that convergence in $\mathscr{L}(w, z)$ implies convergence in $\mathscr{S}$ if and only if $w \geq 0$ and $z \leq 0$. What does this imply about the dual spaces?

3.3. The Two-Sided Laplace Transformation

We shall call f a *Laplace-transformable generalized function* if it possesses the following properties:

(i) f is a functional on some domain $d(f)$ of conventional functions.

(ii) f is additive in the sense that, if ϕ, θ, and $\phi + \theta$ are all members of $d(f)$, then $\langle f, \theta + \phi \rangle = \langle f, \theta \rangle + \langle f, \phi \rangle$.

(iii) $\mathscr{L}_{a,b} \subset d(f)$ for at least one pair of real numbers a and b with $a < b$.

(iv) For every $\mathscr{L}_{c,d} \subset d(f)$, the restriction of f to $\mathscr{L}_{c,d}$ is in $\mathscr{L}'_{c,d}$.

For each such f there exists a unique set Λ_f in $\mathscr{R}^1$ defined as follows: A point σ is in Λ_f if and only if there exist two real numbers a_σ and b_σ depending on σ such that $a_\sigma < \sigma < b_\sigma$ and $\mathscr{L}_{a_\sigma, b_\sigma} \subset d(f)$. According to this definition, the open interval (a_σ, b_σ) is contained in Λ_f; moreover, Λ_f is the union of all such open intervals (a_σ, b_σ) arising as σ varies through Λ_f and, for each σ, (a_σ, b_σ) varies through all choices for which $\mathscr{L}_{a_\sigma, b_\sigma} \subset d(f)$. Consequently, Λ_f is an open set.

Let σ_1 be the infimum of Λ_f and σ_2 the supremum of Λ_f. Here, $\sigma_1 = -\infty$ and $\sigma_2 = +\infty$ are possible. Choose two sequences $\{c_\nu\}$ and $\{d_\nu\}$ such that $c_\nu \to \sigma_1 +, d_\nu \to \sigma_2 -, c_\nu \in \Lambda_f, d_\nu \in \Lambda_f, c_\nu < d_\nu$. By condition (iv) and the paragraph before Example 3.2-1, $f \in \mathscr{L}'_{c_\nu, c_\nu}$ and $f \in \mathscr{L}'_{d_\nu, d_\nu}$ for every ν. Next, choose λ as in Sec. 3.2, note *IV* and use Sec. 3.2, Eq. (4) to extend f onto every $\mathscr{L}_{c_\nu, d_\nu}$ and thereby onto $\mathscr{L}(\sigma_1, \sigma_2)$. The extended functional f_1 is defined on $\mathscr{L}(\sigma_1, \sigma_2)$ by

$$\langle f_1, \phi \rangle = \langle f, \lambda\phi \rangle + \langle f, (1 - \lambda)\phi \rangle$$

where $\lambda(t)$ is a fixed smooth function on $\mathscr{R}^1$ such that $\lambda(t) = 0$ for $t < -1$ and $\lambda(t) = 1$ for $t > 1$. According to Lemma 3.2-2, f_1 is continuous and linear on $\mathscr{L}(\sigma_1, \sigma_2)$. Moreover, if ϕ is a member of both $d(f)$ and $\mathscr{L}(\sigma_1, \sigma_2)$, the value $\langle f_1, \phi \rangle$ obtained through this extension of f coincides with the original value $\langle f, \phi \rangle$. Indeed, for the given ϕ there is some ν for which

$\lambda\phi \in \mathscr{L}_{c_\nu, c_\nu} \subset d(f)$ and $(1-\lambda)\phi \in \mathscr{L}_{d_\nu, d_\nu} \subset d(f)$, and therefore the right-hand side of the last equation is equal to $\langle f, \phi \rangle$ according to the additivity property (ii).

Consequently, f has been extended into a functional f_1 on $d(f) \cup \mathscr{L}(\sigma_1, \sigma_2)$ having the following two properties:

(*A*) *The restriction of f_1 to $\mathscr{L}(\sigma_1, \sigma_2)$ is a member of $\mathscr{L}'(\sigma_1, \sigma_2)$.*

(*B*) *The restriction of f_1 to $d(f)$ coincides with f.*

By Lemma 3.2-2 again, there is no other functional on $d(f) \cup \mathscr{L}(\sigma_1, \sigma_2)$ having these two properties.

Henceforth, it is understood that every Laplace-transformable generalized function f has been extended into the functional f_1, which we shall also denote by f. Under this convention, we have the following result: *To every Laplace-transformable generalized function f there corresponds a unique nonvoid open interval (σ_1, σ_2) such that $f \in \mathscr{L}'(\sigma_1, \sigma_2)$ and in addition $f \notin \mathscr{L}'(w, z)$ if either $w < \sigma_1$ or $z > \sigma_2$ (more precisely, f has a continuous linear restriction to $\mathscr{L}(\sigma_1, \sigma_2)$ and is not defined on all of $\mathscr{L}(w, z)$ if either $w < \sigma_1$ or $z > \sigma_2$).*

As an illustration of these ideas, consider

EXAMPLE 3.3-1. Let f be a Laplace-transformable generalized function defined only on $\mathscr{L}_{0,1} \cup \mathscr{L}_{2,3}$ and undefined otherwise. Then, Λ_f is the union of the open intervals $(0, 1)$ and $(2, 3)$. On the other hand, the corresponding f_1 is a member of $\mathscr{L}'(0, 3)$ but is not defined on all of $\mathscr{L}(w, z)$ if either $w < 0$ or $z > 3$. When we apply the Laplace-transformation, we shall implicitly extend f into f_1 but will continue to use the symbol f.

We are at last ready to define our *generalized two-sided Laplace transformation*, which we denote by $\mathfrak{L}$. For a given Laplace-transformable generalized function f, we let Ω_f denote the *open* strip in the complex s plane:

$$\Omega_f = \{s: \sigma_1 < \operatorname{Re} s < \sigma_2\}$$

where σ_1 and σ_2 are defined as the infimum and supremum of Λ_f, as before. Then, the *Laplace transform* F of f is defined as a conventional function on Ω_f by

$$F(s) \triangleq (\mathfrak{L}f)(s) \triangleq \langle f(t), e^{-st} \rangle \qquad s \in \Omega_f \tag{1}$$

For any fixed $s \in \Omega_f$ the right-hand side has a meaning as the application of $f \in \mathscr{L}'(\sigma_1, \sigma_2)$ to $e^{-st} \in \mathscr{L}(\sigma_1, \sigma_2)$ (or equivalently, as the application of $f \in \mathscr{L}'_{a,b}$ to $e^{-st} \in \mathscr{L}_{a,b}$ for any a and b such that $\sigma_1 < a \leq \operatorname{Re} s \leq b < \sigma_2$). We call Ω_f the *region* (or *strip*) *of definition for* $\mathfrak{L}f$, and σ_1 and σ_2 the *abscissas of definition*. We will refer to the operation $\mathfrak{L}$ that maps f into F

as the *Laplace transformation* and will reserve the name "Laplace transform" for the function $F(s)$. Whenever we write $\mathfrak{L}f$, it is understood that f is a Laplace-transformable generalized function that has been extended as explained above.

In certain cases it may be possible to assign a meaning to (1) on the boundary of Ω_f. For example, set $1_+(t)$ equal to 1 for $t > 0$, $\frac{1}{2}$ for $t = 0$, and 0 for $t < 0$. Then, $1_+(t)/(1 + t^2) \in \mathscr{L}'_{0,b}$ for every b. Thus, we could define its Laplace transform for purely imaginary values of $s = i\omega$ by

$$F(i\omega) = \left\langle \frac{1_+(t)}{1+t^2}, e^{-i\omega t} \right\rangle$$

since $e^{-i\omega t} \in \mathscr{L}_{0,b}$ if $b \geq 0$. However, we choose not to consider such boundary points as part of the region of definition; instead, we shall always restrict the definition (1) to the interior of the strip $\sigma_1 \leq \operatorname{Re} s \leq \sigma_2$.

The reason for this is the following. If f is a Laplace-transformable generalized function and its transform has the strip of definition $\sigma_1 < \operatorname{Re} s < \sigma_2$, then a knowledge of f on any space $\mathscr{L}_{c,d}$, where $\sigma_1 < c$ and $d < \sigma_2$, (or even on $\mathscr{D}$ alone) will uniquely determine f on $\mathscr{L}(\sigma_1, \sigma_2)$. (See Sec. 3.2, note *I*.) Therefore, $(\mathfrak{L}f)(s)$ will also be determined throughout Ω_f. This may not be the case however on the boundary of Ω_f. For instance, if σ_1 and σ_2 are finite, f may assign nonzero values to certain $\phi \in \mathscr{L}_{\sigma_1, \sigma_2}$ and yet be the zero functional on $\mathscr{L}(\sigma_1, \sigma_2)$. (See Example 3.2-1) In this case, (1) will be identically equal to zero throughout Ω_f, and at the same time it may be defined and not identically equal to zero on the boundary of Ω_f. In order to avoid this complication we have restricted our definition of $\mathfrak{L}f$ to the interior of $\{s\colon \sigma_1 \leq \operatorname{Re} s \leq \sigma_2\}$.

If $f(t)$ is a locally integrable function such that $e^{-\sigma t}f(t)$ is bounded on $-\infty < t < \infty$ for each real number σ in some interval $\sigma_1 < \sigma < \sigma_2$, then its conventional Laplace transform

$$\int_{-\infty}^{\infty} f(t)e^{-st}\,dt \tag{2}$$

exists for at least $\sigma_1 < \operatorname{Re} s < \sigma_2$ and can be identified with our generalized Laplace transform (1). This is an immediate consequence of Sec. 3.2, note *V*.

An important fact should be pointed out here. When specifying the Laplace transform $F(s)$ of a generalized function f, it is equally important to state the strip of definition Ω_f. This is because the same function $F(s)$ may be generated by two entirely different generalized functions. What will differ will be the strips of definition. For example, the function $1/s$ is the Laplace transform of both $1_+(t)$ and $-1_+(-t)$. However, the corresponding strips of definition are the half-planes $\operatorname{Re} s > 0$ and $\operatorname{Re} s < 0$, respectively.

Moreover, it is even possible for one strip of definition to be contained within the other. More specifically, let $\mathfrak{L}f = F(s)$ for $s \in \Omega_f$, and let $a < \text{Re } s < b$ be any proper substrip of Ω_f. Then, another Laplace-transformable generalized function h can always be constructed such that $\mathfrak{L}h = F(s)$ with the strip of definition Ω_h for $\mathfrak{L}h$ being precisely $a < \text{Re } s < b$. As an illustration of this, consider

EXAMPLE 3.3-2. With δ denoting as usual the delta functional, we have $\mathfrak{L}\delta = 1$ for $-\infty < \text{Re } s < \infty$. Now choose arbitrarily two real numbers a and b with $a < b$. Let $h = \delta + f_1$, where f_1 is the generalized function f_1 that was discussed in Example 3.2-1. As was indicated in that example, f_1 is the zero member of $\mathscr{L}'_{c,d}$, if $a < c$ and $d < b$, and therefore is the zero member of $\mathscr{L}'(a, b)$. On the other hand, f_1 does not exist on all of $\mathscr{L}_{c,d}$ if either $c < a$ or $b < d$. Thus, $\mathfrak{L}f_1 = 0$ and $\Omega_{f_1} = \{s: a < \text{Re } s < b\}$. Consequently, $\mathfrak{L}h = \mathfrak{L}(\delta + f_1) = 1$ for $a < \text{Re } s < b$, but $\mathfrak{L}h$ does not exist for $-\infty < \text{Re } s < a$ and $b < \text{Re } s < \infty$.

In Sec. 3.5 (Theorem 3.5-2) we shall see that a specification of $F(s)$ and the strip of definition $\Omega_f = \{s: \sigma_1 < \text{Re } s < \sigma_2\}$ uniquely determines that member f of $\mathscr{L}'(\sigma_1, \sigma_2)$ for which $\mathfrak{L}f = F(\text{s})$ for $s \in \Omega_f$.

As in the classical case, a generalized Laplace transform is an analytic function at least within its strip of definition. More precisely, we can state

THEOREM 3.3-1 (*The Analyticity Theorem*). *If* $\mathfrak{L}f = F(s)$ *for* $s \in \Omega_f$, *then* $F(s)$ *is analytic on* Ω_f *and*

$$DF(s) = \langle f(t), -te^{-st} \rangle \qquad s \in \Omega_f. \tag{3}$$

PROOF. Let s be an arbitrary but fixed point in $\Omega_f = \{s: \sigma_1 < \text{Re } s < \sigma_2\}$. Choose the real positive numbers a, b, and r such that $\sigma_1 < a < \text{Re } s - r < \text{Re } s + r < b < \sigma_2$. Also, let Δs be a complex increment such that $|\Delta s| < r$. For $\Delta s \neq 0$, we invoke the definition (1) of $F(\text{s})$ to write

$$\frac{F(s + \Delta s) - F(s)}{\Delta s} - \langle f(t), \frac{\partial}{\partial s} e^{-st} \rangle = \langle f(t), \psi_{\Delta s}(t) \rangle \tag{4}$$

where

$$\psi_{\Delta s}(t) = \frac{1}{\Delta s} [e^{-(s+\Delta s)t} - e^{-st}] - \frac{\partial}{\partial s} e^{-st}$$

Note that $\psi_{\Delta s}(t) \in \mathscr{L}_{a,b}$ so that (4) has a sense. We shall show that, as $|\Delta s| \to 0$, $\psi_{\Delta s}(t)$ converges in $\mathscr{L}_{a,b}$ to zero. Because $f \in \mathscr{L}'_{a,b}$, this will imply that $\langle f, \psi_{\Delta s} \rangle \to 0$. In view of (4) and our freedom to choose a arbitrarily close to σ_1 and b arbitrarily close to σ_2, (3) and therefore the analyticity of $F(s)$ on Ω_f will thereby be established.

To proceed, let C denote the circle with center at s and radius r_1 where $0 < r < r_1 < \min(\operatorname{Re} s - a, b - \operatorname{Re} s)$. We may interchange differentiation on s with differentiation on t and invoke Cauchy's integral formulas (Churchill [1], p. 120) to write

$$(-D_t)^k \psi_{\Delta s}(t) = \frac{1}{\Delta s}[(s + \Delta s)^k e^{-(s+\Delta s)t} - s^k e^{-st}] - \frac{\partial}{\partial s} s^k e^{-st}$$

$$= \frac{1}{2\pi i \Delta s}\int_C \left(\frac{1}{\zeta - s - \Delta s} - \frac{1}{\zeta - s}\right)\zeta^k e^{-\zeta t}\, d\zeta$$

$$- \frac{1}{2\pi i}\int_C \frac{\zeta^k e^{-\zeta t}}{(\zeta - s)^2}\, d\zeta = \frac{\Delta s}{2\pi i}\int_C \frac{\zeta^k e^{-\zeta t}}{(\zeta - s - \Delta s)(\zeta - s)^2}\, d\zeta$$

Now, for all $\zeta \in C$ and $-\infty < t < \infty$, $|\kappa_{a,b}(t)\zeta^k e^{-\zeta t}| \leq K$, where K is a constant independent of ζ and t. Moreover, $|\zeta - s - \Delta s| > r_1 - r > 0$ and $|\zeta - s| = r_1$. Consequently.

$$|\kappa_{a,b}(t) D_t{}^k \psi_{\Delta s}(t)| \leq \frac{|\Delta s|}{2\pi}\int_C \frac{K}{(r_1 - r)r_1{}^2}\,|d\zeta|$$

$$\leq \frac{|\Delta s| K}{(r_1 - r)r_1}$$

The right-hand side is independent of t and converges to zero as $|\Delta s| \to 0$. This shows that $\psi_{\Delta s}(t)$ converges to zero in $\mathscr{L}_{a,b}$ as $|\Delta s| \to 0$, which ends the proof.

PROBLEM 3.3-1. Under the hypothesis of Theorem 3.3-1, prove that

$$D^k F(s) = \langle f(t), (-t)^k e^{-st}\rangle \qquad s \in \Omega_f, \qquad k = 1, 2, 3, \dots \tag{5}$$

through the following inductive argument: Assume that (5) is true for k replaced by $k - 1$. It is true for $k = 0$ by definition. Then, show that (5) also holds by following the proof of Theorem 3.3-1. (Another proof of (5) is indicated in the next section.)

PROBLEM 3.3-2. Let $\mathfrak{L}f = F(s)$ for $\Omega_f = \{s\colon \sigma_1 < \operatorname{Re} s < \sigma_2\}$. Define $e^{-\sigma t}f(t)$ as a functional on $\mathscr{S}$ by

$$\langle e^{-\sigma t}f, \phi\rangle \triangleq \langle f, e^{-\sigma t}\phi\rangle \qquad \phi \in \mathscr{S}.$$

Prove that $e^{-\sigma t}f \in \mathscr{S}'$ for $\sigma_1 < \sigma < \sigma_2$. (Actually, the converse is also true; namely, if $e^{-\sigma t}f \in \mathscr{S}'$ for $\sigma_1 < \sigma < \sigma_2$, then $f \in \mathscr{L}'(\sigma_1, \sigma_2)$ so that $(\mathfrak{L}f)(s)$ exists for at least $\sigma_1 < \operatorname{Re} s < \sigma_2$. See Zemanian [2].)

PROBLEM 3.3-3. The quantity

$$f(t) = \sum_{\nu=-\infty}^{\infty} e^{-|t|}\,\delta(t - \nu)$$

is defined as a functional on any function $\phi(t)$ on $-\infty < t < \infty$ by

$$\langle f, \phi \rangle \triangleq \sum_{\nu=-\infty}^{\infty} e^{-|\nu|}\phi(\nu)$$

if the right-hand side exists. Show that

$$(\mathfrak{L}f)(s) = \frac{1 - e^{-2}}{(1 - e^{s-1})(1 - e^{-s-1})} \qquad -1 < \operatorname{Re} s < 1.$$

PROBLEM 3.3-4. A generalized two-sided Laplace transformation that is defined on every member of $\mathscr{D}'$ can be constructed by adapting the results indicated in Zemanian [1], Secs. 7.6, 7.7, and 7.8, as follows:

(*a*) Let $\phi \in \mathscr{D}$, and let g be a locally integrable function such that, for every real number, c, $e^{-ct}g(t)$ is absolutely integrable on $-\infty < t < \infty$. Also, let

$$\Phi(s) = (\mathfrak{L}\phi)(s) = \int_{-\infty}^{\infty} \phi(t)e^{-st}\,dt$$

$$G(s) = \int_{-\infty}^{\infty} g(t)e^{-st}\,dt.$$

Show that, for any real value of c,

$$\int_{c-i\infty}^{c+i\infty} G(s)\Phi(s)\,ds = 2\pi i \int_{-\infty}^{\infty} g(t)\phi(-t)\,dt. \tag{6}$$

This suggests the following definition of $\mathfrak{L}f$ for any $f \in \mathscr{D}'$.

$$\langle \mathfrak{L}f, \mathfrak{L}\phi \rangle = 2\pi i \langle f, \check{\phi} \rangle \tag{7}$$

Here, $\check{\phi}(t) \triangleq \phi(-t)$. Thus, $\mathfrak{L}$ on $\mathscr{D}'$ is the adjoint of the mapping $\mathfrak{L}\phi \mapsto 2\pi i\check{\phi}$ where ϕ traverses $\mathscr{D}$. Note that, when $f = g$, the left-hand side of (7) is to be interpreted as an integral on a vertical line in the complex s plane, namely, as the left-hand side of (6).

(*b*) Let n be a positive integer. Let $\mathscr{G}_n$ be the space of all entire functions Φ such that

$$v_{n,k}(\Phi) \triangleq \sup_{s\in\mathscr{C}^1} |e^{-n|\sigma|}s^k\Phi(s)| < \infty \qquad \sigma = \operatorname{Re} s; \qquad k = 0, 1, 2, \ldots$$

Assign to $\mathscr{G}_n$ the topology generated by the multinorm $\{v_{n,k}\}_{k=0}^{\infty}$. Verify that $\mathscr{G}_n$ is a complete countably multinormed space and that $\Phi \triangleq \mathfrak{L}\phi \mapsto 2\pi i\check{\phi}$ is an isomorphism from $\mathscr{G}_n$ onto $\mathscr{D}_K$ where K is the interval $-n \leq t \leq n$.

(*c*) Show that $\mathscr{G} \triangleq \bigcup_{n=1}^{\infty} \mathscr{G}_n$ can be defined as a strict countable-union space in the natural way and that $\Phi \triangleq \mathfrak{L}\phi \mapsto 2\pi i\check{\phi}$ is an isomorphism from $\mathscr{G}$ onto $\mathscr{D}$. This verifies that, when $\mathfrak{L}f$ is defined for $f \in \mathscr{D}'$ as indicated in part (a), $f \leftrightarrow \mathfrak{L}f$ is an isomorphism from $\mathscr{D}'$ onto $\mathscr{G}'$, where $\mathscr{G}'$ is the dual of $\mathscr{G}$.

3.4. Operation-Transform Formulas

In this section we define a number of operations that are applicable to the members of any space $\mathscr{L}'_{a,b}$ or $\mathscr{L}'(w, z)$ and then show how these operations are transformed under the Laplace transformation. Throughout this section we assume that $F(s) = \mathfrak{L}f$ for $s \in \Omega_f$ and that σ_1 and σ_2 denote the abscissas of definition for $\mathfrak{L}f$.

Differentiation: Given any $\phi \in \mathscr{L}_{a,b}$, we see from Sec. 3.2, Eq. (1) that $\gamma_k(-D\phi) = \gamma_{k+1}(\phi)$. Therefore, $-D$ (i.e., differentiation and then multiplication by -1) is a continuous linear mapping of $\mathscr{L}_{a,b}$ into itself, according to Lemma 1.10-1. By Theorem 1.10-1, the adjoint operator D, which is now generalized differentiation (see Sec. 2.5), is a continuous linear mapping of $\mathscr{L}'_{a,b}$ into itself. Consequently, $-D$ is also a continuous linear mapping of any countable-union space $\mathscr{L}(w, z)$ into itself, and the generalized operator D is a continuous linear mapping of $\mathscr{L}'(w, z)$ into itself.

This leads to the following operation-transform formula:

$$\mathfrak{L}D^k f(t) = s^k F(s) \qquad s \in \Omega_f, \qquad k = 1, 2, 3, \ldots \tag{1}$$

Indeed, we have that $f \in \mathscr{L}'(\sigma_1, \sigma_2)$, $e^{-st} \in \mathscr{L}(\sigma_1, \sigma_2)$, and

$$\langle D^k f(t), e^{-st}\rangle = \langle f(t), (-D_t)^k e^{-st}\rangle = \langle f(t), s^k e^{-st}\rangle = s^k F(s)$$

Formula (1) represents that property of the Laplace transformation which makes it so useful as an operational tool for solving differential equations with constant coefficients. We discuss this matter at the end of Sec. 3.6.

Multiplication by a function in $\mathscr{O}_M$: $\mathscr{O}_M$ is a space of smooth functions defined as follows: $\theta(t)$ is in $\mathscr{O}_M$ if and only if it is smooth on $-\infty < t < \infty$ and for each nonnegative integer k there exists an integer N_k for which $(1 + t^2)^{-N_k} D^k \theta(t)$ is bounded on $-\infty < t < \infty$.

Let ϕ be an arbitrary member of $\mathscr{L}_{c,d}$, and let a and b be arbitrary real numbers such that $a < c$ and $d < b$. Then, for any $\theta \in \mathscr{O}_M$ the operation $\phi \mapsto \theta\phi$ is a continuous linear mapping of $\mathscr{L}_{c,d}$ into $\mathscr{L}_{a,b}$. To show this, we write

$$\kappa_{a,b} D^k(\theta\phi) = \sum_{\nu=0}^{k} \binom{k}{\nu} \left(\frac{\kappa_{a,b}}{\kappa_{c,d}} D^{k-\nu}\theta\right)(\kappa_{c,d} D^\nu \phi)$$

and note that a constant B_k can be chosen such that

$$\left|\frac{\kappa_{a,b}}{\kappa_{c,d}} D^{k-\nu}\theta\right| < B_k \qquad -\infty < t < \infty, \qquad \nu = 0, 1, \ldots, k$$

Hence,

$$\gamma_{a,b,k}(\theta\phi) \le B_k \sum_{\nu=0}^{k} \binom{k}{\nu} \gamma_{c,d,\nu}(\phi)$$

Thus, $\theta\phi$ is in $\mathscr{L}_{a,b}$ whenever ϕ is in $\mathscr{L}_{c,d}$. The linearity of the mapping $\phi \mapsto \theta\phi$ is obvious, and its continuity is implied by Lemma 1.10-1.

It follows now that, for every choice of w and z and any $\theta \in \mathcal{O}_M$, $\phi \mapsto \theta\phi$ is a continuous linear mapping of $\mathscr{L}(w, z)$ into itself (i.e., θ is a multiplier for $\mathscr{L}(w, z)$). Indeed, for any given sequence $\{\phi_\nu\}$ that converges in $\mathscr{L}(w, z)$, we can choose the real numbers a, b, c, and d such that $w < a < c$, $d < b < z$, with $\{\phi_\nu\}$ converging in $\mathscr{L}_{c,d}$. Our previous result then shows that $\{\theta\phi_\nu\}$ converges in $\mathscr{L}_{a,b}$ and hence in $\mathscr{L}(w, z)$. Thus, the mapping is truly continuous, its linearity being obvious.

Using the notation described in Sec. 2.5, we now conclude that $f \mapsto \theta f$ is a continuous linear mapping of $\mathscr{L}'(w, z)$ into itself. This result and the fact that $t^k \in \mathcal{O}_M$ for every $k = 0, 1, 2, \ldots$ leads to another operation-transform formula.

For any fixed s such that $\sigma_1 < \operatorname{Re} s < \sigma_2$,

$$\langle tf(t), e^{-st}\rangle = \langle f(t), te^{-st}\rangle \tag{2}$$

where each side of this equation has a sense as the application of a member of $\mathscr{L}'(\sigma_1, \sigma_2)$ to a member of $\mathscr{L}(\sigma_1, \sigma_2)$. In view of Sec. 3.3, Eq. (3), (2) can be rewritten as

$$\mathfrak{L}tf(t) = -DF(s) \qquad s \in \Omega_f. \tag{3}$$

For any positive integer k and fixed (but arbitrary) point s in Ω_f, we can justify the following manipulations in a similar way.

$$\begin{aligned}\langle t^k f(t), e^{-st}\rangle &= \langle t^{k-1} f(t), te^{-st}\rangle = -D_s\langle t^{k-1}f(t), e^{-st}\rangle \\ &= -D_s\langle t^{k-2}f(t), te^{-st}\rangle = (-D_s)^2\langle t^{k-2}f(t), e^{-st}\rangle \\ &= \cdots = (-D_s)^k\langle f(t), e^{-st}\rangle\end{aligned} \tag{4}$$

Thus, we have obtained the operation-transform formula:

$$\mathfrak{L}t^k f(t) = \langle f(t), t^k e^{-st}\rangle = (-D)^k F(s) \qquad s \in \Omega_f, \quad k = 1, 2, 3, \ldots \tag{5}$$

Multiplication by an exponential function: Let α be a fixed complex number and $r = \operatorname{Re} \alpha$. We shall first prove that $\phi(t) \mapsto e^{-\alpha t}\phi(t)$ is an isomorphism from $\mathscr{L}_{a-r,b-r}$ onto $\mathscr{L}_{a,b}$. Indeed,

$$\kappa_{a,b}(t)D^k e^{-\alpha t}\phi(t) = \sum_{\nu=0}^{k}\binom{k}{\nu}(-\alpha)^{k-\nu}\frac{\kappa_{a,b}(t)e^{-\alpha t}}{\kappa_{a-r,b-r}(t)}\kappa_{a-r,b-r}(t)D^\nu\phi(t)$$

The function $\kappa_{a,b}(t)e^{-\alpha t}/\kappa_{a-r,b-r}(t)$ is bounded on $-\infty < t < \infty$ by 1. Consequently,

$$\gamma_{a,b,k}(e^{-\alpha t}\phi) \leq \sum_{\nu=0}^{k}\binom{k}{\nu}|\alpha|^{k-\nu}\gamma_{a-r,b-r,\nu}(\phi)$$

and therefore $e^{-\alpha t}\phi \in \mathscr{L}_{a,b}$ whenever $\phi \in \mathscr{L}_{a-r,b-r}$. Since the mapping is clearly linear, this also shows that it is continuous from $\mathscr{L}_{a-r,b-r}$ into $\mathscr{L}_{a,b}$ (Lemma 1.10-1).

On the other hand, multiplication by $e^{\alpha t}$ is the unique inverse mapping of $\phi \mapsto e^{-\alpha t}\phi$, and by a similar argument it is a continuous linear mapping of all of $\mathscr{L}_{a,b}$ into $\mathscr{L}_{a-r,b-r}$. So truly, $\phi \mapsto e^{-\alpha t}\phi$ is an isomorphism from $\mathscr{L}_{a-r,b-r}$ onto $\mathscr{L}_{a,b}$. Consequently, $\phi \mapsto e^{-\alpha t}\phi$ is also an isomorphism from $\mathscr{L}(w-r, z-r)$ onto $\mathscr{L}(w, z)$; here, we use the convention $-\infty - r = -\infty$ when $w = -\infty$ and $\infty - r = \infty$ when $z = \infty$.

In accordance with Sec. 2.5, it now follows that $f \mapsto e^{-\alpha t}f$ is an isomorphism from $\mathscr{L}'_{a,b}$ onto $\mathscr{L}'_{a-r,b-r}$, as well as from $\mathscr{L}'(w, z)$ onto $\mathscr{L}'(w-r, z-r)$. Therefore, if $\mathfrak{L}F = F(s)$ for $s \in \Omega_f$, the equation

$$\langle e^{-\alpha t}f(t), e^{-st}\rangle = \langle f(t), e^{-(s+\alpha)t}\rangle \qquad s+\alpha \in \Omega_f \tag{6}$$

has a sense. Indeed, we have $f(t) \in \mathscr{L}'(\sigma_1, \sigma_2)$, $e^{-(s+\alpha)t} \in \mathscr{L}(\sigma_1, \sigma_2)$, $e^{-\alpha t}f(t) \in \mathscr{L}'(\sigma_1 - r, \sigma_2 - r)$, and $e^{-st} \in \mathscr{L}(\sigma_1 - r, \sigma_2 - r)$. Equation (6) is the same as

$$\mathfrak{L}e^{-\alpha t}f(t) = F(s+\alpha) \qquad s+\alpha \in \Omega_f \tag{7}$$

Shifting (or translation): Let τ be a fixed real number. The mapping $\phi(t) \mapsto \phi(t+\tau)$ is readily seen to be a continuous linear mapping of $\mathscr{L}_{a,b}$ into $\mathscr{L}_{a,b}$ because

$$\kappa_{a,b}(t)D_t{}^k\phi(t+\tau) = \frac{\kappa_{a,b}(t)}{\kappa_{a,b}(t+\tau)}\kappa_{a,b}(t+\tau)D_{t+\tau}^k\phi(t+\tau)$$

and $\kappa_{a,b}(t)/\kappa_{a,b}(t+\tau)$ is bounded on $-\infty < t < \infty$. The unique inverse mapping is $\phi(t) \mapsto \phi(t-\tau)$, and it maps all of $\mathscr{L}_{a,b}$ into $\mathscr{L}_{a,b}$. Hence, $\phi(\mathrm{t}) \mapsto \phi(t+\tau)$ is an isomorphism from $\mathscr{L}_{a,b}$ onto $\mathscr{L}_{a,b}$ (i.e., an automorphism on $\mathscr{L}_{a,b}$). Therefore, it is also an automorphism on $\mathscr{L}(w, z)$.

We denote the adjoint of the mapping $\phi(t) \mapsto \phi(t+\tau)$ by $f(t) \mapsto f(t-\tau)$, since this is what we would have if f were a conventional function, and we write

$$\langle f(t-\tau), \phi(t)\rangle = \langle f(t), \phi(t+\tau)\rangle \tag{8}$$

By Theorem 1.10-2, $f(t) \mapsto f(t+\tau)$ is an automorphism on $\mathscr{L}'_{a,b}$, as well as on $\mathscr{L}'(w, z)$ for every a, b, w, and z. As a consequence of all this, we have

$$\mathfrak{L}f(t-\tau) = e^{-s\tau}F(s) \qquad s \in \Omega_f \tag{9}$$

PROBLEM 3.4-1. Let $\mathfrak{L}f = F(s)$ for $s \in \Omega_f$. First show that $\phi(t) \mapsto \phi(-t)$ is an isomorphism from $\mathscr{L}_{-b,-a}$ onto $\mathscr{L}_{a,b}$. Then, for $f \in \mathscr{L}'_{a,b}$ define the mapping $f(t) \mapsto f(-t)$ by $\langle f(-t), \phi(t)\rangle \triangleq \langle f(t), \phi(-t)\rangle$. Finally, show that $\mathfrak{L}f(-t) = F(-s)$ for $-s \in \Omega_f$.

PROBLEM 3.4-2. Let τ be a fixed positive number. Also, let $\mathfrak{L}f = F(s)$ for $s \in \Omega_f$. Show that $\phi(t) \mapsto \phi(t/\tau)$ is an isomorphism from $\mathscr{L}_{\tau a, \tau b}$ onto $\mathscr{L}_{a,b}$. For $f \in \mathscr{L}'_{a,b}$ define the mapping $f(t) \mapsto f(\tau t)$ by $\langle f(\tau t), \phi(t)\rangle \triangleq \langle f(t), \tau^{-1}\phi(t/\tau)\rangle$. Prove that $\mathfrak{L}f(\tau t) = \tau^{-1}F(s/\tau)$ for $s/\tau \in \Omega_f$.

PROBLEM 3.4-3. Let

$$\psi_n(t) = e^{-t/2} \sum_{m=0}^{n} \binom{n}{m} \frac{(-t)^m}{m!} = e^{t/2} D^n \left(\frac{e^{-t}t^n}{n!}\right) \qquad n = 0, 1, 2, \ldots$$

These are the *Laguerre functions* [Erdelyi (Ed.) [1], Vol. II, pp. 188–189]. Show that

$$\mathfrak{L}\psi_n(t)1_+(t) = \int_0^\infty \psi_n(t)e^{-st}\, dt = \frac{(s-\frac{1}{2})^n}{(s+\frac{1}{2})^{n+1}} \qquad \operatorname{Re} s > -\tfrac{1}{2} \tag{10}$$

3.5. Inversion and Uniqueness

We shall now derive an inversion formula for our Laplace transformation. This in turn will imply a uniqueness property; namely, two Laplace-transformable generalized functions having the same strip of definition $\sigma_1 < \operatorname{Re} s < \sigma_2$ and the same transform must be identical on $\mathscr{L}(\sigma_1, \sigma_2)$. The proof of the inversion formula requires two lemmas.

LEMMA 3.5-1. *Let $\mathfrak{L}f = F(s)$ for $\sigma_1 < \operatorname{Re} s < \sigma_2$, let $\phi \in \mathscr{D}$, and set $\Psi(s) = \int_{-\infty}^{\infty} \phi(t)e^{st}\, dt$. Then, for any fixed real number r with $0 < r < \infty$,*

$$\int_{-r}^{r} \langle f(\tau), e^{-s\tau}\rangle \Psi(s)\, d\omega = \left\langle f(\tau), \int_{-r}^{r} e^{-s\tau}\Psi(s)\, d\omega \right\rangle$$

where $s = \sigma + i\omega$ and σ is fixed with $\sigma_1 < \sigma < \sigma_2$.

PROOF. There is nothing to prove if $\phi(t) \equiv 0$. So, assume $\phi(t) \not\equiv 0$. Now, note that $F(s)$ is analytic for $\sigma_1 < \operatorname{Re} s < \sigma_2$, and $\Psi(s)$ is an entire function. Therefore, the above integrals certainly exist. Moreover,

$$\left| D_\tau^{\,k} \int_{-r}^{r} e^{-s\tau}\Psi(s)\, d\omega \right| \le e^{-\sigma\tau} \int_{-r}^{r} |s^k \Psi(s)|\, d\omega$$

so that $\int_{-r}^{r} e^{-s\tau}\Psi(s)\, d\omega$ is a member of $\mathscr{L}(\sigma_1, \sigma_2)$.

Next, partition the path of integration on the straight line from $s = \sigma - ir$ to $s = \sigma + ir$ into m intervals, each of length $2r/m$, and let $s_\nu = \sigma + i\omega_\nu$ be any point in the νth interval. Consider

$$\theta_m(\tau) \triangleq \sum_{\nu=1}^{m} e^{-s_\nu \tau}\Psi(s_\nu)\frac{2r}{m} \tag{1}$$

By applying $f(\tau)$ to (1) term by term, we get

$$\langle f(\tau), \theta_m(\tau)\rangle = \sum_{\nu=1}^{m} \langle f(\tau), e^{-s_\nu\tau}\rangle \Psi(s_\nu)\frac{2r}{m}$$

$$\to \int_{-r}^{r} \langle f(\tau), e^{-s\tau}\rangle \Psi(s)\, d\omega \qquad m \to \infty$$

in view of the fact that $\langle f(\tau), e^{-s\tau}\rangle\Psi(s)$ is a continuous function of ω.

Next, choose a and b such that $\sigma_1 < a < \sigma < b < \sigma_2$. Since $f \in \mathscr{L}'_{a,b}$, all that remains to be proven is that $\theta_m(\tau)$ converges in $\mathscr{L}_{a,b}$ to $\int_{-r}^{r} e^{-s\tau}\Psi(s)\, d\omega$. So, we need merely show that, for each fixed k the following quantity converges uniformly to zero on $-\infty < \tau < \infty$ as $m \to \infty$.

$$A(\tau, m) \triangleq \kappa_{a,b}(\tau) D_\tau^k \left[\theta_m(\tau) - \int_{-r}^{r} e^{-s\tau}\Psi(s)\, d\omega\right]$$

$$= (-1)^k \kappa_{a,b}(\tau) \sum_{\nu=1}^{m} s_\nu^k e^{-s_\nu\tau}\Psi(s_\nu)\frac{2r}{m} \tag{2}$$

$$- (1)^k \kappa_{a,b}(\tau) \int_{-r}^{r} s^k e^{-s\tau}\Psi(s)\, d\omega$$

Now,

$$|\kappa_{a,b}(\tau)e^{-s\tau}| = \kappa_{a,b}(\tau)e^{-\sigma\tau} \to 0$$

as $|\tau| \to \infty$ because $a < \sigma < b$. So, given any $\varepsilon > 0$, we can choose T so large that for all $|\tau| > T$

$$|\kappa_{a,b}(\tau)e^{-s\tau}| < \frac{\varepsilon}{3}\left[\int_{-r}^{r} |s^k\Psi(s)|\, d\omega\right]^{-1}$$

Since $\phi(t) \not\equiv 0$, the right-hand side is finite. (Prove this.) Now, for all $|\tau| > T$, the magnitude of the second term on the right-hand side of (2) is bounded by $\varepsilon/3$. Moreover, again for $|\tau| > T$, the magnitude of the first term on the right-hand side of (2) is bounded by

$$\frac{\varepsilon}{3}\left[\int_{-r}^{r} |s^k\Psi(s)|\, d\omega\right]^{-1} \sum_{\nu=1}^{m} |s_\nu^k\Psi(s_\nu)|\frac{2r}{m}$$

We can now choose m_0 so large that for all $m > m_0$ the last expression is less than $2\varepsilon/3$. Therefore, for all $|\tau| > T$ and all $m > m_0$, $|A(\tau, m)| < \varepsilon$.

Finally, $\kappa_{a,b}(\tau)s^k e^{-s\tau}\Psi(s)$ is a uniformly continuous function of (τ, ω) on the domain $-T \leq \tau \leq T$, $-r \leq \omega \leq r$. Therefore, in view of (2), there exists an m_1 such that, for all $m > m_1$, $|A(\tau, m)| < \varepsilon$ on $-T \leq \tau \leq T$ as well. Thus, when $m > \max(m_0, m_1)$, $|A(\tau, m)| < \varepsilon$ on $-\infty < \tau < \infty$. Q.E.D.

LEMMA 3.5-2. *Let a, b, σ, and r be real numbers with $a < \sigma < b$. Also, let $\phi \in \mathcal{D}$. Then,*

$$\frac{1}{\pi}\int_{-\infty}^{\infty} \phi(t+\tau)e^{\sigma t}\,\frac{\sin rt}{t}\,dt$$

converges in $\mathcal{L}_{a,b}$ to $\phi(\tau)$ as $r \to \infty$.

PROOF. In the following assume that $r > 0$. It is a fact that

$$\int_{-\infty}^{\infty} \frac{\sin rt}{t}\,dt = \pi.$$

Thus, our objective is to prove that, for each $k = 0, 1, 2, \ldots,$

$$B_r(\tau) \triangleq \frac{1}{\pi}\,\kappa_{a,b}(\tau)D_\tau^{\,k}\int_{-\infty}^{\infty}[\phi(t+\tau)e^{\sigma t} - \phi(\tau)]\,\frac{\sin rt}{t}\,dt$$

converges uniformly to zero on $-\infty < \tau < \infty$ as $r \to \infty$. Since ϕ is smooth and of bounded support, we may differentiate under the integral sign:

$$\begin{aligned} B_r(\tau) &= \frac{\kappa_{a,b}(\tau)}{\pi}\int_{-\infty}^{\infty}[e^{\sigma t}D_\tau^{\,k}\phi(t+\tau) - D_\tau^{\,k}\phi(\tau)]\,\frac{\sin rt}{t}\,dt \\ &= \frac{\kappa_{a,b}(\tau)}{\pi}\left[\int_{-\infty}^{-\delta} + \int_{-\delta}^{\delta} + \int_{\delta}^{\infty}\right] \\ &= I_{1,r}(\tau) + I_{2,r}(\tau) + I_{3,r}(\tau) \end{aligned}$$

Here, $I_{1,r}(\tau)$, $I_{2,r}(\tau)$, and $I_{3,r}(\tau)$ denote the quantities obtained by integrating over the intervals $-\infty < t < -\delta$, $-\delta < t < \delta$, and $\delta < t < \infty$, respectively, where $\delta > 0$.

First, consider $I_{2,r}(\tau)$. The function

$$H(t, \tau) \triangleq \kappa_{a,b}(\tau)t^{-1}[e^{\sigma t}D_\tau^{\,k}\phi(t+\tau) - D_\tau^{\,k}\phi(\tau)] \tag{3}$$

is a continuous function of (t, τ) for all τ and all $t \neq 0$. Moreover, since ϕ is smooth, (3) tends to

$$\kappa_{a,b}(\tau)D_t[e^{\sigma t}D_\tau^{\,k}\phi(t+\tau)]\big|_{t=0} \tag{4}$$

as $t \to 0$. Upon assigning the value (4) to $H(0, \tau)$, we obtain a function $H(t, \tau)$ that is continuous everywhere on the (t, τ) plane. Since ϕ is of bounded support, $H(t, \tau)$ is bounded on the domain $\{(t, \tau): -\delta < t < \delta, -\infty < \tau < \infty\}$ by, say, the constant M. Thus, given an $\varepsilon > 0$, we can choose δ so small that

$$|I_{2,r}(\tau)| = \left| \frac{1}{\pi} \int_{-\delta}^{\delta} H(t, \tau) \sin rt \, dt \right| \leq \frac{2M\delta}{\pi} < \varepsilon \qquad -\infty < \tau < \infty$$

Fix δ this way.

Next, consider $I_{1,r}(\tau)$. Set $I_{1,r}(\tau) = J_{1,r}(\tau) - J_{2,r}(\tau)$ where

$$J_{1,r}(\tau) = \frac{1}{\pi} \int_{-\infty}^{-\delta} \kappa_{a,b}(\tau) e^{\sigma t} D_\tau^k \phi(t + \tau) \frac{\sin rt}{t} \, dt$$

$$J_{2,r}(\tau) = \frac{1}{\pi} \kappa_{a,b}(\tau) D_\tau^k \phi(\tau) \int_{-\infty}^{-r\delta} \frac{\sin z}{z} \, dz$$

Since $\kappa_{a,b}(\tau) D_\tau^k \phi(\tau)$ is continuous and of bounded support, it is bounded on $-\infty < \tau < \infty$. By the convergence of the improper integral $\int_{-\infty}^{0} z^{-1} \sin z \, dz$, it follows therefore that $J_{2,r}(\tau)$ tends uniformly to zero on $-\infty < \tau < \infty$ as $r \to \infty$.

To show that $J_{1,r}(\tau)$ does the same, first integrate by parts and use the fact that $\phi(\tau)$ is of bounded support to obtain

$$\begin{aligned} J_{1,r}(\tau) = {} & \frac{e^{-\sigma\delta} \cos r\delta}{\pi r \delta} \kappa_{a,b}(\tau) D_\tau^k \phi(\tau - \delta) \\ & + \frac{1}{\pi r} \int_{-\infty}^{-\delta} \cos rt \, \kappa_{a,b}(\tau) D_t \left[\frac{e^{\sigma t}}{t} D_\tau^k \phi(t + \tau) \right] dt \end{aligned} \tag{5}$$

The first term on the right-hand side tends uniformly to zero on $-\infty < \tau < \infty$ as $r \to \infty$ because σ and δ are fixed and $\kappa_{a,b}(\tau) D_\tau^k \phi(\tau - \delta)$ is a bounded function of τ.

Moreover,

$$\begin{aligned} \kappa_{a,b}(\tau) D_t \left[\frac{e^{\sigma t}}{t} D_\tau^k \phi(t + \tau) \right] = {} & \kappa_{a,b}(\tau) e^{\sigma t} \left(\frac{\sigma}{t} - \frac{1}{t^2} \right) D_\tau^k \phi(t + \tau) \\ & + \kappa_{a,b}(\tau) \frac{e^{\sigma t}}{t} D_\tau^{k+1} \phi(t + \tau) \end{aligned} \tag{6}$$

But, for every k, $\kappa_{a,b}(\tau) e^{\sigma t} D_\tau^k \phi(t + \tau)$ is bounded on the (t, τ) plane. This is because $D_\tau^k \phi(t + \tau)$ is bounded and has its support contained in the strip $\{(t, \tau): |t + \tau| < A\}$ where A is a sufficiently large number, whereas $\kappa_{a,b}(\tau) e^{\sigma t}$ is bounded on this strip by virtue of the inequality $a < \sigma < b$.

Thus, (6) is bounded on the domain $\{(t, \tau): -\infty < t < -\delta, -\infty < \tau < \infty\}$ by, say, the constant N. This result and the assumption that the support of $\phi(\tau)$ is contained in the interval $-A \leq \tau \leq A$ implies that the second term on the right-hand side of (5) is bounded by $2NA/\pi r$, which tends to zero as $r \to \infty$.

So truly, $J_{1,r}(\tau)$ and therefore $I_{1,r}(\tau)$ tend uniformly to zero on $-\infty < \tau < \infty$ as $r \to \infty$. A similar argument shows that $I_{3,r}(\tau)$ also tends uniformly to zero on $-\infty < \tau < \infty$ as $r \to \infty$. Thus, we have established that

$$\overline{\lim_{r\to\infty}} |B_r(\tau)| \leq \varepsilon$$

and since $\varepsilon > 0$ is arbitrary, our proof is complete.

We are now ready to state and prove the inversion formula.

THEOREM 3.5-1. *Let $\mathfrak{L}f = F(s)$ for $\sigma_1 < \operatorname{Re} s < \sigma_2$. Also, let r be a real variable. Then, in the sense of convergence in $\mathscr{D}'$,*

$$f(t) = \lim_{r\to\infty} \frac{1}{2\pi i} \int_{\sigma - ir}^{\sigma + ir} F(s)e^{st}\,ds \tag{7}$$

where σ is any fixed real number such that $\sigma_1 < \sigma < \sigma_2$.

Note: This formula is true even in the sense of convergence in $\mathscr{L}'(\sigma_1, \sigma_2)$ (Zemanian [3]). We won't need this result however.

PROOF. The idea of this proof is to transfer the inversion formula onto a transform of $\phi \in \mathscr{D}$, and to use the fact that the resulting expression converges to ϕ with respect to the topology of the testing-function space $\mathscr{L}_{a,b}$. Indeed, we will employ this scheme of proof several times when proving the inversion formulas for some of the other generalized integral transformations considered subsequently.

Let $\phi \in \mathscr{D}$, and choose the real numbers a and b such that $\sigma_1 < a < \sigma < b < \sigma_2$. We have to show that

$$\lim_{r\to\infty} \left\langle \frac{1}{2\pi i} \int_{\sigma - ir}^{\sigma + ir} F(s)e^{st}\,ds, \phi(t) \right\rangle = \langle f, \phi \rangle \tag{8}$$

Now, the integral on s is a continuous function of t, and therefore the left-hand side without the limit notation can be rewritten as

$$\frac{1}{2\pi} \int_{-\infty}^{\infty} \phi(t) \int_{-r}^{r} F(s)e^{st}\,d\omega\,dt \qquad s = \sigma + i\omega, \qquad r > 0$$

Since $\phi(t)$ is of bounded support and the integrand is a continuous function of (t, ω), the order of integration may be changed. This yields

$$\frac{1}{2\pi} \int_{-r}^{r} \langle f(\tau), e^{-s\tau} \rangle \int_{-\infty}^{\infty} \phi(t)e^{st}\,dt\,d\omega$$

which by Lemma 3.5-1 is the same as

$$\left\langle f(\tau), \frac{1}{2\pi}\int_{-r}^{r} e^{-s\tau}\int_{-\infty}^{\infty}\phi(t)e^{st}\,dt\,d\omega\right\rangle$$

The order of integration for the repeated integral herein may be changed because again ϕ is of bounded support and the integrand is a continuous function of (t, ω). Upon doing this, we obtain

$$\left\langle f(\tau), \frac{1}{2\pi}\int_{-\infty}^{\infty}\phi(t)\int_{-r}^{r} e^{s(t-\tau)}\,d\omega\,dt\right\rangle$$
$$= \left\langle f(\tau), \frac{1}{\pi}\int_{-\infty}^{\infty}\phi(t+\tau)e^{\sigma t}\frac{\sin rt}{t}\,dt\right\rangle$$

The last expression tends to $\langle f(\tau), \phi(\tau)\rangle$ as $r \to \infty$ because $f \in \mathscr{L}'_{a,b}$ and, according to Lemma 3.5-2, the testing function in the last expression converges in $\mathscr{L}_{a,b}$ to $\phi(\tau)$. This completes the proof.

By using Theorem 3.5-1, we can readily prove the following uniqueness property for the Laplace transformation

THEOREM 3.5-2 (*The Uniqueness Theorem*). *If* $\mathfrak{L}f = F(s)$ *for* $s \in \Omega_f$ *and* $\mathfrak{L}h = H(s)$ *for* $s \in \Omega_h$, *if* $\Omega_f \cap \Omega_h$ *is not empty, and if* $F(s) = H(s)$ *for* $s \in \Omega_f \cap \Omega_h$, *then* $f = h$ *in the sense of equality in* $\mathscr{L}'(w, z)$, *where the interval* $w < \sigma < z$ *is the intersection of* $\Omega_f \cap \Omega_h$ *with the real axis.*

PROOF. f and h must assign the same value to each $\phi \in \mathscr{D}$. Indeed, choose σ such that $w < \sigma < z$. Then, by invoking Theorem 3.5-1 and equating $F(\sigma + i\omega)$ to $H(\sigma + i\omega)$ in (8), we immediately obtain $\langle f, \phi\rangle = \langle h, \phi\rangle$.

Furthermore, $\mathscr{D}$ is dense in $\mathscr{L}(w, z)$, and f and h are both members of $\mathscr{L}'(w, z)$. Therefore, $\langle f, \theta\rangle = \langle h, \theta\rangle$ for every $\theta \in \mathscr{L}(w, z)$ as well. Q.E.D.

PROBLEM 3.5-1. Show that Theorem 3.5-1 remains true when (7) is rewritten as

$$f(t) = \lim_{r, r' \to \infty} \frac{1}{2\pi i}\int_{\sigma - ir}^{\sigma + ir'} F(s)e^{st}\,ds \qquad \sigma_1 < \sigma < \sigma_2$$

Here, r and r' tend to infinity independently.

PROBLEM 3.5-2. Let $\phi \in \mathscr{D}$ be such that $\phi(t) \not\equiv 0$ on $-\infty < t < \infty$. Define $\Psi(s)$ as in Lemma 3.5-1, and let r be a fixed positive number. Prove that

$$\int_{-r}^{r} |s^k\Psi(s)|\,d\omega > 0 \qquad s = \sigma + i\omega; \qquad k = 0, 1, 2, \ldots$$

This fact was used in the proof of Lemma 3.5-1. Be sure not to use any results of this section that would lead to circular reasoning.

3.6. Characterization of Laplace Transforms and an Operational Calculus

First of all, we shall show that Laplace transforms can be characterized in the following way.

THEOREM 3.6-1. *Necessary and sufficient conditions for a function $F(s)$ to be the Laplace transform of a generalized function f (according to our definition in Sec. 3.3, Eq. (1)) and for the corresponding strip of definition to be $\Omega_f = \{s: \sigma_1 < \operatorname{Re} s < \sigma_2\}$ are that $F(s)$ be analytic on Ω_f and, for each closed substrip $\{s: a \leq \operatorname{Re} s \leq b\}$ of Ω_f $(\sigma_1 < a < b < \sigma_2)$, there be a polynomial P such that $|F(s)| \leq P(|s|)$ for $a \leq \operatorname{Re} s \leq b$. The polynomial P will depend in general on the choices of a and b.*

PROOF. *Necessity*: The analyticity of $F(s)$ has already been established in Theorem 3.3-1. Now, by our definition of the Laplace transformation, f is a member of $\mathscr{L}'_{a,b}$ where $\sigma_1 < a < b < \sigma_2$. So, by Sec. 3.2, note *III*, there exist a constant C and a nonnegative integer r such that for $a \leq \operatorname{Re} s \leq b$

$$|F(s)| = |\langle f(t), e^{-st}\rangle| \leq C \max_{0 \leq k \leq r} \sup_t |\kappa_{a,b}(t) D_t^k e^{-st}|$$

$$= C \max_{0 \leq k \leq r} |s|^k \sup_t |\kappa_{a,b}(t) e^{-st}| \leq P(|s|)$$

That $P(|s|)$ depends in general on the choices of a and b follows from the fact that, for some Laplace transforms at least, the boundary of Ω_f contains poles of $F(s)$. For example, $\mathfrak{L}1_+(t) = 1/s$ for $0 < \operatorname{Re} s < \infty$.

Sufficiency: We first prove

LEMMA 3.6-1. *If, on the strip $\{s: a < \operatorname{Re} s < b\}$, $G(s)$ is an analytic function that satisfies $|G(s)| \leq K|s|^{-2}$, where K is a constant, and if*

$$(1) \qquad g(t) = \frac{1}{2\pi i} \int_{\sigma - i\infty}^{\sigma + i\infty} G(s) e^{st}\, ds \qquad a < \sigma < b$$

then $g(t)$ is a continuous function that does not depend upon the choice of σ, and $g(t)$ generates a regular generalized function in $\mathscr{L}'(a, b)$. Moreover, $\mathfrak{L}g = G(s)$ for at least $a < \operatorname{Re} s < b$.

PROOF. That $g(t)$ does not depend on σ follows directly from Cauchy's theorem and the bound on $G(s)$. Next, the expression (1) can be rewritten as

$$(2) \qquad e^{-\sigma t} g(t) = \frac{1}{2\pi} \int_{-\infty}^{\infty} G(\sigma + i\omega) e^{i\omega t}\, d\omega \qquad a < \sigma < b$$

Since $G(s)$ is analytic on $s = \sigma + i\omega$ and $|G(s)| \le K|s|^{-2}$,

$$\int_{-\infty}^{\infty} |G(\sigma + i\omega)e^{i\omega t}|\, d\omega \le \int_{-\infty}^{\infty} |G(\sigma + i\omega)|\, d\omega < \infty \tag{3}$$

Thus, the integral in (2) converges uniformly for all t, which implies the continuity of $g(t)$ (Apostol [1], p. 438 and p. 441). Moreover, it is a standard result from the theory of the conventional Laplace transformation (Widder [1], p. 265) that

$$G(s) = \int_{-\infty}^{\infty} g(t)e^{-st}\, dt \tag{4}$$

for at least $a < \operatorname{Re} s < b$.

Finally, (2) and (3) demonstrate that, for each choice of σ in the interval $a < \sigma < b$, $e^{-\sigma t}g(t)$ is bounded on $-\infty < t < \infty$. In view of Sec. 3.2, note V, the conventional function $g(t)$ generates a regular generalized function g in $\mathscr{L}'(a, b)$. Hence, according to our definition of $\mathfrak{L}$, $\mathfrak{L}g$ exists and is equal to (4) for at least $a < \operatorname{Re} s < b$ (see Sec. 3.3, Eq. (2), and the associated discussion). The proof of Lemma 3.6-1 is complete.

Turning to the proof of the sufficiency part of Theorem 3.6-1, we let the strip $a \le \operatorname{Re} s \le b$ be a fixed but arbitrary closed substrip of Ω_f. We also let $Q(s)$ be a polynomial that is different from zero on the strip $a \le \operatorname{Re} s \le b$ and satisfies

$$\left|\frac{F(s)}{Q(s)}\right| \le \frac{K}{|s|^2} \qquad a < \operatorname{Re} s < b \tag{5}$$

where K is a constant. Let $G(s) = F(s)/Q(s)$. Then, $g(t)$, as determined by (1), possesses the properties stated in the conclusion of Lemma 3.6-1. Set $f(t) = Q(D)g(t)$ where D represents generalized differentiation in $\mathscr{L}'(a, b)$. Then, f is also in $\mathscr{L}'(a, b)$ because $\mathscr{L}'(a, b)$ is closed under differentiation. By Sec. 3.4, Eq. (1), $(\mathfrak{L}f)(s) = Q(s)G(s) = F(s)$ for at least $a < \operatorname{Re} s < b$.

Next, let $\{a_\nu\}$ and $\{b_\nu\}$ be two monotonic sequences of real numbers such that $a_\nu \to \sigma_1+$ and $b_\nu \to \sigma_2-$. We have proven in the preceding paragraph that, for each pair of numbers a_ν and b_ν, there exists an $f_\nu \in \mathscr{L}'(a_\nu, b_\nu)$ such that $\mathfrak{L}f_\nu = F(s)$ on $a_\nu < \operatorname{Re} s < b_\nu$. According to the uniqueness theorem (Theorem 3.5-2), for $\mu < \nu$, the restriction of f_ν to $\mathscr{L}(a_\mu, b_\mu)$ must coincide with f_μ. Consequently, there is a member $f \in \mathscr{L}'(\sigma_1, \sigma_2)$ whose restriction to each $\mathscr{L}(a_\nu, b_\nu)$ coincides with f_ν and for which $\mathfrak{L}f = F(s)$ for $s \in \Omega_f$. That f can be so specified that the tube of definition for $\mathfrak{L}f$ is precisely Ω_f follows from the discussion concerning Example 3.3-2. This completes the proof of Theorem 3.6-1.

By virtue of this theorem and the uniqueness theorem, we can now conclude that for any choice of σ_1 and σ_2 ($\sigma_1 < \sigma_2$) the Laplace transformation is a one-to-one mapping from $\mathscr{L}'(\sigma_1, \sigma_2)$ onto the space of analytic functions on the strip $\sigma_1 < \text{Re } s < \sigma_2$ which satisfy polynomial growth conditions as stated in Theorem 3.6-1. These growth conditions will of course be different for different members of $\mathscr{L}'(\sigma_1, \sigma_2)$.

Our proof of the sufficiency part of Theorem 3.6-1 and the uniqueness theorem furnish another inversion formula (see (6) below) for our Laplace transformation. This formula is at times a convenient one to use when a specific inverse transform is to be determined.

COROLLARY 3.6-1a. *Let* $\mathfrak{L}f = F(s)$ *for* $s \in \Omega_f$. *Choose three fixed real numbers* a, σ, *and* b *in* Ω_f *such that* $a < \sigma < b$, *and choose a polynomial* $Q(s)$ *that has no zeros for* $a \leq \text{Re } s \leq b$ *and satisfies* (5). *Then, in the sense of equality in* $\mathscr{L}'(a, b)$,

$$f(t) = Q(D_t) \frac{1}{2\pi i} \int_{\sigma - i\infty}^{\sigma + i\infty} \frac{F(s)}{Q(s)} e^{st} \, ds \qquad a < \sigma < b \tag{6}$$

where D_t *represents generalized differentiation in* $\mathscr{L}'(a, b)$ *and the integral converges in the conventional sense to a continuous function that generates a regular member of* $\mathscr{L}'(a, b)$.

There are some other inversion formulas for our generalized Laplace transformation (see Zemanian [3]).

Before leaving this section, we discuss an operational calculus that is generated by the Laplace transformation. Consider the linear differential equation

$$Lu(t) \triangleq (a_n D^n + a_{n-1} D^{n-1} + \cdots + a_0) u(t) = g(t) \tag{7}$$

where the a_ν are constants, $a_n \neq 0$, and $g(t)$ is a given Laplace-transformable generalized function. We can find a solution $u(t)$ by applying the Laplace transformation to (7) and invoking Sec. 3.4, Eq. (1) to obtain

$$B(s) U(s) = G(s)$$

where

$$\begin{aligned} B(s) &= a_n s^n + a_{n-1} s^{n-1} + \cdots + a_0 \\ U(s) &= \mathfrak{L}u \\ G(s) &= \mathfrak{L}g \qquad s \in \Omega_g = \{s: \sigma_{g_1} < \text{Re } s < \sigma_{g_2}\} \end{aligned}$$

If $B(s)$ has no zeros on Ω_g, then by virtue of Theorem 3.6-1 there exists a generalized function $u(t)$ whose Laplace transform is equal to $G(s)/B(s)$ on Ω_g. According to Theorem 3.5-2 and Sec. 3.4, Eq. (1), $u(t)$ is unique as a member of $\mathscr{L}'(\sigma_{g_1}, \sigma_{g_2})$ and satisfies (7) in the sense of differentiation and equality in $\mathscr{L}'(\sigma_{g_1}, \sigma_{g_2})$.

On the other hand, if $B(s)$ does possess zeros in Ω_g, these will be finite in number, and there will be a set of m adjacent substrips:

$$\sigma_{g_1} = \sigma_0 < \text{Re } s < \sigma_1, \; \sigma_1 < \text{Re } s < \sigma_2, \; \ldots, \; \sigma_{m-1} < \text{Re } s < \sigma_m = \sigma_{g_2}$$

on which $G(s)/B(s)$ is analytic and satisfies the growth condition of Theorem 3.6-1. Consequently, for any given substrip, say, $\sigma_\nu < \text{Re } s < \sigma_{\nu+1}$ there exists a unique member $u(t)$ of $\mathscr{L}'(\sigma_\nu, \sigma_{\nu+1})$ which satisfies (7) in the space $\mathscr{L}'(\sigma_\nu, \sigma_{\nu+1})$ and whose Laplace transform is equal to $G(s)/B(s)$ on $\sigma_\nu < \text{Re } s < \sigma_{\nu+1}$. For any other choice of substrip, there will be a different solution. (It is a fact that the difference between any two such solutions is a smooth function on $-\infty < t < \infty$ that satisfies the homogeneous equation $Lu = 0$ in the conventional sense.)

PROBLEM 3.6-1. Determine the generalized function whose Laplace transform is $-s + s \log Cs$ with the region of definition being $\text{Re } s > 0$. Here, $C = e^\gamma$, and γ is Euler's constant $(0.5772\cdots)$. Use the fact that

$$\mathfrak{L}1_+(t) \log t = -\frac{\log Cs}{s} \qquad \text{Re } s > 0$$

PROBLEM 3.6-2. We have seen that the operational calculus generated by our Laplace transformation yields a unique solution to any differential equation such as (7) once the strip of definition is fixed. On the other hand, one usually assigns initial conditions to such a differential equation in order to obtain a unique solution. Why is there no contradiction here?

PROBLEM 3.6-3. Find all possible Laplace-transformable solutions to the differential equation:

$$(D^2 - 1)u = D^2\delta(t)$$

PROBLEM 3.6-4. Explain how the operational calculus discussed in this section can be extended to simultaneous linear differential equations with constant coefficients.

3.7. Convolution

The convolution of generalized functions arises in a variety of mathematical problems and is related to the behavior of many physical systems. (See, for example, Zemanian [1], Chapters 5, 6, and 10.) This section is devoted to a discussion of this process in the case where the generalized functions are members of some space $\mathscr{L}'_{a,b}$ where $a \le b$. Moreover, the results obtained will be extended to the spaces $\mathscr{L}'(w, z)$ where $w < z$.

The convolution product $f * g$ of two generalized functions f and g in $\mathscr{L}'_{a,b}$ $(a \le b)$ is defined by the expression:

$$(1) \qquad \langle f * g, \phi \rangle \triangleq \langle f(t), \langle g(\tau), \phi(t + \tau) \rangle \rangle \qquad \phi \in \mathscr{L}_{a,b}$$

In order to make sense out of this, we shall have to investigate the expression

$$\psi(t) \triangleq \langle g(\tau), \phi(t+\tau)\rangle \tag{2}$$

Since $\mathscr{L}_{a,b}$ is closed under the shifting operation (Sec. 3.4), the right-hand side of (2) exists for each choice of t and defines $\psi(t)$ as a conventional function on $\mathscr{R}^1$.

LEMMA 3.7-1. *Assume that $g \in \mathscr{L}'_{a,b}$, $\phi \in \mathscr{L}_{a,b}$, and ψ is defined by* (2). *Then, $\psi(t)$ is smooth, and*

$$D^k\psi(t) = \langle g(\tau), D_t^k\phi(t+\tau)\rangle \qquad k = 1, 2, 3, \ldots \tag{3}$$

PROOF. For t fixed and $\Delta t \neq 0$,

$$\frac{1}{\Delta t}[\psi(t+\Delta t) - \psi(t)] - \langle g(\tau), D_t\phi(t+\tau)\rangle = \langle g(\tau), \theta_{\Delta t}(\tau)\rangle \tag{4}$$

where

$$\theta_{\Delta t}(\tau) = \frac{1}{\Delta t}[\phi(t+\Delta t+\tau) - \phi(t+\tau)] - D_t\phi(t+\tau)$$

We shall show that $\theta_{\Delta t}(\tau)$ converges in $\mathscr{L}_{a,b}$ to the zero function as $\Delta t \to 0$. Since g is a continuous linear functional on $\mathscr{L}_{a,b}$, this will show that the right-hand side of (4) converges to zero, and (3) will thereby be established for the case $k = 1$.

Assume that t and τ are fixed, and let $\phi^{(p)}(x)$ denote $D_x{}^p\phi(x)$. By using Taylor's formula with exact remainder where Δt is treated as the independent variable, we may write (Widder [2], p. 43)

$$\phi^{(p)}(t+\tau+\Delta t) = \phi^{(p)}(t+\tau) + \Delta t\phi^{(p+1)}(t+\tau) + \int_0^{\Delta t}(\Delta t - y)\phi^{(p+2)}(t+\tau+y)\,dy \qquad p = 0, 1, 2, \ldots$$

Therefore,

$$\theta_{\Delta t}{}^{(p)}(\tau) = \frac{1}{\Delta t}\int_0^{\Delta t}(\Delta t - y)\phi^{(p+2)}(t+\tau+y)\,dy \tag{5}$$

Moreover, for a fixed t, for all τ, and for $|\Delta t| < 1$,

$$\kappa_{a,b}(\tau) \sup_{|y|<|\Delta t|} |\phi^{(p+2)}(t+\tau+y)| \tag{6}$$

is bounded by a constant, say, B. Hence,

$$|\kappa_{a,b}(\tau)\theta_{\Delta t}^{(p)}(\tau)| \leq \frac{B}{|\Delta t|}\int_0^{\Delta t}(\Delta t - y)\,dy = \tfrac{1}{2}|\Delta t|B$$

This proves that $\theta_{\Delta t}(\tau)$ converges in $\mathscr{L}_{a,b}$ to zero as $\Delta t \to 0$. As mentioned above, (3) is hereby established when $k = 1$.

Since $\mathscr{L}_{a,b}$ is closed under differentiation (Sec. 3.4), we may repeatedly apply this result to write

$$D^k\psi(t) = D^{k-1}\langle g(\tau), D_t\phi(t+\tau)\rangle = D^{k-2}\langle g(\tau), D_t^2\phi(t+\tau)\rangle$$
$$= \cdots = \langle g(\tau), D_t^k\phi(t+\tau)\rangle$$

Our proof is complete.

LEMMA 3.7-2. *In addition to the hypothesis of Lemma* 3.7-1, *assume that* $a \le b$. *Then*, $\psi \in \mathscr{L}_{a,b}$.

PROOF. In view of the preceding lemma, we need merely show that, for each k, $\kappa_{a,b}(t)D^k\psi(t)$ is bounded for $-\infty < t < \infty$. By Lemma 3.7-1 and Sec. 3.2, note *III*, there exist a constant C and a nonnegative integer r such that

$$|\kappa_{a,b}(t)D^k\psi(t)| = |\kappa_{a,b}(t)\langle g(\tau), D_t^k\phi(t+\tau)\rangle|$$
$$\le C \max_{0\le p\le r} \sup_\tau |\kappa_{a,b}(t)\kappa_{a,b}(\tau)D_\tau^p D_t^k\phi(t+\tau)|$$
$$= C \max_{0\le p\le r} \sup_\tau \left| \frac{\kappa_{a,b}(t)\kappa_{a,b}(\tau)}{\kappa_{a,b}(t+\tau)} \kappa_{a,b}(t+\tau)\phi^{(p+k)}(t+\tau)\right| \tag{7}$$
$$\le C\left[\sup_\tau \left|\frac{\kappa_{a,b}(t)\kappa_{a,b}(\tau)}{\kappa_{a,b}(t+\tau)}\right|\right] \max_{0\le p\le r} \gamma_{a,b,p+k}(\phi)$$

Consequently, our lemma will be proven if we can show that the positive function

$$K(t,\tau) \triangleq \frac{\kappa_{a,b}(t)\kappa_{a,b}(\tau)}{\kappa_{a,b}(t+\tau)} \tag{8}$$

is bounded on the (t, τ) plane.

For $t \ge 0$ and $\tau \ge 0$, $K(t,\tau) = 1$. For $t \ge 0$ and $\tau \le 0$, we consider two cases: If $t + \tau \ge 0$, $K(t,\tau) = e^{(b-a)\tau} \le 1$. If $t + \tau \le 0$, $K(t,\tau) = e^{(a-b)t} \le 1$. Thus, $K(t,\tau)$ is bounded by 1 for $t \ge 0$ and $-\infty < \tau < \infty$. Similar arguments show that $K(t,\tau)$ is bounded on the rest of the (t,τ) plane. Q.E.D.

LEMMA 3.7-3. *If* $a \le b$, $g \in \mathscr{L}'_{a,b}$, $\{\phi_\nu\}_{\nu=1}^\infty$ *converges in* $\mathscr{L}_{a,b}$ *to zero, and* $\psi_\nu(t) \triangleq \langle g(\tau), \phi_\nu(t+\tau)\rangle$, *then* $\{\psi_\nu\}_{\nu=1}^\infty$ *also converges in* $\mathscr{L}_{a,b}$ *to zero.*

PROOF. Since (8) is a bounded function, our conclusion follows directly from (7).

We are finally ready to define convolution. Let f and g be arbitrary members of $\mathscr{L}'_{a,b}$ where $a \le b$. The *convolution product* $f * g$ is defined as a functional on $\mathscr{L}_{a,b}$ by (1). The right-hand side of (1) has a sense since, by Lemma 3.7-2, $\langle g(\tau), \phi(t+\tau)\rangle$ is a member of $\mathscr{L}_{a,b}$ whenever ϕ is. This functional is obviously linear on $\mathscr{L}_{a,b}$. That it is continuous on $\mathscr{L}_{a,b}$ is a direct consequence of Lemma 3.7-3. Thus, we have

THEOREM 3.7-1. *If f and g are members of $\mathscr{L}'_{a,b}$ where $a \leq b$,* then *$f * g$, as defined by* (1), *is also a member of $\mathscr{L}'_{a,b}$.*

Convolution is the operation that maps any ordered pair (f, g) of elements $f \in \mathscr{L}'_{a,b}$ and $g \in \mathscr{L}'_{a,b} (a \leq b)$ into the element $f * g$ of $\mathscr{L}'_{a,b}$. It is a *bilinear* operation in the following sense: If $f, g, h \in \mathscr{L}'_{a,b}$ and if α and β are any complex numbers, then

$$f * (\alpha g + \beta h) = \alpha(f * g) + \beta(f * h)$$

and

$$(\alpha g + \beta h) * f = \alpha(g * f) + \beta(h * f)$$

Under certain restrictions, our convolution of generalized functions corresponds to the customary convolution of conventional functions. Assume again that $a \leq b$, and let f and g be locally integrable functions such that $f/\kappa_{a,b}$ and $g/\kappa_{a,b}$ are absolutely integrable on $-\infty < t < \infty$. By Sec. 3.2, note V, f and g generate regular members of $\mathscr{L}'_{a,b}$, which we also denote by f and g respectively. Then, for any $\phi \in \mathscr{L}_{a,b}$

$$\langle f * g, \phi\rangle = \langle f(t), \langle g(\tau), \phi(t+\tau)\rangle\rangle$$

$$= \int_{-\infty}^{\infty} dt \int_{-\infty}^{\infty} f(t) g(\tau) \phi(t+\tau)\, d\tau \tag{9}$$

The integrand in the right-hand side is locally integrable as a function of (t, τ). It is also absolutely integrable on the (t, τ) plane. Indeed,

$$|f(t)g(\tau)\phi(t+\tau)| = \left|\frac{f(t)g(\tau)}{\kappa_{a,b}(t)\kappa_{a,b}(\tau)}\right| \left|\frac{\kappa_{a,b}(t)\kappa_{a,b}(\tau)}{\kappa_{a,b}(t+\tau)}\right| |\kappa_{a,b}(t+\tau)\phi(t+\tau)|$$

The first factor on the right-hand side is integrable on the (t, τ) plane, whereas the second and third factors are continuous and bounded. (See the proof of Lemma 3.7-2.) Hence, by Fubini's theorem (Williamson [1], Theorem 4.2b), we may convert the repeated integral in (9) into a double integral on the (t, τ) plane. Upon applying the change of variables $x = t$ and $y = t + \tau$ (Williamson [1], Theorem 5.1f) and noting that the Jacobian is equal to 1, we can equate (9) to

$$\int_{-\infty}^{\infty} \int_{-\infty}^{\infty} f(x) g(y - x) \phi(y)\, dx\, dy$$

By Fubini's theorem again, this becomes

(10) $$\int_{-\infty}^{\infty} \phi(y) \int_{-\infty}^{\infty} f(x)g(y-x)\,dx\,dy$$

The integral

(11) $$\int_{-\infty}^{\infty} f(x)g(y-x)\,dx$$

is the *conventional convolution product* of f and g. It is a locally integrable function of y because, by Fubini's theorem once more, $\phi(y)\int_{-\infty}^{\infty} f(x)g(y-x)\,dx$ is locally integrable and we are free to choose $\phi(y) \equiv 1$ on any finite interval. Moreover, since $\phi(y)$ is an arbitrary member of $\mathscr{L}_{a,b}$, we can conclude that $\kappa_{a,b}^{-1}(y)\int_{-\infty}^{\infty} f(x)\,g(y-x)\,dx$ is absolutely integrable on $-\infty < y < \infty$. By Sec. 3.2, note V again, (11) is a regular member of $\mathscr{L}'_{a,b}$. In summary, the conventional convolution product (11) of f and g generates a regular member of $\mathscr{L}'_{a,b}$, which, in view of (10), is equal to $f * g$ in the sense of equality in $\mathscr{L}'_{a,b}$.

The convolution process can be readily extended to the space $\mathscr{L}'(w, z)$ where $w < z$. In particular, if f and g are members of $\mathscr{L}'(w, z)$, then their convolution product $f * g$ is defined as that member of $\mathscr{L}'(w, z)$ whose restriction to each space $\mathscr{L}_{a,b}$, where $w < a \le b < z$, coincides with the convolution product of the restrictions of f and g to $\mathscr{L}_{a,b}$. In other words, we again have the definition (1) where now ϕ is an arbitrary member of $\mathscr{L}(w, z)$. It follows that convolution is a bilinear operation that maps any ordered pair of elements in $\mathscr{L}'(w, z)$ into another member of $\mathscr{L}'(w, z)$. Moreover, the convolution product of regular members of $\mathscr{L}'(w, z)$ agrees with the conventional convolution product (11) of the corresponding conventional functions.

We end this section by listing some rather easily established formulas: As usual, let δ be the delta functional, k a positive integer, and τ a fixed real number. Then, in the sense of convolution in either $\mathscr{L}'_{a,b}$ or $\mathscr{L}'(w, z)$, where $a \le b$ or $w < z$,

(12) $$\delta * f = f$$

(13) $$(D^k\delta) * f = D^k f$$

(14) $$\delta(t-\tau) * f(t) = f(t-\tau)$$

PROBLEM 3.7-1. Show that, if $w < z$, if $\{f_\nu\}_{\nu=1}^{\infty}$ converges in $\mathscr{L}'(w, z)$ to f, and if $g \in \mathscr{L}'(w, z)$, then $\{f_\nu * g\}_{\nu=1}^{\infty}$ converges in $\mathscr{L}'(w, z)$ to $f * g$.

PROBLEM 3.7-2. Prove formulas (12), (13), and (14).

3.8. The Laplace Transformation of Convolution

Here, we shall derive the so-called *exchange formula* (see (1) below), which states that convolution is converted by the Laplace transformation into multiplication. This result will then be used to establish some other properties of convolution.

THEOREM 3.8-1. *Let* $\mathfrak{L}f = F(s)$ *for* $s \in \Omega_f$ *and* $\mathfrak{L}g = G(s)$ *for* $s \in \Omega_g$. *If* $\Omega_f \cap \Omega_g$ *is not empty, then* $f * g$ *exists in the sense of convolution in* $\mathscr{L}'(w, z)$ *where the interval* $w < \sigma < z$ *is the intersection of* $\Omega_f \cap \Omega_g$ *with the real axis. Moreover,*

$$(1) \qquad \mathfrak{L}(f * g) = F(s)G(s) \qquad s \in \Omega_f \cap \Omega_g$$

PROOF. By the definition of Laplace-transformable generalized functions and by Sec. 3.2, the restrictions of f and g to $\mathscr{L}(w, z)$ are members of $\mathscr{L}'(w, z)$. Therefore, $f * g$ exists in the sense of convolution in $\mathscr{L}'(w, z)$. Moreover, for any s such that $w < \operatorname{Re} s < z$, we can invoke Sec. 3.7, Eq. (1), to write

$$(2) \qquad \begin{aligned} \mathfrak{L}(f * g) &= \langle f(t), \langle g(\tau), e^{-s(t+\tau)} \rangle \rangle \\ &= \langle f(t), e^{-st} \rangle \langle g(\tau), e^{-s\tau} \rangle = F(s)G(s) \qquad \text{Q.E.D.} \end{aligned}$$

The exchange formula shows that the convolution of Laplace-transformable generalized functions is commutative and associative. More specifically, we have

THEOREM 3.8-2. *Let* $\mathfrak{L}f = F(s)$ *for* $s \in \Omega_f$ *and* $\mathfrak{L}g = G(s)$ *for* $s \in \Omega_g$. *If* $\Omega_f \cap \Omega_g$ *is not empty, then* $f * g = g * f$ *(commutativity) in the sense of equality in* $\mathscr{L}'(w, z)$ *where the interval* $w < \sigma < z$ *is the intersection of* $\Omega_f \cap \Omega_g$ *with the real axis. If, in addition,* $\mathfrak{L}h = H(s)$ *for* $s \in \Omega_h$ *and* $\Omega_f \cap \Omega_g \cap \Omega_h$ *is not empty, then,* $f * (g * h) = (f * g) * h$ *(associativity) in the sense of equality in* $\mathscr{L}'(w', z')$ *where the interval* $w' < \sigma < z'$ *is the intersection of* $\Omega_f \cap \Omega_g \cap \Omega_h$ *with the real axis.*

PROOF. The manipulation (2) shows that $\mathfrak{L}(f * g) = F(s)G(s) = G(s)F(s) = \mathfrak{L}(g * f)$ for $s \in \Omega_f \cap \Omega_g$. This in turn implies our first conclusion by virtue of the uniqueness theorem (Theorem 3.5-2). The second conclusion is proven in the same way.

We can differentiate or shift the convolution product of two Laplace-transformable generalized functions f and g by differentiating or shifting either f or g within the product. More specifically, let f and g satisfy the hypothesis of Theorem 3.8-1, let τ be a fixed real number, and let S_τ denote the shifting operator: $S_\tau f(t) = f(t - \tau)$. Then,

(3) $$D(f * g) = (Df) * g = f * (Dg)$$

and

(4) $$S_\tau(f * g) = (S_\tau f) * g = f * (S_\tau g)$$

where again these equalities are understood to be in the space $\mathscr{L}'(w, z)$ specified in Theorem 3.8-1. Indeed, to show (3), we invoke Sec. 3.4, Eq. (1), and the exchange formula to write

$$\begin{aligned}\mathfrak{L}D(f * g) &= s[F(s)G(s)] = [sF(s)]G(s) = \mathfrak{L}[(Df) * g] \\ &= F(s)[sG(s)] = \mathfrak{L}(f * Dg)\end{aligned}$$

and then invoke the uniqueness theorem. We can get (4) in the same way by making use of Sec. 3.4, Eq. (9).

EXAMPLE 3.8-1. Many physical systems set up a correspondence between an input variable f and an output variable v which is specified by a convolution operator $w*$. That is, there exists a fixed generalized function w, the so-called *unit impulse response*, such that the output v to any permissible input f is given by the convolution product $v = w * f$. (See, for example, Zemanian [1], Chapters 5, 6, and 10.) Under certain circumstances we can employ the Laplace transformation and the exchange formula to determine a possible input f from a knowledge of w and the output v.

As a simple illustration, assume that w is the regular generalized function:

$$w(t) = 1_+(t) \sin t$$

Then, a simple computation shows that w is Laplace-transformable with $\Omega_w = \{s\colon \operatorname{Re} s > 0\}$, and

$$\mathfrak{L}w = \frac{1}{s^2 + 1}$$

Assuming that f is also Laplace-transformable with $\Omega_w \cap \Omega_f$ being not empty, we may apply the exchange formula to $v = w * f$ to get

$$(\mathfrak{L}v)(s) = \frac{1}{s^2 + 1} (\mathfrak{L}f)(s)$$

Consequently, for at least $s \in \Omega_w \cap \Omega_f$ we have

$$(\mathfrak{L}f)(s) = (s^2 + 1)(\mathfrak{L}v)(s)$$

and, by Sec. 3.4, Eq. (1) and the uniqueness theorem,

$$f = D^2 v + v.$$

When v is a regular generalized function whose second derivative is not regular, this solution has a sense in terms of generalized functions but not in terms of conventional functions, despite the fact that the given data, namely, w and v, can be interpreted as conventional functions.

PROBLEM 3.8-1. Let f and g be members of $\mathscr{L}'(w, z)$ for some w and z with $w < z$. Show that, if $f * g = 0$ in $\mathscr{L}'(w, z)$, then either $f = 0$ or $g = 0$ (or both) in $\mathscr{L}'(w, z)$.

PROBLEM 3.8-2. Show that the space $\mathscr{L}'(w, z)$, where $w < z$, is a commutative convolution algebra. (See Zemanian [1], pp. 149–150 for the definition of such an algebra.) Does it have a unit element? (It contains no divisors of zero according to the preceding problem.)

PROBLEM 3.8-3. Let

$$f(t) = \sum_{v=-\infty}^{\infty} e^{-|t|} \delta(t - \nu)$$

as defined in Problem 3.3-3. Show that

$$(f * f)(t) = \sum_{v=-\infty}^{\infty} a_\nu \delta(t - \nu)$$

where

$$a_\nu = \sum_{\mu=-\infty}^{\infty} e^{-|\mu|} e^{-|\nu-\mu|}$$

Then, determine the Laplace transform of $f * f$.

3.9. The Cauchy Problem for the Wave Equation in One-Dimensional Space

As an example of the use of our generalized Laplace transformation, we shall employ it to solve the Cauchy problem for the wave equation in one-dimensional space where the initial conditions are generalized functions. Let x and t be one-dimensional real variables with $-\infty < x < \infty$ and $0 < t < \infty$. Consider the partial differential equation:

(1) $$c^2 D_x^2 u = D_t^2 u$$

This is the homogeneous wave equation. Here, x and t are customarily interpreted as the space and time variables respectively, and c is a real positive number which represents the speed of the wave.

We shall seek a solution to (1) of the form $u = u_t(x)$ where $u_t(x)$ is a generalized function on $-\infty < x < \infty$ depending on the parameter t. (The subscript t on u_t does *not* denote differentiation with respect to t.)

In (1), D_x^2 represents generalized differentiation (Sec. 2.5), whereas D_t^2 represents parametric differentiation (Sec. 2.6). Moreover, we shall only require that (1) hold in the sense of equality in $\mathscr{D}'$. Finally, we impose the following initial conditions: As $t \to 0+$, $u_t(x)$ converges in $\mathscr{D}'$ to $f(x)$, and $D_t u_t(x)$ converges in $\mathscr{D}'$ to $g(x)$, where f and g are given members of $\mathscr{D}'$.

By assuming that f and g are Laplace-transformable generalized functions whose strips of definition intersect, we shall formally derive the solution through the use of the Laplace transformation. After that, it will be shown that the solution truly satisfies the differential equation (1) and the stated initial conditions. When applying the Laplace transformation $\mathfrak{L}$, we treat t as a parameter and x as the independent variable. Thus, we write $\mathfrak{L}[u_t(x)] = U_t(s)$. Here, $U_t(s)$ is a conventional function of both t and s for $0 < t < \infty$ and s restricted to an appropriate strip of definition which we assume is not empty.

Under the assumption that $\mathfrak{L}$ and D_t commute, (1) is transformed into

$$c^2 s^2 U_t(s) = D_t^2 U_t(s)$$

Consequently,

$$U_t(s) = A(s)e^{-cst} + B(s)e^{cst} \tag{2}$$

Next, set $\mathfrak{L}[f(x)] = F(s)$ for $s \in \Omega_f$ and $\mathfrak{L}[g(x)] = G(s)$ for $s \in \Omega_g$. By assuming furthermore that the limiting process $t \to 0+$ can be interchanged with $\mathfrak{L}$, we obtain for $s \in \Omega_f \cap \Omega_g$

$$F(s) = U_t(s)|_{t\to 0+} = A(s) + B(s)$$

and

$$G(s) = D_t U_t(s)|_{t \to 0+} = -csA(s) + csB(s)$$

These equations may be solved for $A(s)$ and $B(s)$ and the result substituted into (2) to obtain

$$U_t(s) = \frac{1}{2}\left[F(s) - \frac{G(s)}{cs}\right]e^{-cst} + \frac{1}{2}\left[F(s) + \frac{G(s)}{cs}\right]e^{cst} \tag{3}$$

The terms $F(s)e^{-cst}$ and $F(s)e^{cst}$ are Laplace transforms on the strip Ω_f, in view of Theorem 3.6-1. However, the factor s^{-1} in the other terms of (3) implies a possible restriction on the tube of definition Ω_u for $\mathfrak{L}[u_t(x)]$. Let R_+ and R_- denote the half-planes Re $s > 0$ and Re $s < 0$, respectively. If $\Omega_f \cap \Omega_g$ and R_+ intersect, we choose $\Omega_u = \Omega_f \cap \Omega_g \cap R_+$. Also, observe that $\mathfrak{L}[1_+(x)] = s^{-1}$ for $s \in R_+$. Thus, the inverse Laplace transform of $s^{-1}G(s)$ in this case is determined by the equation $\mathfrak{L}[1_+(x) * \mathrm{g(x)}] = s^{-1}G(s)$ for $s \in \Omega_g \cap R_+$. On the other hand, if $\Omega_f \cap \Omega_g$ and R_+ do not intersect, then $\Omega_f \cap \Omega_g$ must lie entirely within R_- because these are open sets. In this case we choose $\Omega_u = \Omega_f \cap \Omega_g$. Since $\mathfrak{L}[-1_+(-x)] = s^{-1}$ for $s \in R_-$, the inverse Laplace transform of $s^{-1}G(s)$ is now determined by the equation $\mathfrak{L}[-1_+(-x) * g(x)] = s^{-1}G(s)$ for $s \in \Omega_g \cap R_-$.

Set $h(x) = 1_+(x) * g(x)$ in the first case and $\mathrm{h(x)} = -1_+(-x) * g(x)$ in the second case. By invoking Sec. 3.4, Eq. (9), we finally determine the inverse Laplace transform of (3):

$$u_t(x) = \tfrac{1}{2}[f(x-ct) - c^{-1}h(x-ct) + f(x+ct) + c^{-1}h(x+ct)] \tag{4}$$

For each fixed t this is a specific generalized function on $-\infty < x < \infty$, so that it truly has a sense as a generalized function on $-\infty < x < \infty$ depending on the parameter t.

A *primitive* of a generalized function q is another generalized function whose derivative is equal to q. In the present situation h is a primitive of g. Indeed, it is readily shown that $D1_+(x) = -D1_+(-x) = \delta(x)$. Then, when $h = 1_+ * g$, we have from Sec. 3.7, Eq. (12) and Sec. 3.8, Eq. (3) that $Dh = (D1_+) * g = \delta * g = g$, and similarly when $h(x) = -1_+(-x) * g(x)$.

It is also true that we may replace h in (4) by any other primitive of g without altering our solution. This is because any two primitives of a given distribution differ by no more than a constant (Zemanian [1], Sec. 2.6), and this constant will cancel out of (4) when h is replaced by another primitive of g.

Having formally developed our solution (4), we shall now prove that it satisfies both the differential equation (1) and the stated initial conditions.

Let $f^{(\nu)}$ denote the generalized νth derivative of f. Now, $D_x^2 f(x-ct) = f^{(2)}(x-ct)$ when equality is taken in $\mathscr{D}'$ because for any $\phi \in \mathscr{D}$

$$\langle D_x^2 f(x-ct), \phi(x)\rangle = \langle f(x-ct), D_x^2\phi(x)\rangle = \langle f(x), D_{x+ct}^2\,\phi(x+ct)\rangle$$

$$= \langle f(x), D_x^2\phi(x+ct)\rangle = \langle f^{(2)}(x), \phi(x+ct)\rangle = \langle f^{(2)}(x-ct), \phi(x)\rangle$$

On the other hand, in the definition of parametric differentiation (Sec. 2.6, Eq. (1)) let us take the limit in the sense of convergence in $\mathscr{D}'$. Then, for any $\phi \in \mathscr{D}$ and $\Delta t \to 0$, the quantity

$$-c\left\langle \frac{1}{-c\Delta t}[f(x-ct-c\Delta t) - f(x-ct)], \phi(x)\right\rangle$$

converges to $\langle -cf^{(1)}(x-ct), \phi(x)\rangle$ according to Zemanian [1] pp. 53–54, so that $D_t f(x-ct) = -cf^{(1)}(x-ct)$ in $\mathscr{D}'$. A repetition of this argument with f replaced by $f^{(1)}$ shows that $D_t^2 f(x-ct) = c^2 f^{(2)}(x-ct)$ in $\mathscr{D}'$. Consequently, $D_t^2 f(x-ct) = c^2 D_x^2 f(x-ct)$ in $\mathscr{D}'$. Similar equalities can also be stated when $f(x-ct)$ is replaced by either $f(x+ct)$, $h(x-ct)$ or $h(x+ct)$. Thus, (4) certainly satisfies the differential equation (1) in the required sense.

We turn now to the initial conditions. For any $\phi \in \mathscr{D}$ and in the sense of convergence in $\mathscr{D}$,

$$\tfrac{1}{2}[\phi(x - ct) + \phi(x + ct)] \to \phi(x) \tag{5}$$

and

$$\tfrac{1}{2}[\phi(x - ct) - \phi(x + ct)] \to 0$$

as $t \to 0^+$. This assertion follows from the fact that ϕ is smooth and of bounded support, so that $\phi(x)$ and each of its derivatives are uniformly continuous on $-\infty < x < \infty$. Next, the definition of shifting (Sec. 3.4 Eq. (8)) also holds when ϕ is restricted to $\mathscr{D}$; moreover, shifting is an automorphism on $\mathscr{D}'$. Noting that f and h are continuous linear functionals on $\mathscr{D}$, we can therefore write

$$\langle u_t(x), \phi(x) \rangle = \tfrac{1}{2} \langle f(x), \phi(x - ct) + \phi(x + ct) \rangle + \frac{1}{2c} \langle h(x), \phi(x - ct)$$
$$- \phi(x + ct) \rangle \to \langle f(x), \phi(x) \rangle + 0$$

for $t \to 0^+$. Thus, the first initial condition is satisfied.

To verify the second initial condition, we employ the equations in $\mathscr{D}'$:

$$D_t f(x - ct) = -cf^{(1)}(x - ct), \; D_t f(x + ct) = cf^{(1)}(x + ct)$$
$$D_t h(x - ct) = -cg(x - ct), \; D_t h(x + ct) = cg(x + ct)$$

and proceed as above.

A final comment is in order here. In verifying that (4) is truly a solution to our problem, we made no use of the fact that (4) is a Laplace-transformable generalized function. Actually, neither the initial conditions nor (4) need be Laplace-transformable. In fact, if f and g are arbitrary distributions and h any distributional primitive of g, (4) still satisfies (1) and the initial conditions in the stated sense. This illustrates the fact that the transform method of solving partial differential equations may be useful in indicating what the solution is, even though the given data, such as the initial or boundary conditions or the inhomogeneous term in the differential equation, may not be transformable. In such a situation, one may attempt to find the solution by assuming that the given data is transformable and proceeding formally as above to obtain a possible solution. A final argument, not employing the transformation in question, is then required to establish whether one truly has a solution.

PROBLEM 3.9-1. Prove that (4) satisfies the second initial condition.

PROBLEM 3.9-2. In a formal way, find a solution $u = u(x, t)$ on the domain $\{(x, t)\colon -\infty < x < \infty, 0 < t < \infty\}$ to the differential equation:

$$D_x^2 u = D_t u$$

under the following initial condition: As $t \to 0+$, $u(x, t)$ converges in some sense to a Laplace-transformable generalized function $f(x)$. This is the Cauchy problem for the (normalized) heat equation for one-dimensional flow. It will be discussed in Sec. 7.5 under weaker conditions on $f(x)$. *Hint*:

$$(4\pi t)^{-1/2} \int_{-\infty}^{\infty} e^{-x^2/4t} e^{-xs}\, dx = e^{s^2 t} \qquad -\infty < \operatorname{Re} s < \infty, \qquad 0 < t < \infty$$

3.10. The Right-Sided Laplace Transformation

A special case of the conventional two-side Laplace transformation, namely, the right-sided (or one-sided) Laplace transformation, arises when the conventional function $f(t)$ under consideration is equal to zero almost everywhere on an interval of the form $-\infty < t < T_f$, where T_f depends in general on f. In this case the lower limit on the transformation integral can be replaced by T_f, and we obtain

$$F(s) = \int_{T_f}^{\infty} f(t) e^{-st}\, dt \tag{1}$$

Moreover, if this integral converges for one value of s, it will converge for every s in some half-plane $\sigma_f < \operatorname{Re} s < \infty$, or perhaps in the entire s plane (Widder [1], p. 37).

In this section we shall show how these classical results can be extended to generalized functions. The basic idea is to construct a collection of testing-function spaces by combining the techniques used in constructing the spaces $\mathscr{L}_{a,b}$ and $\mathscr{E}$, the technique for $\mathscr{L}_{a,b}$ being used to provide the proper exponential order for the resulting generalized functions as $t \to \infty$ and the technique for $\mathscr{E}$ being used to insure that these generalized functions are concentrated on intervals of the form $T \le t < \infty$ $(T > -\infty)$.

Let's be specific. For each choice of the real number a, we define a testing-function space $\mathscr{L}_a$ as follows: $\phi(t)$ is a member of $\mathscr{L}_a$ if and only if $\phi(t)$ is smooth on $-\infty < t < \infty$ and, for each nonnegative integer k and each real (finite) number T, it satisfies

$$\rho_{a,T,k}(\phi) \triangleq \sup_{T<t<\infty} |e^{at} D^k \phi(t)| < \infty$$

Thus, ϕ and all its derivatives are of the order $O(e^{-at})$ as $t \to \infty$ but have no restriction on their rate of growth as $t \to -\infty$. $\mathscr{L}_a$ is a linear space. Moreover, each $\rho_{a,T,k}$ is a seminorm on $\mathscr{L}_a$, and the collection of all such seminorms separates $\mathscr{L}_a$ since the assumption that $\rho_{a,T,0}(\phi) = 0$ for every T implies that $\phi(t) \equiv 0$ on $-\infty < t < \infty$.

We assign to $\mathscr{L}_a$ the topology generated by the multinorm $\{\rho_{a, T, k}\}_{T, k}$ where T traverses the real numbers and k traverses the nonnegative integers. By an argument similar to the one given in Sec. 2.3, we see that the same topology is generated in $\mathscr{L}_a$ by the countable multinorm $\{\rho_{a, T_p, k}\}_{p, k=0}^{\infty}$ where T_p is restricted to a sequence of points tending to $-\infty$. Consequently, $\mathscr{L}_a$ is a countably multinormed space. $\mathscr{L}_a$ is a complete space (show this), and therefore its dual $\mathscr{L}_a'$ is also complete according to Theorem 1.8-3. Moreover, $\mathscr{L}_a$ is a testing-function space, and $\mathscr{L}_a'$ is a generalized-function space.

An obvious but crucial property of $\mathscr{L}_a$ is that e^{-st} is a member of $\mathscr{L}_a$ if and only if Re $s \geq a$. On the other hand, for each positive integer k, $t^k\phi(t)$ is in $\mathscr{L}_a$ if and only if Re $s > a$.

Let $\phi \in \mathscr{L}_{a, b}$ where b is arbitrary. Then,

$$\rho_{a, T, k}(\phi) = \sup_{T<t<\infty} \left| \frac{e^{at}}{\kappa_{a, b}(t)} \kappa_{a, b}(t) D^k\phi(t) \right| \leq C\gamma_{a, b, k}(\phi)$$

where $C = \sup_{T<t<\infty}|e^{at}/\kappa_{a, b}(t)|$. Consequently, for every choice of b, $\mathscr{L}_{a, b}$ is a subspace of $\mathscr{L}_a$, and the topology of $\mathscr{L}_{a, b}$ is stronger than that induced on it by $\mathscr{L}_a$. It follows that the restriction of any $f \in \mathscr{L}_a'$ to $\mathscr{L}_{a, b}$ is a member of $\mathscr{L}_{a, b}'$, no matter what the choice of b is.

It can be shown in a similar way that, if $a < c$, then $\mathscr{L}_c \subset \mathscr{L}_a$, and the topology of $\mathscr{L}_c$ is stronger than that induced on it by $\mathscr{L}_a$. Hence, the restriction of any $f \in \mathscr{L}_a'$ to $\mathscr{L}_c$ is in $\mathscr{L}_c'$; speaking more loosely once again, we shall simply say in this case that f is also in $\mathscr{L}_c'$.

Next, let w denote either a real number or $-\infty$. Choose a sequence $\{a_\nu\}_{\nu=1}^{\infty}$ of real numbers with $a_\nu > w$ which converges monotonically to $w+$ as $\nu \to \infty$. Then, $\mathscr{L}(w)$ is defined as the countable-union space $\bigcup_{\nu=1}^{\infty} \mathscr{L}_{a_\nu}$; a sequence converges in $\mathscr{L}(w)$ if and only if it converges in one of the spaces $\mathscr{L}_{a_\nu}$. Moreover, $\mathscr{L}(w)$ does not depend on the choice of $\{a_\nu\}$. Both $\mathscr{L}(w)$ and its dual $\mathscr{L}'(w)$ are complete spaces. If $w < u$, then $\mathscr{L}(u) \subset \mathscr{L}(w)$, and convergence in $\mathscr{L}(u)$ implies convergence in $\mathscr{L}(w)$; thus, the restriction of any $f \in \mathscr{L}'(w)$ to $\mathscr{L}(u)$ is in $\mathscr{L}'(u)$. Similarly, the restriction of $f \in \mathscr{L}'(w)$ to $\mathscr{L}(w, z)$ is a member of $\mathscr{L}'(w, z)$ for every z. Furthermore, $\mathscr{D}$ is dense in $\mathscr{L}(w)$ (show this), even though it is not dense in $\mathscr{L}_a$ for every a. It follows that the members of $\mathscr{L}'(w)$ are distributions, but this is not so for the members of $\mathscr{L}_a'$. It also follows that the values that any $f \in \mathscr{L}'(w)$ assigns to $\mathscr{D}$ uniquely determine f on $\mathscr{L}(w)$. Finally, $\mathscr{E}'$ is a subspace of $\mathscr{L}'(w)$. (Show this too.)

In view of the foregoing remarks, f certainly possesses a Laplace transform in the sense of Sec. 3.3 if $f \in \mathscr{L}'(w)$ for some w because f will then be a member of $\mathscr{L}'(w, \infty)$. Moreover, there exists a real number σ_1 (possibly

$\sigma_1 = -\infty$) such that $f \in \mathscr{L}'(\sigma_1)$ and $f \notin \mathscr{L}'(w)$ if $w < \sigma_1$. The Laplace transform $\mathfrak{L}f$ of f is defined again by

$$(2) \qquad F(s) \triangleq (\mathfrak{L}f)(s) \triangleq \langle f(t), e^{-st} \rangle \qquad s \in \Omega_f$$

but now the region of definition is a half-plane extending infinitely to the right: $\Omega_f = \{s: \sigma_1 < \operatorname{Re} s < \infty\}$. The right-hand side of (2) also has a sense as the application of $f \in \mathscr{L}'(\sigma_1)$ to $e^{-st} \in \mathscr{L}(\sigma_1)$ where $\operatorname{Re} s > \sigma_1$. Under these circumstances we call f a *right-sided Laplace-transformable generalized function* and (2) a *right-sided Laplace transform.*

The next several paragraphs show that this terminology is consistent with that for the conventional right-sided Laplace transformation.

THEOREM 3.10-1. *Every right-sided Laplace-transformable generalized function is concentrated on a semi-infinite interval of the form $T_f \leq t < \infty$, where T_f is finite and depends on f* (i.e., *$\langle f, \phi \rangle = 0$ for every smooth function ϕ that is identically equal to zero on a neighborhood of $T_f \leq t < \infty$*).

PROOF. This proof is quite similar to the necessity part of the argument of Theorem 2.3-1. Let Λ_n denote the semi-infinite interval $-\mathrm{n} \leq t < \infty$. Suppose that f is not concentrated on any interval of the form $T_f \leq t < \infty$ This means that for every n there exists a smooth function ϕ_n which vanishes on Λ_n and for which $\langle f, \phi_n \rangle \neq 0$. Note that $\phi_n \in \mathscr{L}_a$ for every a. Set $\theta_n = \phi_n / \langle f, \phi_n \rangle$. Since $f \in \mathscr{L}_a'$ whenever $a > \sigma_1$, we may apply f to θ_n. Hence, $\langle f, \theta_n \rangle = 1$ for every n. Now, for each choice of a the sequence $\{\theta_n\}_{n=1}^{\infty}$ converges in $\mathscr{L}_a$ to the identically zero function because for any given interval $T \leq t < \infty$ $(T > -\infty)$ all members of $\{\theta_n\}$, except possibly a finite number of them, will be identically equal to zero on $T \leq t < \infty$. Consequently, $\langle f, \theta_n \rangle \to 0$ as $n \to \infty$. This contradicts the previous statement that $\langle f, \theta_n \rangle = 1$ for every n. Q.E.D.

If $f(t)$ is a locally integrable function such that $f(t) = 0$ almost everywhere on $-\infty < t < T_f$ and $f(t)e^{-at}$ is absolutely integrable on $-\infty < t < \infty$, then $f(t)$ generates a regular member of $\mathscr{L}_a'$ (which we again denote by f) through the definition:

$$(3) \qquad \langle f, \phi \rangle \triangleq \int_{-\infty}^{\infty} f(t)\phi(t)\, dt \qquad \phi \in \mathscr{L}_a$$

Similarly, f is a regular member of $\mathscr{L}'(w)$ if the above conditions hold for every $a > w$. The proof of these assertions is straightforward.

Conversely, if f is a regular member of $\mathscr{L}'(w)$, then, by the definition of such generalized functions (see Sec. 2.4, Eq. (1)), f is generated by (3) where the right-hand side converges in the Lebesgue sense for every $\phi \in \mathscr{L}(w)$. In this case, the conventional function $f(t)$ must be equal to zero almost everywhere on some interval $-\infty < t < T_f$. Indeed, by Theorem

3.10-1, there exists a real number T_f such that $\int_{-\infty}^{\infty} f(t)\phi(t)\, dt = 0$ for every $\phi \in \mathscr{D}$ whose support is contained in $-\infty < t < T_f$. Now, choose arbitrarily two real numbers x and y such that $-\infty < x < y < T_f$. Set

$$\phi(t) = \begin{cases} 0 & t \leq x \quad \text{and} \quad t \geq y \\ \exp\left(-\dfrac{1}{t-x} - \dfrac{1}{y-t}\right) & x < t < y \end{cases}$$

Then, for any positive integer n, $[\phi(t)]^{1/n}$ is also in $\mathscr{D}$, and we have

$$\langle f, \phi^{1/n} \rangle = \int_x^y f(t)[\phi(t)]^{1/n}\, dt = 0$$

Since, for all n, $[\phi(t)]^{1/n}$ is uniformly bounded by some constant M, we may invoke Lebesgue's theorem of dominated convergence (Williamson [1], p. 60) and write

$$0 = \lim_{n \to \infty} \langle f, \phi^{1/n} \rangle = \int_x^y f(t) \lim_{n \to \infty} [\phi(t)]^{1/n}\, dt$$

$$= \int_x^y f(t)\, dt$$

In view of the arbitrariness of x and y, we can conclude that $f(t) = 0$ almost everywhere on $-\infty < t < T_f$ (Williamson [1], p. 87). This confirms our assertion.

It now follows that the Laplace transform $\mathfrak{L}f$ of a regular generalized function f in $\mathscr{L}'(\sigma_1)$ is equal to

$$F(s) = \int_{T_f}^{\infty} f(t)e^{-st}\, dt \qquad \operatorname{Re} s > \sigma_1$$

Thus, our present theory is truly an extension to generalized functions of the conventional right-sided Laplace transformation.

Since the right-sided Laplace transformation is a special case of the two-sided one, the results of the preceding section can be carried over to the present case. For example, the analyticity theorem (Theorem 3.3-1), the uniqueness theorem (Theorem 3.5-2), and the inversion formulas (Theorem 3.5-1 and Corollary 3.6-1a) remain valid statements, where now σ_2 is equal to ∞. Moreover, we can characterize Laplace transforms in the present case as follows.

THEOREM 3.10-2. *Necessary and sufficient conditions for $F(s)$ to be the Laplace transform of a right-sided Laplace-transformable generalized function which is concentrated on $T \leq t < \infty$ $(T > -\infty)$ are that there be a half-*

plane Re $s \geq a$ *on which* $F(s)$ *is analytic and there be a polynomial* $P(|s|)$ *for which*

$$|F(s)| \leq e^{-\text{Re}\, sT} P(|s|) \qquad \text{Re}\, s \geq a$$

PROBLEM 3.10-1. Show that, if $a < c$, then the restriction of any $f \in \mathscr{L}_a'$ to $\mathscr{L}_c$ is in $\mathscr{L}_c'$.

PROBLEM 3.10-2. Prove that $\mathscr{L}_a$ is a complete space.

PROBLEM 3.10-3. Prove that $\mathscr{D}$ is dense in $\mathscr{L}(w)$.

PROBLEM 3.10-4. Show that $\mathscr{E}'$ is a subspace of $\mathscr{L}'(w)$.

PROBLEM 3.10-5. Let $f(t)$ be a locally integrable function such that $f(t) = 0$ for $-\infty < t < T$. Show that $f(t)$ generates a generalized function in $\mathscr{L}'(w)$ through the definition (3), if $e^{-at}f(t)$ is absolutely integrable on $-\infty < t < \infty$ for each $a > w$.

PROBLEM 3.10-6. Show by example that the converse to the statement, "every right-sided Laplace transform has a region of definition extending infinitely to the right," is not true in general. That is, find a Laplace-transformable generalized function that is not right-sided but nevertheless possesses a two-sided transform with a region of definition of the form $\{s: \sigma_1 < \text{Re}\, s < \infty\}$.

PROBLEM 3.10-7. Let $\mathscr{S}$ denote the space of testing functions of rapid descent on $-\infty < t < \infty$, and let $\mathscr{S}'$ be its dual, the space of distributions of slow growth. (These spaces were described in Problems 1.6-4, 1.7-3, 1.8-4, and 1.9-3.) Also, let $\lambda_T(t)$ be a smooth function such that $\lambda_T(t) \equiv 0$ for $t < T - 1$ and $\lambda_T(t) \equiv 1$ for $t > T$. Prove the following assertions.

(*a*) If $\{\phi_\nu\}_{\nu=1}^\infty$ converges in $\mathscr{L}_c$ to zero, then, for every $a < c$, $\{\lambda_T(t) e^{at}\phi_\nu(t)\}_{\nu=1}^\infty$ converges in $\mathscr{S}$ to zero.

(*b*) Let $f \in \mathscr{D}'$ have its support bounded on the left at $T_f < -\infty$, let $T < T_f$, and let $e^{-at}f \in \mathscr{S}'$ for every $a > \sigma_1$. Define f as a functional on $\mathscr{L}_c$ for every $c > a$ by

$$\langle f, \phi \rangle \triangleq \langle e^{-at}f, \lambda_T e^{at}\phi \rangle \qquad \phi \in \mathscr{L}_c$$

Then, $f \in \mathscr{L}_c'$. Moreover, since c may be chosen arbitrarily close to a,

$$\langle f, e^{-st} \rangle = \langle e^{-at}f, \lambda_T e^{(a-s)t} \rangle \qquad \text{Re}\, s > a > \sigma_1$$

(*c*) Conversely, if $\{\phi_\nu\}_{\nu=1}^\infty$ converges in $\mathscr{S}$ to zero, then $\{e^{-at}\phi_\nu\}_{\nu=1}^\infty$ converges in $\mathscr{L}_a$ to zero.

(*d*) Let $f \in \mathscr{L}_a'$ for every $a > \sigma_1$. Define $e^{-at}f(t)$ as a functional on $\mathscr{S}$ by

$$\langle e^{-at}f, \phi \rangle \triangleq \langle f, e^{-at}\phi \rangle \qquad \phi \in \mathscr{S}$$

Then, $e^{-at}f \in \mathscr{S}'$. Also, choose T to the left of an interval $T_f \leq t < \infty$ on which f is concentrated. Since a may be chosen arbitrarily close to σ_1,

$$\langle e^{-at}f, \lambda_T e^{(a-s)t}\rangle = \langle f, e^{-st}\rangle \qquad \text{Re } s > \sigma_1$$

These results show that the definition of the right-sided generalized Laplace transformation given in Zemanian [1], Sec. 8.3 is equivalent to the definition of this section.

PROBLEM 3.10-8. State why the operation-transform formulas, given by Sec. 3.4, Eqs. (1), (5), (7), and (9) also hold for the right-sided Laplace transformation.

PROBLEM 3.10-9. Show that, when f and h are right-sided Laplace-transformable generalized functions, the conclusion of the uniqueness theorem (Theorem 3.5-2) can be strengthened by stating that "$f = h$ in $\mathscr{L}'(w)$."

PROBLEM 3.10-10. Prove Theorem 3.10-2. *Hint*: For the necessity part, use the fact that, for Re $s > w > 0$ and $f \in \mathscr{L}'(w)$.

$$\langle f(t), e^{-st}\rangle = \langle f(t), \lambda[|s|(t - T)]e^{-st}\rangle$$

where λ is a smooth function such that $\lambda(t) \equiv 0$ for $t < -1$ and $\lambda(t) \equiv 1$ for $t > -\frac{1}{2}$. Also, invoke Theorem 1.8-1. For the sufficiency part, make use of the following classical fact (Zemanian [1], Theorem 8.2-3 and Sec. 8.3, Eq. (9)): If, on the half-plane Re $s \geq a$, $G(s)$ is analytic and satisfies $|G(s)| \leq K|s|^{-2}e^{-\text{Re } sT}$. where K is a constant, and if

$$g(t) \triangleq \frac{1}{2\pi i}\int_{c-i\infty}^{c+i\infty} G(s)e^{st}\, ds \qquad c \geq a$$

then $g(t)$ is a continuous function for all t, $g(t) = 0$ for $t < T$ and

$$G(s) = \int_T^\infty g(t)e^{-st}\, dt$$

for at least Re $s > a$, where $g(t)e^{-at}$ is bounded on $-\infty < t < \infty$.

PROBLEM 3.10-11. Following the development of Sec. 3.7, derive a convolution operation on the generalized functions in $\mathscr{L}_a'$. Then, prove the exchange formula for the right-sided Laplace transformation. In this development work in terms of the spaces $\mathscr{L}_a$, $\mathscr{L}(w)$, and their duals, instead of $\mathscr{L}_{a,b}$, $\mathscr{L}(w, z)$, and their duals.

PROBLEM 3.10-12. The regular generalized function $1_+(t)$ is a member of $\mathscr{L}'(0)$. For $f \in \mathscr{L}'(0)$ the convolution $1_+(t) * f(t)$ can be interpreted as a generalized integral: $\int_{-\infty}^t f(x)\, dx$. Why? Derive an operation-transform formula for the mapping $f \mapsto \int_{-\infty}^t f(x)\, dx$.

PROBLEM 3.10-13. It is more customary to base the conventional one-side Laplace transformation on the integral:

$$\int_0^\infty f(t)e^{-st}\,dt \tag{4}$$

where, in distinction to (1), the lower limit is fixed at $t = 0$. Extend this form of the Laplace transformation to generalized functions as follows: Let I be the open interval $(0, \infty)$, and restrict t to I. $\mathscr{L}_{+,a}$ denotes the space of all smooth functions $\phi(t)$ on I such that

$$\lambda_{a,k}(\phi) \triangleq \sup_{0<t<\infty} |e^{at}D^k\phi(t)| < \infty \qquad k = 0, 1, 2, \ldots$$

and its topology is generated by $\{\lambda_{a,k}\}_{k=0}^\infty$.

(a) Show that $\mathscr{L}_{+,a}$ is a complete countably multinormed space.

(b) Let $\{a_\nu\}_{\nu=1}^\infty$ be a monotonic sequence of real numbers such that $a_\nu \to w^+$, where w is a real number or $-\infty$. Show that $\mathscr{L}_+(w) = \bigcup_{\nu=1}^\infty \mathscr{L}_{+,a_\nu}$ can be defined as a countable-union space.

(c) Call the generalized function f $\mathfrak{L}_+$-transformable if $f \in \mathscr{L}_+'(w)$ for some w, where $\mathscr{L}_+'(w)$ is the dual of $\mathscr{L}_+(w)$. Let σ_f be the infimum of all such w. Define

$$F(s) \triangleq (\mathfrak{L}_+ f)(s) \triangleq \langle f(t), e^{-st}\rangle \qquad \operatorname{Re} s > \sigma_f \tag{5}$$

Show that $F(s)$ is analytic for $\operatorname{Re} s > \sigma_f$.

(d) Under what conditions on the locally integrable function $f(t)$ can the integral (4) be considered as a special case of (5)?

3.11. The n-Dimensional Laplace Transformation

In this section it is understood that $t = \{t_1, t_2, \ldots, t_n\} \in \mathscr{R}^n$, and similarly $a, b, \sigma, \omega \in \mathscr{R}^n$. Also, $s = \sigma + i\omega = \{s_1, s_2, \ldots, s_n\} \in \mathscr{C}^n$. As is customary, we use the notations:

$$e^{-st} = \exp(-s_1t_1 - s_2t_2 - \cdots - s_nt_n)$$

$$s^t = s_1^{t_1}s_2^{t_2}\cdots s_n^{t_n}$$

Let $f(t)$ be a function from $\mathscr{R}^n$ into $\mathscr{C}^1$. The *conventional n-dimensional Laplace transformation* maps a suitably restricted function of this sort into another function $F(s)$, where $F(S)$ maps $\mathscr{C}^n$ into $\mathscr{C}^1$ by means of the integral:

$$F(s) = \int_{\mathscr{R}^n} f(t)e^{-st}\,dt \tag{1}$$

We shall indicate how this transformation can be extended to certain generalized functions on $\mathcal{R}^n$. Our discussion will parallel that of Secs. 3.2 and 3.3. At the end of this section we merely quote some properties of the n-dimensional generalized Laplace transformation and very briefly indicate how the generalized convolution process can also be extended to the n-dimensional case; for proofs of these latter results see Zemanian [2], An application of the two-dimensional Laplace transformation is given in the next section.

Let t_ν, a_ν, and b_ν be the νth components of t, a, and b respectively. Set

$$\kappa_{a,b}(t) = \prod_{\nu=1}^{n} \kappa_{a_\nu, b_\nu}(t_\nu)$$

where

$$\kappa_{a_\nu, b_\nu}(t_\nu) = \begin{cases} \exp a_\nu t_\nu & 0 \le t_\nu < \infty \\ \exp b_\nu t_\nu & -\infty < t_\nu < 0 \end{cases}$$

We now use the symbol $\mathcal{L}_{a,b}$ to denote the space of all complex-valued smooth functions $\phi(t)$ from $\mathcal{R}^n$ into $\mathcal{C}^1$ such that, for every integer $k \ge 0$ ($k \subset \mathcal{R}^n$),

$$\gamma_k(\phi) \triangleq \gamma_{a,b,k}(\phi) \triangleq \sup_{t \in \mathcal{R}^n} |\kappa_{a,b}(t) D^k \phi(t)| < \infty$$

As before, under the topology generated by the multinorm $\{\gamma_k\}_{k \ge 0}$, $\mathcal{L}_{a,b}$ is a complete countably multinormed space (i.e., a Fréchet space). Here again, $e^{-st} \in \mathcal{L}_{a,b}$ if and only if $a \le \operatorname{Re} s \le b$; also, $t^k e^{-st} \in \mathcal{L}_{a,b}$ ($k = 1, 2, 3, \ldots$) if and only if $a < \operatorname{Re} s < b$. The dual $\mathcal{L}'_{a,b}$ of $\mathcal{L}_{a,b}$ is a linear space that is complete under its customary weak topology.

As in the one-dimensional case, if $a \le c$ and $d \le b$, then $\mathcal{L}_{c,d} \subset \mathcal{L}_{a,b}$, and the topology of $\mathcal{L}_{c,d}$ is stronger than the topology induced on $\mathcal{L}_{c,d}$ by $\mathcal{L}_{a,b}$. Consequently, the restriction of any $f \in \mathcal{L}'_{a,b}$ to $\mathcal{L}_{c,d}$ is a member of $\mathcal{L}_{c,d}$. Moreover, notes *I*, *II*, *III*, and *V* of Sec. 3.2 can be immediately extended to the n-dimensional case, at least so far as they apply to the $\mathcal{L}_{a,b}$ spaces. However, the concept of the countable-union space $\mathcal{L}(w, z)$ cannot be applied in the way it was in the one-dimensional case without some loss of generality; we shall simply do without it. A more crucial deviation from the one-dimensional case is that Lemma 3.2-2 is no longer true; that is, when f is a member of $\mathcal{L}'_{a,a}$ and also of $\mathcal{L}'_{b,b}$, where $a < b$, it does not necessarily follow that f is or can be extended into a member of $\mathcal{L}'_{a,b}$. As an illustration of this, we offer

EXAMPLE 3.11-1. In the two-dimensional case where $t = \{t_1, t_2\}$, consider the regular generalized function:

$$e^{-|t|} = \exp\left(-\sqrt{t_1^2 + t_2^2}\right)$$

For $a = \{a_1, 0\}$ and $-1 < a_1 < 1$, $e^{-|t|}$ is a member of $\mathscr{L}'_{a,a}$ because $e^{-|t|}/\kappa_{a,a}(t)$ is absolutely integrable on the t plane. Similarly, for $b = \{0, b_2\}$ and $-1 < b_2 < 1$, $e^{-|t|}$ is a member of $\mathscr{L}'_{b,b}$. Next, let $\phi(t) = \exp(-a_1 t_1 - b_2 t_2)$. Then, for $a_1 < 0$ and $b_2 > 0$ and for $a = \{a_1, 0\}$ and $b = \{0, b_2\}$ as above, ϕ is a member of $\mathscr{L}_{a,b}$ because, with $k = \{k_1, k_2\}$ being a non-negative integer in $\mathscr{R}^2$,

$$\kappa_{a,b}(t) D^k \phi(t) = (-a_1)^{k_1} (-b_2)^{k_2} \kappa_{a_1,0}(t_1) e^{-a_1 t_1} \kappa_{0,b_2}(t_2) e^{-b_2 t_2}$$

and the right-hand side is bounded on the t plane.

Finally, choose $a_1 = -9/10$ and $b_2 = 9/10$. Then, as above, $e^{-|t|} \in \mathscr{L}'_{a,a}$ and $e^{-|t|} \in \mathscr{L}'_{b,b}$. However, $e^{-|t|}$ is not a member of $\mathscr{L}'_{a,b}$ because

$$\langle e^{-|t|}, \phi(t) \rangle \triangleq \int_{-\infty}^{\infty} \int_{-\infty}^{\infty} \exp\left[-\sqrt{t_1^2 + t_2^2} + \frac{9}{10}(t_1 - t_2) \right] dt_1 \, dt_2$$

and this integral is divergent. For instance, on the line $t_2 = -t_1$ the integrand is equal to

$$\exp\left(-\sqrt{2}|t_1| + \frac{9}{5} t_1\right)$$

which tends to infinity as $t_1 \to \infty$. This implies that, in some angular sector whose sides are the radial lines $t_2 = -(1 \pm \varepsilon) t_1$ where $t_1 > 0$ and ε is a small positive number, the integrand remains greater than 1, so that the above integral must certainly diverge.

In the one-dimensional case, our definition of a Laplace-transformable generalized function depended on Lemma 3.2-2. Since this lemma does not hold in n-dimensions, we must now proceed in a more general fashion.

LEMMA 3.11-1. *Let $\ell \in \mathscr{R}^1$ with $0 \leq \ell \leq 1$. Set $\sigma = \ell a + (1 - \ell) b$ (i.e., σ lies on the straight line connecting the points a and b). Then,*

$$\phi(t) \mapsto \frac{e^{-at}}{e^{-at} + e^{-bt}} \phi(t)$$

is a continuous linear mapping of $\mathscr{L}_{\sigma,\sigma}$ into $\mathscr{L}_{a,a}$. Similarly,

$$\phi(t) \mapsto \frac{e^{-bt}}{e^{-at} + e^{-bt}} \phi(t)$$

is a continuous linear mapping of $\mathscr{L}_{\sigma,\sigma}$ into $\mathscr{L}_{b,b}$.

PROOF. Leibniz's rule for the differentiation of a product holds also in n-dimensions:

$$D^k(\psi\phi) = \sum_{0 \leq p \leq k} \binom{k}{p} (D^{k-p}\psi)(D^p\phi)$$

Here, p ranges over all integers in $\mathscr{R}^n$ satisfying $0 \leq p \leq k$, and

$$\binom{k}{p} \triangleq \binom{k_1}{p_1}\binom{k_2}{p_2} \cdots \binom{k_n}{p_n}$$

Therefore, we may write

$$\begin{aligned} &\kappa_{a,\,a}(t) D^k \frac{e^{-at}\phi(t)}{e^{-at}+e^{-bt}} \\ (2) \qquad &= \sum_{0 \leq p \leq k} \binom{k}{p} \frac{\kappa_{a,\,a}(t)}{\kappa_{\sigma,\,\sigma}(t)} \left[D^{k-p} \frac{e^{-at}}{e^{-at}+e^{-bt}} \right] \kappa_{\sigma,\,\sigma}(t) D^p \phi(t) \end{aligned}$$

Some computation shows that for each $k - p$ there is a constant B_{k-p} such that

$$(3) \qquad \left| \frac{\kappa_{a,\,a}(t)}{\kappa_{\sigma,\,\sigma}(t)} D^{k-p} \frac{e^{-at}}{e^{-at}+e^{-bt}} \right| < B_{k-p} \qquad t \in \mathscr{R}^n$$

Consequently, (2) yields

$$\gamma_{a,\,a,\,k}\left[\frac{e^{-at}\phi(t)}{e^{-at}+e^{-bt}}\right] \leq \sum_{0 \leq p \leq k} \binom{k}{p} B_{k-p} \gamma_{\sigma,\,\sigma,\,p}[\phi(t)]$$

This implies our first conclusion. The second conclusion is obtained by interchanging the roles of a and b.

A functional f on an arbitrary domain $d(f)$ of conventional functions will be called *additive* if, for every finite set $\{\phi_\nu\}$ such that $\phi_\nu \in d(f)$ and $\sum_\nu \phi_\nu \in d(f)$, we have $\langle f, \sum_\nu \phi_\nu \rangle = \sum_\nu \langle f, \phi_\nu \rangle$. In the following we at times make the statement that $f \in \mathscr{L}'_{a,\,a}$, $f \in \mathscr{L}'_{b,\,b}, \ldots,$ and $f \in \mathscr{L}'_{z,\,z}$. By this we mean that f is an additive functional on $\mathscr{L}_{a,\,a} \cup \mathscr{L}_{b,\,b} \cup \cdots \cup \mathscr{L}_{z,\,z}$ whose restriction to each space in this union is linear and continuous.

Lemma 3.11-1 implies

LEMMA 3.11-2. *Let $f \in \mathscr{L}'_{a,\,a}$ and $f \in \mathscr{L}'_{b,\,b}$ where $a \neq b$. Set $\sigma = \ell a + (1 - \ell) b$ where $\ell \in \mathscr{R}^1$, $0 \leq \ell \leq 1$. Then, f can be extended into a member of $\mathscr{L}'_{\sigma,\,\sigma}$ by defining f as a functional on $\mathscr{L}_{\sigma,\,\sigma}$ through*

$$(4) \qquad \langle f, \phi \rangle \triangleq \left\langle f(t), \frac{e^{-at}\phi(t)}{e^{-at}+e^{-bt}} \right\rangle + \left\langle f(t), \frac{e^{-bt}\phi(t)}{e^{-at}+e^{-bt}} \right\rangle \qquad \phi \in \mathscr{L}_{\sigma,\,\sigma}$$

The definition (4) does not alter the values of f on $\mathscr{L}_{a,\,a}$ or $\mathscr{L}_{b,\,b}$ when $\ell = 0$ or $\ell = 1$ respectively; see Problem 3.11-2. On the other hand, this extension of f is additive on the domain $\mathscr{L}_{a,\,a} \cup \mathscr{L}_{\sigma,\,\sigma} \cup \mathscr{L}_{b,\,b}$. Moreover, f is unique: There is no other additive functional on the domain

$\mathscr{L}_{a,a} \cup \mathscr{L}_{\sigma,\sigma} \cup \mathscr{L}_{b,b}$ whose restrictions to $\mathscr{L}_{a,a}$ and $\mathscr{L}_{b,b}$ coincide with those of f. Indeed, assume that g is such a functional. Any $\phi \in \mathscr{L}_{\sigma,\sigma}$ may be decomposed into $\phi = \phi_a + \phi_b$ where, according to Lemma 3.11-1,

$$\phi_a \triangleq \frac{e^{-at}\phi(t)}{e^{-at} + e^{-bt}} \in \mathscr{L}_{a,a}$$

and

$$\phi_b \triangleq \frac{e^{-bt}\phi(t)}{e^{-at} + e^{-bt}} \in \mathscr{L}_{b,b}$$

Then,

$$\langle g, \phi \rangle = \langle g, \phi_a \rangle + \langle g, \phi_b \rangle = \langle f, \phi_a \rangle + \langle f, \phi_b \rangle = \langle f, \phi \rangle$$

Hence, g coincides with f on $\mathscr{L}_{\sigma,\sigma}$ as well.

As in the one-dimensional case, we shall call f a *Laplace-transformable generalized function* if it possesses the following four properties:

(i) f is a functional on some domain $d(f)$ of conventional functions.

(ii) If $\{\phi_\nu\}$ and $\{\psi_\nu\}$ are finite sets whose members are in $d(f)$ and if $\sum_\nu \phi_\nu = \sum_\nu \psi_\nu$(here, we do not require that $\sum_\nu \phi_\nu \in d(f)$), then $\sum_\nu \langle f, \phi_\nu \rangle = \sum_\nu \langle f, \psi_\nu \rangle$. If, in addition, $\sum_\nu \phi_\nu \in d(f)$, then $\langle f, \sum_\nu \phi_\nu \rangle = \sum_\nu \langle f, \phi_\nu \rangle$. (This is an extension of the additivity property used in the one-dimensional case.)

(iii) $\mathscr{L}_{a,b} \subset d(f)$ for at least one pair of points a and b, $a \in \mathscr{R}^n$, $b \in \mathscr{R}^n$, with $a < b$.

(iv) For every $\mathscr{L}_{c,d} \subset d(f)$, the restriction of f to $\mathscr{L}_{c,d}$ is in $\mathscr{L}'_{c,d}$.

By the n-dimensional analog to Sec. 3.2, note *IV*, if $a < b$ and $\mathscr{L}_{a,b} \subset d(f)$, then $f \in \mathscr{L}'_{c,c}$ for every c such that $a \leq c \leq b$. Let $\Lambda_f{}^0$ be the set of all $c \in \mathscr{R}^n$ for which there exists a pair $a_c, b_c \in \mathscr{R}^n$ with $a_c < c < b_c$ such that $\mathscr{L}_{a_c, b_c} \subset d(f)$. $\Lambda_f{}^0$ is an open set since it is the union of all such open sets $\{c: a_c < c < b_c\}$.

According to Lemma 3.11-2, f can be extended by means of (4) into a member of $\mathscr{L}'_{\sigma,\sigma}$ for every σ that lies on a straight line segment whose endpoints lie in $\Lambda_f{}^0$. By property (ii) of f, this extension of f into a member of $\mathscr{L}'_{\sigma,\sigma}$ does not depend on the choice of the straight line segment. (Prove this.) Let $\Lambda_f{}^1$ denote the set of all such σ. $\Lambda_f{}^1$ contains $\Lambda_f{}^0$.

$\Lambda_f{}^1$ is also an open set. Indeed, an arbitrary $\sigma \in \Lambda_f{}^1$ lies on a straight-line segment L having end points c_1 and c_2 in $\Lambda_f{}^0$. Since $\Lambda_f{}^0$ is open, there is an $r \in \mathscr{R}^1$ with $r > 0$ such that both of the spheres, $S_1 \triangleq \{c: c \in \mathscr{R}^n, |c - c_1| < r\}$ and $S_2 \triangleq \{c: c \in \mathscr{R}^n, |c - c_2| < r\}$, lie entirely within $\Lambda_f{}^0$. But then, every straight-line segment that is parallel to L and has one end point in S_1 and the other in S_2 lies entirely in $\Lambda_f{}^1$. This implies that the sphere $\{c: c \in \mathscr{R}^n, |c - \sigma| < r\}$ is also contained in Λ_f^1. Hence, every point of Λ_f^1 is an interior point; that is, Λ_f^1 is open.

Next, we can extend f by means of (4) again into a member of $\mathscr{L}'_{\sigma,\sigma}$ for every $\sigma \in \mathscr{R}^n$ lying on a straight-line segment whose end points lie in Λ_f^1. The second property of f again implies that this extension is independent of the choice of the straight-line segment. Let Λ_f^2 be the set of all such σ. Λ_f^2 is also an open set and contains Λ_f^1. Continues this process indefinitely. Let Ξ_f be the union of all such sets: $\Xi_f = \Lambda_f^0 \cup \Lambda_f^1 \cup \Lambda_f^2 \cup \cdots$. Consequently, Ξ_f is also open. Moreover, $\sigma \in \Xi_f$ if and only if f can be extended into a member of $\mathscr{L}'_{\sigma,\sigma}$ by the said procedure after a finite number of steps. (Actually, this procedure terminates in the sense that after the nth step the subsequent spaces Λ_f^m get no larger; that is, $\Lambda_f^m = \Lambda_f^n$ for every $m > n$. Moreover, Ξ_f is the convex hull of Λ_f^0; see Wilansky [1], p. 27. However, we do not need these facts.) The resulting functional f is additive on the domain $\bigcup_{\sigma \in \Xi_f} \mathscr{L}_{\sigma,\sigma}$ and unique in the sense that there is no other additive functional on the said domain whose restriction to each $\mathscr{L}_{\sigma,\sigma}$, where $\sigma \in \Lambda_f^0$, coincides with f. (Prove this also.)

Henceforth, we shall always assume that every Laplace-transformable generalized function f has been extended in this way onto the domain $\bigcup_{\sigma \in \Xi_f} \mathscr{L}_{\sigma,\sigma}$.

A set Θ in $\mathscr{R}^n$ or $\mathscr{C}^n$ is called *convex* if every straight-line segment with endpoints in Θ is contained in Θ. In other words, Θ is convex if and only if, for $\ell \in \mathscr{R}^1$ and $0 < \ell < 1$, $\ell a + (1 - \ell)b$ is contained in Θ whenever a and b are members of Θ.

THEOREM 3.11-1. *Ξ_f is a convex set.*

PROOF. Let a and b be arbitrary points in Ξ_f, and let $\ell \in \mathscr{R}^1$ with $0 < \ell < 1$. Set $\sigma = \ell a + (1 - \ell)b$. By the definition of Ξ_f, $a \in \Lambda_f^p$ and $b \in \Lambda_f^q$ for some p and q. But then, $\sigma \in \Lambda_f^s$ where $s = 1 + \max(p, q)$. Therefore, $\sigma \in \Xi_f$. Q.E.D.

LEMMA 3.11-3. *Let f be a Laplace-transformable generalized function. If the set $\Psi = \{\sigma: a \le \sigma \le b\}$ is contained in Ξ_f, then f can be extended into a member of $\mathscr{L}'_{a,b}$. This extension is unique in that only one member of $\mathscr{L}'_{a,b}$ has restrictions to all $\mathscr{L}_{\sigma,\sigma} (\sigma \in \Psi)$ that coincide with f on these $\mathscr{L}_{\sigma,\sigma}$.*

PROOF. Let λ be a smooth function on $\mathscr{R}^1$ such that $\lambda(\xi) = 0$ for $\xi < -1$ and $\lambda(\xi) = 1$ for $\xi > 1$. Then, for every $\phi \in \mathscr{L}_{a,b}$,

$$\phi(t) = \phi(t) \prod_{\nu=1}^{n} \{\lambda(t_\nu) + [1 - \lambda(t_\nu)]\} \tag{5}$$

By expanding the product on the right-hand side, we obtain a finite sum of terms, a typical term being

$$\phi(t)\lambda(t_1)[1 - \lambda(t_2)] \cdots \lambda(t_n) = \phi\lambda_{+,-,\ldots,+} \tag{6}$$

In the notation $\lambda_{+,-,\ldots,+}$ the first subscript $+$ represents the factor $\lambda(t_1)$, the second subscript $-$ represents the factor $1-\lambda(t_2)$, etc.

Set $a=\{a_1,\ldots,a_n\}$ and $b=\{b_1,\ldots,b_n\}$. We shall now show that the function (6) is a member of $\mathscr{L}_{\sigma,\sigma}$ where $\sigma=\{a_1,b_2,\ldots,a_n\}$, the component a_ν (or b_ν) being chosen if the νth subscript in the right-hand side of (6) is a plus sign (or respectively a minus sign). The function

$$\kappa_{\sigma,\sigma}D^k(\phi\lambda_{+,-,\ldots,+}) \tag{7}$$

is equal to zero everywhere on $\mathscr{R}^n$ except on the sector $t_1>-1$, $t_2<1,\cdots$, $t_n>-1$. On that sector $\kappa_{\sigma,\sigma}/\kappa_{a,b}$ is bounded by a constant, say, B. Thus, the magnitude of (7) is bounded by

$$B\kappa_{a,b}|D^k(\phi\lambda_{+,-,\ldots,+})|\le B\sum_{0\le p\le k}\binom{k}{p}\gamma_{a,b,p}(\phi)|D^{k-p}\lambda_{+,-,\ldots,+}|$$

Since every partial derivative of $\lambda_{+,-,\ldots,+}$ is bounded on $\mathscr{R}^n$, this verifies our assertion and shows moreover that $\phi\mapsto\phi\lambda_{+,-,\ldots,+}$ is a continuous linear mapping of $\mathscr{L}_{a,b}$ into $\mathscr{L}_{\sigma,\sigma}$.

By using the expansion of (5) into a sum of terms like (6), we extend f into a linear functional on $\mathscr{L}_{a,b}$ through the definition:

$$\langle f,\phi\rangle\triangleq\langle f,\phi\lambda_{+,+,\ldots,+}\rangle+\cdots+\langle f,\phi\lambda_{-,-,\ldots,-}\rangle\qquad \phi\in\mathscr{L}_{a,b} \tag{8}$$

We have shown in the preceding paragraph that $\phi\mapsto\phi\lambda_{+,+,\ldots,+}$ is a continuous linear mapping of $\mathscr{L}_{a,b}$ into $\mathscr{L}_{\sigma,\sigma}$ where $\sigma=\{a_1,a_2,\cdots,a_n\}\in\Xi_f$. Moreover, f is by hypothesis a member of $\mathscr{L}'_{\sigma,\sigma}$ for every $\sigma\in\Xi_f$. Hence, $\langle f,\phi_\nu\lambda_{+,+,\ldots,+}\rangle\to 0$ as $\nu\to\infty$ whenever $\phi_\nu\to 0$ in $\mathscr{L}_{a,b}$. A similar situation holds for each of the subsequent terms in the right-hand side of (8). Consequently, by Lemma 1.8-2, (8) extends f into a member of $\mathscr{L}'_{a,b}$. That this extension is unique follows from the expansion (5) and (6) and the fact that every $f\in\mathscr{L}'_{a,b}$ is additive on $\mathscr{L}_{a,b}$. This completes the proof.

Once again, we shall always assume that every Laplace-transformable generalized function f has been extended into a member of $\mathscr{L}'_{a,b}$ for every a,b such that the set $\{\sigma\colon a\le\sigma\le b\}$ is contained in Ξ_f.

We are now ready to define our *n-dimensional generalized Laplace transformation* $\mathfrak{L}$. A *tube* in $\mathscr{C}^n$ is any set $\{s\}$ of points s where Re s is restricted to some set in $\mathscr{R}^n$ and Im s ranges throughout all of $\mathscr{R}^n$. Thus, if $s^0=\sigma^0+i\omega^0$ is a member of a tube, then $s=\sigma^0+i\omega$ is too for every $\omega\in\mathscr{R}^n$.

As before, let f be a Laplace-transformable generalized function. The *tube of definition* for the Laplace transform of f is the set $\Omega_f=\{s\colon s\in\mathscr{C}^n,$ Re $s\in\Xi_f\}$, where Ξ_f is defined as above. Since Ξ_f is open and convex,

Ω_f is too. We define the *Laplace transform* $\mathfrak{L}f$ of f as a conventional function on Ω_f by

$$F(s) \triangleq (\mathfrak{L}f)(s) \triangleq \langle f(t), e^{-st} \rangle \qquad s \in \Omega_f \tag{9}$$

The right-hand side has a sense as the application of $f \in \mathscr{L}'_{\sigma,\sigma}$ to $e^{-st} \in \mathscr{L}_{\sigma,\sigma}$ where $\sigma = \text{Re } s \in \Xi_f$. Moreover, since Ω_f is open, for each $s \in \Omega_f$ we can always choose $a, b \in \Xi_f$ such that the tube $\{s: a < \text{Re } s < b\}$ is contained in Ω_f. In view of Lemma 3.11-3 and our convention about the extension of every Laplace-transformable generalized function, the right-hand side of (9) also has a sense as the application of $f \in \mathscr{L}'_{a,b}$ to $e^{-st} \in \mathscr{L}_{a,b}$. As before, whenever we write "$\mathfrak{L}f = F(s)$ for $s \in \Omega_f$", it is understood that f is a Laplace-transformable generalized function, which has been extended as stated, and that Ω_f is the tube of definition for $\mathfrak{L}f$ with Ξ_f being defined as above.

We close our discussion of the n-dimensional Laplace transformation by merely stating some of its properties; see Zemanian [2] for a more detailed discussion.

If $f(t)$ is a locally integrable function on $\mathscr{R}^n$ such that, for all $\sigma = \text{Re } s$ in some open convex subset Ξ of $\mathscr{R}^n$, the integral

$$\int_{\mathscr{R}^n} f(t)e^{-st}\, dt \tag{10}$$

converges absolutely, then $f(t)$ generates a regular generalized function whose Laplace transform coincides with (10) whenever $\text{Re } s \in \Xi$.

The operation-transform formulas of Sec. 3.4 continue to hold in the n-dimensional case. But now, the notation t^k is understood to mean $t_1^{k_1} t_1^{k_2} \cdots t_n^{k_n}$, and similarly for s^k.

THEOREM 3.11-2 (*The Analyticity Theorem*). *If $\mathfrak{L}f = F(s)$ for $s \in \Omega_f$, then $F(s)$ is analytic on Ω_f, and*

$$D^k F(s) = \langle f(t), (-1)^{|k|} t^k e^{-st} \rangle \qquad s \in \Omega_f$$

where $|k| = k_1 + k_2 + \cdots + k_n$.

In proving this theorem, we invoke Hartog's theorem (Bochner and Martin [1], p. 140); otherwise, the proof is just like that for the one-dimensional case.

THEOREM 3.11-3 (*The Uniqueness Theorem*). *If $\mathfrak{L}f = F(s)$ for $s \in \Omega_f$ and $\mathfrak{L}h = H(s)$ for $s \in \Omega_h$, if $\Omega_f \cap \Omega_h$ is not empty, and if $F(s) = H(s)$ for $s \in \Omega_f \cap \Omega_h$, then $f = h$ in the sense of equality in every $\mathscr{L}'_{a,b}$ for which the tube $\{s: a \leq \text{Re } s \leq b\}$ is contained in $\Omega_f \cap \Omega_h$.*

THEOREM 3.11-4. *Necessary and sufficient conditions for a function $F(s)$ to be the Laplace transform of a generalized function (according to the definition (9)) and for the correspondiny tube of definition to contain the closed tube $\Theta = \{s: a \le \operatorname{Re} s \le b\}$ are that $F(s)$ be analytic on Θ and that there be a polynomial P such that $|F(s)| \le P(|s|)$ on Θ. The polynomial will depend in general on Θ.*

THEOREM 3.11-5. *Let $\mathfrak{L}f = F(s)$ for $s \in \Omega_f$. Choose a closed subtube Θ of Ω_f of the form $\Theta = \{s: a \le \operatorname{Re} s \le b\} \subset \Omega_f$. Also, let $Q(s)$ be a polynomial in the components of s which is different from zero everywhere on Θ and such that*

$$\left|\frac{F(s)}{Q(s)}\right| \le \frac{K}{|s|^{n+1}} \qquad s \in \Theta$$

where K is a constant and n is the dimension of the complex euclidean space $\mathscr{C}^n$ in which s varies. Then, in the sense of equality in every $\mathscr{L}'_{c,d}$ for which $c > a$ and $d < b$, we have

$$f(t) = Q(D_t)\frac{1}{(2\pi i)^n}\int_{\sigma-i\infty}^{\sigma+i\infty}\frac{F(s)}{Q(s)}e^{st}\,ds \qquad a < \sigma < b$$

Here, the integration is on the domain in $\mathscr{C}^n$ traversed by $s = \sigma + i\omega$ as $\sigma \in \mathscr{R}^n$ remains fixed and ω varies throughout $\mathscr{R}^n$. Moreover, D_t represents generalized differentiation in $\mathscr{L}'_{c,d}$.

The discussion of convolution given in Secs. 3.7 and 3.8 can also be extended to the n-dimensional case with only a few changes. First, we show that for $g \in \mathscr{L}'_{a,b}$ and $\phi_\nu \in \mathscr{L}_{a,b}$, where $a, b \in \mathscr{R}^n$, $a \le b$, and $\nu = 1, 2, 3, \ldots$, the functions

$$\psi_\nu(t) = \langle g(\tau), \phi_\nu(t+\tau)\rangle$$

are also testing functions in $\mathscr{L}_{a,b}$, and $\psi_\nu \to 0$ in $\mathscr{L}_{a,b}$ as $\nu \to \infty$ whenever $\phi_\nu \to 0$ in $\mathscr{L}_{a,b}$.

The *convolution product* $f * g$ of $f \in \mathscr{L}'_{a,b}$ and $g \in \mathscr{L}'_{a,b}$, where $a \le b$, is defined as a functional on $\mathscr{L}_{a,b}$, by

$$\langle f * g, \phi\rangle = \langle f(t), \langle g(\tau), \phi(t+\tau)\rangle\rangle \qquad \phi \in \mathscr{L}_{a,b}$$

Moreover, $f * g$ is also a member of $\mathscr{L}'_{a,b}$; this is an immediate consequence of the preceding paragraph.

The exchange formula for the n-dimensional Laplace transformation is given by

THEOREM 3.11-6. *Let $\mathfrak{L}f = F(s)$ for $s \in \Omega_f$ and $\mathfrak{L}g = G(s)$ for $s \in \Omega_g$. If $\Omega_f \cap \Omega_g$ is not empty, then $f * g$ exists in the sense of convolution in every*

$\mathscr{L}'_{a,b}$ *for which* $a \leq b$ *and the tube* $\{s: a \leq \operatorname{Re} s \leq b\}$ *is contained in* $\Omega_f \cap \Omega_g$. *Moreover,*

$$\mathfrak{L}(f * g) = F(s)G(s) \qquad s \in \Omega_f \cap \Omega_g$$

PROBLEM 3.11-1. Prove the existence of the bound indicated in (3).

PROBLEM 3.11-2. (*a*) Let $a, b \in \mathscr{R}^n$ with $a \neq b$. Also, let $\phi \in \mathscr{L}_{a,a}$. Show that

$$\frac{e^{-bt}\phi(t)}{e^{-at} + e^{-bt}}$$

is also in $\mathscr{L}_{a,a}$. This implies that the definition (4) for the extension of f is consistent with the original definition of f on $\mathscr{L}_{a,a}$.

(*b*) Show that this extension of f is an additive functional on $\mathscr{L}_{a,a} \cup \mathscr{L}_{\sigma,\sigma} \cup \mathscr{L}_{b,b}$.

PROBLEM 3.11-3. Prove Lemma 3.11-2.

PROBLEM 3.11-4. Prove that the extension of f into an additive functional on $\bigcup_{\sigma \in \Xi_f} \mathscr{L}_{\sigma,\sigma}$ does not depend on the choices of the straight-line segments used in constructing the extension. Then, prove that there is only one additive functional on $\bigcup_{\sigma \in \Xi_f} \mathscr{L}_{\sigma,\sigma}$ that coincides with f on every $\mathscr{L}_{\sigma,\sigma}$ for which $\sigma \in \Lambda_f^0$.

PROBLEM 3.11-5. Prove the assertion concerning (10).

PROBLEM 3.11-6. If $f \in \mathscr{E}'$, then f is Laplace-transformable, and the corresponding tube of definition is all of $\mathscr{C}^n$. Why is this so?

PROBLEM 3.11-7. Show that under certain circumstances conventional n-dimensional convolution:

$$\int_{\mathscr{R}^n} f(\tau)g(t - \tau)\, d\tau$$

is a special case of our generalized convolution.

PROBLEM 3.11-8. State and prove an n-dimensional analog to Theorem 3.8-2.

PROBLEM 3.11-9. (*a*) Let Ξ be a nonvoid open convex set in $\mathscr{R}^n$. Let $\mathscr{L}(\Xi)$ be the union of all $\mathscr{L}_{a,b}$ spaces ($a < b$) for which the sets $\{\sigma: a \leq \sigma \leq b\}$ are contained in Ξ. Show that $\mathscr{L}(\Xi)$ is not in general a linear space under the usual rule for addition.

(*b*) Assign the following convergence rule to $\mathscr{L}(\Xi)$: A sequence $\{\phi_\nu\}$ converges in $\mathscr{L}(\Xi)$ if and only if it converges in some space $\mathscr{L}_{a,b}$ ($a < b$) that is contained in $\mathscr{L}(\Xi)$. Show that $\mathscr{L}(\Xi)$ is now a sequential-convergence space. Also, show that $\mathscr{D}$ is dense in $\mathscr{L}(\Xi)$ in the sense that for every

$\phi \in \mathscr{L}(\Xi)$ there exists a sequence $\{\phi_m\}_{m=1}^{\infty}$ such that $\phi_m \in \mathscr{D}$ and $\phi_m \to \phi$ in $\mathscr{L}(\Xi)$ as $m \to \infty$.

(c) Finally, let $\mathscr{L}'(\Xi)$ denote the collection of all functionals on $\mathscr{L}(\Xi)$ whose restriction to each $\mathscr{L}_{a,b}$ $(a < b)$ contained in $\mathscr{L}(\Xi)$ is linear and continuous. Define equality, addition, and multiplication by a complex number for the members of $\mathscr{L}'(\Xi)$ in the usual way. Also, assign the following convergence rule to $\mathscr{L}'(\Xi)$: A sequence $\{f_\nu\}$ converges in $\mathscr{L}'(\Xi)$ if and only if every $f_\nu \in \mathscr{L}'(\Xi)$ and there exists an $f \in \mathscr{L}'(\Xi)$ such that for every $\phi \in \mathscr{L}(\Xi)$ we have that $\langle f_\nu, \phi\rangle \to \langle f, \phi\rangle$ as $\nu \to \infty$. Show that $\mathscr{L}'(\Xi)$ is a sequential-convergence* linear space. Also, show that $\mathscr{L}'(\Xi)$ can be identified as a subspace of $\mathscr{D}'$.

By virtue of Theorem 3.11-3, we have that to every Laplace-transformable generalized function f there corresponds a unique nonvoid open convex set $\Xi_f \subset \mathscr{R}^n$ such that $f \in \mathscr{L}'(\Xi_f)$ and $f \notin \mathscr{L}'(\Theta)$ whenever the open convex set Φ is not entirely contained in Ξ_f.

3.12. The Inhomogeneous Wave Equation in One-Dimensional Space

We illustrate the use of the two-dimensional generalized Laplace transformation by solving the inhomogeneous wave equation:

$$(D_x^2 - c^{-2}D_t^2)u(x, t) = g(x, t) \tag{1}$$

where $x \in \mathscr{R}^1$, $t \in \mathscr{R}^1$, and $(x, t) \in \mathscr{R}^2$; also, $g(x, t)$ is a given Laplace-transformable generalized function, $u(x, t)$ is an unknown generalized function, and c is a real positive number, the speed of the wave. We shall first solve the differential equation:

$$(D_x^2 - c^{-2}D_t^2)h(x, t) = \delta(x, t) \tag{2}$$

where $\delta(x, t)$ is the delta functional concentrated on the origin of the (x, t) plane. Any such solution $h(x, t)$ is called an *elementary solution* (and also a *fundamental solution*) of the wave equation, and the solution $u(x, t)$ to (1) is obtained from $h(x, t)$ by forming the convolution product $u = h * g$, as we shall see.

Let $s \in \mathscr{C}^1$, $p \in \mathscr{C}^1$, and take $(s, p) \in \mathscr{C}^2$ as the independent variable in the transform domain. An application of the two-dimensional Laplace transformation to (2) and the two-dimensional analog to Sec. 3.4, Eq. (1) yields

$$\begin{aligned} H(s, p) &= \langle h(x, t), e^{-sx-pt}\rangle \\ &= [s^2 - c^{-2}p^2]^{-1} \end{aligned} \tag{3}$$

Upon choosing the tube of definition as the set $\Omega_h = \{(s, p)\colon \operatorname{Re} p > |\operatorname{Re} sc|\}$, (3) becomes a known Laplace transform, namely,

$$H(s, p) = \mathfrak{L}\left[-\frac{c}{2}\, 1_+(ct - |x|)\right] \qquad (s, p) \in \Omega_h \tag{4}$$

Indeed, that $1_+(ct - |x|)$ is Laplace-transformable with Ω_h as the corresponding tube of definition follows from the fact that

$$1_+(ct - |x|) \exp(-sx - pt)$$

is absolutely integrable on the (x, t) plane for each $(s, p) \in \Omega_h$ and is not integrable on the (x, t) plane otherwise. Moreover, for $\operatorname{Re} p > |\operatorname{Re} sc|$,

$$\begin{aligned}\mathfrak{L}[1_+(ct - |x|)] &= \int_0^\infty dt \int_{-ct}^{ct} e^{-sx-pt}\, dx \\ &= \int_0^\infty e^{-pt}\, \frac{e^{-sct} - e^{sct}}{-s}\, dt \\ &= \frac{2c}{p^2 - s^2c^2}\end{aligned}$$

which verifies (4). Thus,

$$h(x, t) = -\frac{c}{2}\, 1_+(ct - |x|)$$

is an elementary solution to the wave equation in one-dimensional space.

The solution to (1) is now given by

$$u(x, t) = h(x, t) * g(x, t) \tag{5}$$

For, by combining (2) with the two-dimensional analogs to Sec. 3.7, Eq. (12) and Sec. 3.8, Eq. (3), we obtain

$$\begin{aligned}(D_x^2 - c^{-2}D_t^2)u(x, t) &= [(D_x^2 - c^{-2}D_t^2)h(x, t)] * g(x, t) \\ &= \delta(x, t) * g(x, t) = g(x, t)\end{aligned}$$

Here, it has been implicitly assumed that the convolution (5) exists. This will certainly be the case if the tube of definition Ω_g for $\mathfrak{L}g$ intersects Ω_h. The latter is true if, for example, $g(x, t)$ is concentrated on a finite domain of the (x, t) plane because then Ω_g is all of $\mathscr{C}^2$. Moreover, (5) will satisfy (1) in the sense of equality and differentiation in every space $\mathscr{L}'_{a,b}$ for which the tube $\{s\colon a \le \operatorname{Re} s \le b\}$ is contained in $\Omega_g \cap \Omega_h$.

CHAPTER IV

The Mellin Transformation

4.1. Introduction

The *conventional Mellin transformation* maps a suitably restricted function $f(x)$ defined on $0 < x < \infty$ into a function $F(s)$ defined on some strip in the complex s plane by means of the integral:

$$(1) \qquad F(s) = \int_0^\infty f(x) x^{s-1}\, dx$$

Actually, this transformation can be obtained from the conventional two-sided Laplace transformation by replacing t by $-\log x$ and then $f(-\log x)$ by $f(x)$ in Sec. 3.1, Eq. (1). Because of this, many of the properties of the conventional Mellin transformation can be obtained by applying this change of variable to various properties of the Laplace transformation. The same situation occurs for the generalized Mellin and Laplace transformations, and, as a result, the Mellin transformation of a certain type of generalized function $f(x)$ on $0 < x < \infty$ can be defined as the application of $f(x)$ to the kernel x^{s-1}:

$$(2) \qquad F(s) = \langle f(x), x^{s-1} \rangle$$

The first one to discuss the generalized Mellin transformation was apparently Fung Kang [1]. He employed the methods used by Gelfand and Shilov to generalize the Fourier transformation to all distributions on $-\infty < x < \infty$ (Gelfand and Shilov [1], Vol. 1, Chapter II), and consequently obtained an indirect definition of the generalized Mellin transformation based upon a Parseval equation. On the other hand, the theory presented here is not as general but has the virtue of defining the generalized Mellin transformation directly by (2).

Just as the Laplace transformation generates an operational calculus for differential equations with constant coefficients, the Mellin transformation generates one for differential equations of the form $P(xD_x)u(x) = g(x)$, where P is a polynomial. This result is applied in the latter part of this chapter to analyze a certain time-varying electrical network, whose excitation is a generalized function, and also to solve Laplace's equation in an infinite wedge having a generalized-function boundary condition.

Srivastav and Parihar [1] have applied the generalized Mellin transformation discussed here to the theory of dual integral equations.

Finally, we mention that the generalized Mellin transformation can be extended to the n-dimensional case where $x \in \mathcal{R}^n$ and $x > 0$. For a discussion of this, see Zemanian [2].

4.2. The Testing-Function Spaces $\mathcal{M}_{a,b}$ and $\mathcal{M}(w, z)$ and Their Duals

Throughout this chapter, I denotes the positive half-axis $(0, \infty)$, and the real variables $t \in \mathcal{R}^1$ and $x \in I$ will always be related by $x = e^{-t}$, $t = -\log x$. For any $a, b \in \mathcal{R}^1$, let

$$\zeta_{a,b}(x) \triangleq \begin{cases} x^{-a} & 0 < x \leq 1 \\ x^{-b} & 1 < x < \infty \end{cases}$$

Thus, $\zeta_{a,b}(x) = \kappa_{a,b}(t)$, where $\kappa_{a,b}(t)$ is defined in Sec. 3.2.

$\mathcal{M}_{a,b}$ is the space of all smooth complex-valued functions $\theta(x)$ on I such that for each nonnegative integer k

$$\xi_k(\theta) \triangleq \xi_{a,b,k}(\theta) \triangleq \sup_{0<x<\infty} |\zeta_{a,b}(x) x^{k+1} D_x^k \theta(x)| < \infty$$

$\mathcal{M}_{a,b}$ is a linear space under the usual definitions of addition and multiplication by a complex number. The function x^{s-1} is a member of $\mathcal{M}_{a,b}$ if and only if $a \leq \operatorname{Re} s \leq b$. Also, for any positive integer k, $(\log x)^k x^{s-1}$ is a member of $\mathcal{M}_{a,b}$ if the only if $a < \operatorname{Re} s < b$.

The ξ_k are seminorms on $\mathcal{M}_{a,b}$ and ξ_0 is a norm. Moreover, $\mathcal{M}_{a,b}$ is understood to possess the topology generated by the multinorm $\{\xi_k\}_{k=0}^{\infty}$; thus, $\mathcal{M}_{a,b}$ is a countably multinormed space. By an argument similar to the proof of Lemma 3.2-1, it can be shown that $\mathcal{M}_{a,b}$ is complete. (This fact also follows directly from the next theorem and the fact that $\mathcal{L}_{a,b}$ is complete.) $\mathcal{M}'_{a,b}$, the dual of $\mathcal{M}_{a,b}$, is a linear space to which we assign the usual (weak) topology. By Theorem 1.8-3, $\mathcal{M}'_{a,b}$ is also complete.

If $a \leq c$ and $d \leq b$, then $\mathcal{M}_{c,d} \subset \mathcal{M}_{a,b}$, and the topology of $\mathcal{M}_{c,d}$ is stronger than that induced on it by $\mathcal{M}_{a,b}$. Therefore, the restriction of $f \in \mathcal{M}'_{a,b}$ to $\mathcal{M}_{c,d}$ is a member of $\mathcal{M}'_{c,d}$.

Next, let w denote either a finite real number or $-\infty$, and let z denote either a finite real number or $+\infty$, as before. Choose two monotonic sequences $\{a_\nu\}_{\nu=1}^{\infty}$ and $\{b_\nu\}_{\nu=1}^{\infty}$ such that $a_\nu \to w+$ and $b_\nu \to z-$. Let $\mathcal{M}(w, z)$ be the countable-union space of all $\mathcal{M}_{a_\nu, b_\nu}$ spaces; that is,

$\mathscr{M}(w, z) = \bigcup_{\nu=1}^{\infty} \mathscr{M}_{a_\nu, b_\nu}$ and a sequence converges in $\mathscr{M}(w, z)$ if and only if it converges in one of the $\mathscr{M}_{a_\nu, b_\nu}$ spaces. Also, let $\mathscr{M}'(w, z)$ be the dual of $\mathscr{M}(w, z)$. Since the $\mathscr{M}_{a,b}$ spaces are complete, the $\mathscr{M}(w, z)$ and $\mathscr{M}'(w, z)$ spaces are too (see Sec. 1.7 and Theorem 1.9-2). For every $k = 0, 1, 2, \ldots,$ we have that $(\log x)^k x^{s-1}$ is a member of $\mathscr{M}(w, z)$ if and only if $w < \operatorname{Re} s < z$. Finally, $\mathscr{M}_{a,b}$ and $\mathscr{M}(w, z)$ are testing-function spaces, and $\mathscr{M}'_{a,b}$ and $\mathscr{M}'(w, z)$ are generalized-function spaces.

THEOREM 4.2-1 *Let $x = e^{-t}$. The mapping*

$$\theta(x) \leftrightarrow e^{-t}\theta(e^{-t}) \triangleq \phi(t) \tag{1}$$

is an isomorphism from $\mathscr{M}_{a,b}$ onto $\mathscr{L}_{a,b}$. It is also an isomorphism from $\mathscr{M}(w, z)$ onto $\mathscr{L}(w, z)$. The inverse mapping is given by

$$\phi(t) \leftrightarrow x^{-1}\phi(-\log x) = \theta(x) \tag{2}$$

PROOF. That the mappings (1) and (2) are linear and inverses of one another is obvious. Now, assume that $\theta(x) \in \mathscr{M}_{a,b}$. Some computation shows that $D_t^k[e^{-t}\theta(e^{-t})]$ is equal to a finite sum of terms, a typical term being $a_p x^{p+1} D_x^p \theta(x)$ where $0 \le p \le k$ and a_p is a constant. Thus,

$$\kappa_{a,b}(t) D_t^k[e^{-t}\theta(e^{-t})] = \sum_p a_p \zeta_{a,b}(x) x^{p+1} D_x^p \theta(x)$$

so that

$$\gamma_{a,b,k}(\phi) = \gamma_{a,b,k}[e^{-t}\theta(e^{-t})] \le \sum_p |a_p| \xi_{a,b,p}[\theta(x)]$$

Consequently, (1) is also a continuous mapping of $\mathscr{M}_{a,b}$ into $\mathscr{L}_{a,b}$.

In the other direction, assume that $\phi(t) \in \mathscr{L}_{a,b}$. Again, a straightforward computation shows that

$$x^{k+1} D_x^k[x^{-1}\phi(-\log x)] = \sum_p b_p D_t^p \theta(t)$$

where $0 \le p \le k$ and the b_p are constants. Therefore,

$$\xi_{a,b,k}(\theta) = \xi_{a,b,k}[x^{-1}\phi(-\log x)] \le \sum_p |b_p| \gamma_{a,b,p}(\phi)$$

Thus, (2) is a continuous linear mapping of $\mathscr{L}_{a,b}$ into $\mathscr{M}_{a,b}$.

Since the mappings (1) and (2) are one-to-one, we can now conclude that they are also onto $\mathscr{L}_{a,b}$ and $\mathscr{M}_{a,b}$ respectively. The assertion concerning the $\mathscr{M}(w, z)$ and $\mathscr{L}(w, z)$ spaces follows immediately. Our proof is complete.

We can relate $\mathscr{M}'_{a,b}$ to $\mathscr{L}'_{a,b}$ and also $\mathscr{M}'(w, z)$ to $\mathscr{L}'(w, z)$ by employing the analog to the formula for a change of variable in an integral. Indeed, let $\phi(t)$ and $\theta(x)$ be related according to (1) and (2). To each $f(x) \in \mathscr{M}'_{a,b}$ or

$f(x) \in \mathcal{M}'(w, z)$, we can associate a functional on $\mathcal{L}_{a,b}$ or $\mathcal{L}(w, z)$ respecttively, which we denote by $f(e^{-t})$, through the equation:

$$\langle f(e^{-t}), \phi(t)\rangle \triangleq \langle f(x), \theta(x)\rangle \tag{3}$$

Thus, the mapping $f(x) \mapsto f(e^{-t})$ is the adjoint of the mapping $\phi(t) \mapsto \theta(x)$. Our notation $f(e^{-t})$ is motivated by the fact that this is what we would write if f were a conventional function and (3) were an equation between integrals. By Theorems 1.10-2 and 4.2-1, $f(x) \mapsto f(e^{-t})$ is an isomorphism from $\mathcal{M}'_{a,b}$ onto $\mathcal{L}'_{a,b}$ and also an isomorphism from $\mathcal{M}'(w, z)$ onto $\mathcal{L}'(w, z)$.

On the other hand, to each $g(t) \in \mathcal{L}'_{a,b}$ or $g(t) \in \mathcal{L}'(w, z)$ we associate a functional $g(-\log x)$ on $\mathcal{M}_{a,b}$ or $\mathcal{M}(w, z)$ respectively by the definition:

$$\langle g(-\log x), \theta(x)\rangle \triangleq \langle g(t), \phi(t)\rangle \tag{4}$$

the notation $g(-\log x)$ being chosen for the same reason as before. Thus, $g(t) \mapsto g(-\log x)$ is the inverse of the mapping $f(x) \mapsto f(e^{-t})$ and is an isomorphism from $\mathcal{L}'_{a,b}$ onto $\mathcal{M}'_{a,b}$ as well as from $\mathcal{L}'(w, z)$ onto $\mathcal{M}'(w, z)$.

For the sake of easy reference, we restate these results in

THEOREM 4.2-2. *Let θ and ϕ be related by* (1) *and* (2). *The mapping $f(x) \mapsto f(e^{-t})$, which is defined by* (3), *is an isomorphism from $\mathcal{M}'_{a,b}$ onto $\mathcal{L}'_{a,b}$ as well as from $\mathcal{M}'(w, z)$ onto $\mathcal{L}'(w, z)$. The inverse mapping $g(t) \mapsto g(-\log x)$ is defined by* (4).

Other results related to the spaces $\mathcal{M}_{a,b}$, $\mathcal{M}'_{a,b}$, $\mathcal{M}(w, z)$, and $\mathcal{M}'(w, z)$ are listed below. These can be established either directly by modifying the proofs of the corresponding results for the spaces $\mathcal{L}_{a,b}$, $\mathcal{L}'_{a,b}$, $\mathcal{L}(w, z)$, and $\mathcal{L}'(w, z)$ or by invoking Theorems 4.2-1 and 4.2-2 and similar assertions for other spaces such as $\mathcal{D}$, $\mathcal{D}(I)$, $\mathcal{D}'$, and $\mathcal{D}'(I)$.

I. $\mathcal{D}(I)$ is a subspace of both $\mathcal{M}_{a,b}$ and $\mathcal{M}(w, z)$, and convergence in $\mathcal{D}(I)$ implies convergence in $\mathcal{M}_{a,b}$ and also in $\mathcal{M}(w, z)$. Therefore, the restriction of any $f \in \mathcal{M}'_{a,b}$ or $f \in \mathcal{M}'(w, z)$ to $\mathcal{D}(I)$ is in $\mathcal{D}'(I)$. Moreover, $\mathcal{D}(I)$ is dense in $\mathcal{M}(w, z)$. This implies that $\mathcal{M}'(w, z)$ is a subspace of $\mathcal{D}'(I)$; it also implies that the values that $f \in \mathcal{M}'(w, z)$ assigns to $\mathcal{D}(I)$ uniquely determine the values that f assigns to $\mathcal{M}(w, z)$.

II. If $a \leq c$ and $d \leq b$, then $\mathcal{M}_{c,d} \subset \mathcal{M}_{a,b}$, and the topology of $\mathcal{M}_{c,d}$ is stronger than the topology induced on it by $\mathcal{M}_{a,b}$. Therefore, the restriction of any $f \in \mathcal{M}'_{a,b}$ to $\mathcal{M}_{c,d}$ is in $\mathcal{M}'_{c,d}$. (We pointed this out before.) On the other hand, if $w \leq u$ and $v \leq z$, $\mathcal{M}(u, v)$ is a dense subspace of $\mathcal{M}(w, z)$, and convergence in $\mathcal{M}(u, v)$ implies convergence in $\mathcal{M}(w, z)$. Therefore, $\mathcal{M}'(w, z)$ is a subspace of $\mathcal{M}'(u, v)$.

III. For any a and b, $\mathcal{M}_{a,b}$ is a dense subspace of $\mathcal{E}(I)$, and the topology of $\mathcal{M}_{a,b}$ is stronger than the topology induced on $\mathcal{M}_{a,b}$ by $\mathcal{E}(I)$.

Therefore, $\mathscr{E}'(I)$ can be identified with a subspace of $\mathscr{M}'_{a,b}$. Similar considerations show that $\mathscr{E}'(I)$ can also be identified with a subspace of $\mathscr{M}'(w, z)$, whatever be w and z.

IV. For each $f \in \mathscr{M}'_{a,b}$ there exist a positive constant C and a nonnegative integer r such that, for every $\theta \in \mathscr{M}_{a,b}$,

$$|\langle f, \theta \rangle| \le C \max_{0 \le k \le r} \xi_{a,b,k}(\theta)$$

V. If $f(x)$ is a locally integrable function on $0 < x < \infty$ such that $f(x)/\zeta_{a,b}(x)$ is absolutely integrable on $0 < x < \infty$, then $f(x)$ generates a regular member f of $\mathscr{M}'_{a,b}$ through the definition:

$$\text{(5)} \qquad \langle f, \theta \rangle \triangleq \int_0^\infty f(x)\theta(x)\, dx \qquad \theta \in \mathscr{M}_{a,b}$$

Also, if these conditions on $f(x)$ are satisfied for every a and b such that $a > w$ and $b < z$, then (5) also generates a regular member f of $\mathscr{M}'(w, z)$.

PROBLEM 4.2-1. Prove that $\mathscr{M}_{a,b}$ is complete without using Theorem 4.2-1.

PROBLEM 4.2-2. Show that (1) is an isomorphism from $\mathscr{D}(I)$ onto $\mathscr{D}$ and also an isomorphism from $\mathscr{E}(I)$ onto $\mathscr{E}$. This implies that (3) defines an isomorphism between $\mathscr{D}'$ and $\mathscr{D}'(I)$, as well as between $\mathscr{E}'$ and $\mathscr{E}'(I)$.

PROBLEM 4.2-3. Prove notes *I*, *II*, *III*, and *V* directly without using Theorems 4.2-1 and 4.2-2.

PROBLEM 4.2-4. State and prove an analog to Lemma 3.2-2 for the $\mathscr{M}_{a,b}$ spaces.

4.3. The Mellin Transformation

We say that f is a *Mellin-transformable generalized function* if it possesses the following properties:

(i) f is a functional on some domain $d(f)$ of conventional functions.

(ii) f is additive; that is, if ϕ, ψ, and $\phi + \psi$ are all members of $d(f)$, then $\langle f, \phi + \psi \rangle = \langle f, \phi \rangle + \langle f, \psi \rangle$.

(iii) $\mathscr{M}_{a,b} \subset d(f)$ for at least one pair of real numbers a, b with $a < b$.

(iv) For every $\mathscr{M}_{c,d} \subset d(f)$, the restriction of f to $\mathscr{M}_{c,d}$ is in $\mathscr{M}'_{c,d}$.

Let Λ_f be the set of all real numbers σ for which there exist a pair of real numbers a_σ, b_σ depending on σ such that $a_\sigma < \sigma < b_\sigma$ and $\mathscr{M}_{a_\sigma, b_\sigma} \subset d(f)$. Also, let σ_1 and σ_2 be the infimum and supremum of Λ_f. Possibly,

$\sigma_1 = -\infty$ and $\sigma_2 = +\infty$. In view of Theorem 4.2-2 and the first part of Sec. 3.3, f can be extended into a functional f_1 on $d(f) \cup \mathscr{M}(\sigma_1, \sigma_2)$ with the following two properties:

(A) *The restriction of* f_1 *to* $\mathscr{M}(\sigma_1, \sigma_2)$ *is a member of* $\mathscr{M}'(\sigma_1, \sigma_2)$.
(B) *The restriction of* f_1 *to* $d(f)$ *coincides with* f.

Moreover this extension is unique.

We shall always assume that every Mellin-transformable generalized function f has been extended into the functional f_1, but we shall denote f_1 simply by f. Under this convention, we have that *for every Mellin transformable generalized function f there exists a unique nonvoid open interval* (σ_1, σ_2) *such that f has a continuous linear restriction to* $\mathscr{M}(\sigma_1, \sigma_2)$ *and is not defined on all of* $\mathscr{M}(w, z,)$ *if either* $w < \sigma_1$ *or* $z > \sigma_2$.

Given a Mellin-transformable generalized function f we let Ω_f denote the strip in $\mathscr{C}^1$:

$$\Omega_f \triangleq \{s: \sigma_1 < \operatorname{Re} s < \sigma_2\}$$

where σ_1 and σ_2 are defined as above. Then, the *Mellin-transform* $\mathfrak{M}f$ of f is a conventional function $F(s)$ defined on Ω_f by

$$(1) \qquad F(s) \triangleq (\mathfrak{M}f)(s) \triangleq \langle f(x), x^{s-1} \rangle \qquad s \in \Omega_f$$

The right-hand side has a meaning because $f \in \mathscr{M}'(\sigma_1, \sigma_2)$ and $x^{s-1} \in \mathscr{M}(\sigma_1, \sigma_2)$ for any fixed $s \in \Omega_f$. The mapping $\mathfrak{M}$: $f \leftrightarrow F$ is called the *Mellin transformation*. Ω_f is the *strip* (or *region*) *of definition for* $\mathfrak{M}f$, and σ_1 and σ_2 are the *abscissas of definition for* $\mathscr{M}f$. Whenever we write $\mathfrak{M}f$, it is understood that f is a Mellin-transformable generalized function that has been extended as explained above.

Under the mapping defined in Theorem 4.2-1, the functions e^{-st} and x^{s-1} correspond to each other. This fact and Theorem 4.2-2 imply

THEOREM 4.3-1. *$f(x)$ is a Mellin-transformable generalized function if and only if $f(e^{-t})$ is a Laplace-transformable generalized function. In this case the strips of definition of $\mathfrak{M}[f(x)]$ and $\mathfrak{L}[f(e^{-t})]$ coincide, and $\mathfrak{M}[f(x)] = F(s) = \mathfrak{L}[f(e^{-t})]$ for $s \in \Omega_f$.*

Because of Theorems 4.2-1, 4.2-2, and 4.3-1, various properties of the Laplace transformation can be carried directly over to the Mellin transformation by making the substitutions: $t \leftrightarrow -\log x$, $\mathscr{L}(\sigma_1, \sigma_2) \leftrightarrow \mathscr{M}(\sigma_1, \sigma_2)$, $\mathscr{L}'(\sigma_1, \sigma_2) \leftrightarrow \mathscr{M}'(\sigma_1, \sigma_2)$, Laplace $\leftrightarrow$ Mellin, $\mathfrak{L} \leftrightarrow \mathfrak{M}$. For reference purposes we list some of these results here.

From Theorem 3.3-1 and Sec. 3.4, Eq. (5) we get

THEOREM 4.3-2 (*The Analyticity Theorem*). *If $\mathfrak{M}f = F(s)$ for $s \in \Omega_f$, then $F(s)$ is analytic on Ω_f, and*

(2) $D_s^k F(s) = \langle f(x), (\log x)^k x^{s-1} \rangle$

$$= \langle f(x), D_s^k x^{s-1} \rangle \qquad s \in \Omega_f; \qquad k = 1, 2, 3, \ldots$$

Similarly, Theorems 3.5-1 and 3.5-2 can be converted into the following inversion and uniqueness theorems.

THEOREM 4.3-3. *Let $\mathfrak{M}f = F(s)$ for $\sigma_1 < \operatorname{Re} s < \sigma_2$, and let r be a real variable. Then, in the sense of convergence in $\mathscr{D}'(I)$,*

$$f(x) = \lim_{r \to \infty} \frac{1}{2\pi i} \int_{\sigma - ir}^{\sigma + ir} F(s) x^{-s} ds$$

where σ is any fixed real number such that $\sigma_1 < \sigma < \sigma_2$.

THEOREM 4.3-4 (*The Uniqueness Theorem*). *If $\mathfrak{M}f = F(s)$ for $s \in \Omega_f$ and $\mathfrak{M}h = H(s)$ for $s \in \Omega_h$, if $\Omega_f \cap \Omega_h$ is not empty, and if $F(s) = H(s)$ for $s \in \Omega_f \cap \Omega_h$, then $f = h$ in the sense of equality in $\mathscr{M}'(w, z)$ where the interval $w < \sigma < z$ is the intersection of $\Omega_f \cap \Omega_h$ with the real axis.*

Furthermore, Theorem 3.6-1 becomes

THEOREM 4.3-5. *Necessary and sufficient conditions for a function $F(s)$ to be the Mellin transform of a generalized function f (according to the definition (1)) and for the corresponding strip of definition to be $\Omega_f = \{s: \sigma_1 < \operatorname{Re} s < \sigma_2\}$ are that $F(s)$ be analytic on Ω_f and, for each closed substrip $\{s: a \le \operatorname{Re} s \le b\}$ of Ω_f $(\sigma_1 < a < b < \sigma_2)$, there be a polynomial P such that $|F(s)| \le P(|s|)$ for $a \le \operatorname{Re} s \le b$. The polynomial P will depend in general on the choices of a and b.*

Theorems 4.3-4 and 4.3-5 imply that, for any choice of σ_1 and σ_2 with $\sigma_1 < \sigma_2$, the Mellin transformation is a one-to-one mapping from $\mathscr{M}'(\sigma_1, \sigma_2)$ onto the space of analytic functions on the strip $\sigma_1 < \operatorname{Re} s < \sigma_2$ which satisfy the polynomial growth conditions stated in Theorem 4.3-5.

The uniqueness theorem also justifies the use of a table of generalized Mellin transforms to invert a particular transform. (See Laughlin [1] for such a table.)

Corollary 3.6-1a can also be modified for the Mellin transformation, but we delay stating the result until the operator $-xD_x$, into which D_t is converted, is discussed. (See Theorem 4.4-1). Another result we shall need is the analog to Lemma 3.6-1, which we state as

THEOREM 4.3-6. *If, on the strip $\{s: a < \operatorname{Re} s < b\}$, $F(s)$ is an analytic function that satisfies $|F(s)| \le K|s|^{-2}$, where K is some constant, and if*

(3) $$f(x) = \frac{1}{2\pi i} \int_{\sigma - i\infty}^{\sigma + i\infty} F(s) x^{-s}\, ds \qquad a < \sigma < b$$

where σ is fixed, then $f(x)$ is a continuous function on $0 < x < \infty$ that does not depend on the choice of σ, and $f(x)$ generates a regular member of $\mathscr{M}'(a, b)$. Moreover, $\mathfrak{M}f = F(s)$ for at least $a < \operatorname{Re} s < b$.

PROBLEM 4.3-1. Without using Theorem 4.2-2, prove directly that every Mellin-transformable generalized function is or can be extended into a member of $\mathscr{M}'(\sigma_1, \sigma_2)$ as stated in this section. *Hint:* Use the analog to Lemma 3.2-2 that was asked for in Problem 4.2-4.

PROBLEM 4.3-2. Show that, if $f(x)$ is a locally integrable function on $0 < x < \infty$ such that, for each σ in some open interval (σ_1, σ_2), $x^{\sigma-1}f(x)$ is absolutely integrable on $0 < x < \infty$, then the conventional Mellin transform:

$$F(s) = \int_0^\infty f(x)x^{s-1}\,dx$$

is equal to the Mellin transform of the regular generalized function generated by $f(x)$.

PROBLEM. 4.3-3. Find the Mellin transforms and the corresponding regions of definition for the following generalized functions. Here, a is a real positive number, k is a nonnegative integer, and $1_+(t)$ denotes the unit-step function.

(*a*) $\delta^{(k)}(x - a)$
(*b*) $x^k 1_+(x - a)$
(*c*) $x^k 1_+(a - x)$
(*d*) e^{-ax}

PROBLEM 4.3-4. Let a be a real positive number. Show that

$$\sum_{\nu=1}^{\infty} \delta(x - a\nu)$$

is a member of $\mathscr{M}'(-\infty, 0)$. Then, show that

$$\mathfrak{M}\sum_{\nu=1}^{\infty} \delta(x - a\nu) = a^{s-1}\zeta(1 - s) \qquad \operatorname{Re} s < 0$$

where ζ denotes Riemann's zeta function (Erdelyi, Ed., [1], Vol. I, p. 32).

PROBLEM 4.3-5. State some necessary and sufficient conditions on the Mellin transform $F = \mathfrak{M}f$ in order for $f(x)$ to be concentrated on the interval $0 < x \leq X (X < \infty)$. Do the same thing for the case where $f(x)$ is concentrated on the interval $X \leq x < \infty$ $(X > 0)$.

PROBLEM 4.3-6. Devise a generalized Mellin transformation that is applicable to all members of $\mathscr{D}'(I)$. Identify the resulting space of Mellin transforms. *Hint:* See Problem 3.3-4.

4.4. Operation-Transform Formulas

We now present a number of operations that may be applied to Mellin-transformable generalized functions and derive the corresponding operation-transform formulas for the Mellin transformation. Throughout this section we assume that $F(s) = \mathfrak{M}f$ for $s \in \Omega_f$. Moreover, k denotes a positive integer.

Multiplication by $(\log x)^k$: The operation $\theta(x) \mapsto (\log x)^k \theta(x)$ is a continuous linear mapping of $\mathscr{M}(w, z)$ into itself for every choice of w and z. The easiest way to prove this is to invoke Theorem 4.2-1 and the corresponding fact concerning the mapping $\phi(t) \mapsto t^k \phi(t)$ where $\phi(t) = e^{-t}\theta(e^{-t})$. Furthermore, in accordance with Sec. 2.5 and Theorem 1.10-2, the adjoint mapping $f(x) \mapsto (\log x)^k f(x)$ is defined on $\mathscr{M}'(w, z)$ by

$$\langle (\log x)^k f, \theta \rangle = \langle f, (\log x)^k \theta \rangle \qquad \theta \in \mathscr{M}(w, z)$$

and is a continuous linear mapping of $\mathscr{M}'(w, z)$ into itself. Therefore, choosing $f(x) \in \mathscr{M}'(\sigma_1, \sigma_2)$, $\theta(x) = x^{s-1}$, and $\sigma_1 < \operatorname{Re} s < \sigma_2$, we may write

$$\langle (\log x)^k f(x), x^{s-1} \rangle = \langle f(x), (\log x)^k x^{s-1} \rangle$$

In view of Theorem 4.3-2 we have obtained the following operation-transform formula:

$$\mathfrak{M}(\log x)^k f = D_s^k F(s) \qquad s \in \Omega_f \tag{1}$$

Multiplication by a power of x: Let α be a complex number and $r = \operatorname{Re} \alpha$. In accordance with Sec. 2.5, the operator $f \mapsto x^\alpha f$ is defined on $f \in \mathscr{M}'(w, z)$ by

$$\langle x^\alpha f(x), \theta(x) \rangle \triangleq \langle f(x), x^\alpha \theta(x) \rangle \qquad \theta \in \mathscr{M}(w - r, z - r) \tag{2}$$

It is readily shown that $\theta \mapsto x^\alpha \theta$ is an isomorphism from $\mathscr{M}(w - r, z - r)$ onto $\mathscr{M}(w, z)$. Therefore, (2) has a sense, and, by Theorem 1.10-2, the adjoint operator $f \mapsto x^\alpha f$ is an isomorphism from $\mathscr{M}'(w, z)$ onto $\mathscr{M}'(w - r, z - r)$.

Now, assume that $\mathfrak{M}f = F(s)$ for $s \in \Omega_f = \{s\colon \sigma_1 < \operatorname{Re} s < \sigma_2\}$. Consequently, $f \in \mathscr{M}'(\sigma_1, \sigma_2)$ and $x^\alpha f \in \mathscr{M}'(\sigma_1 - r, \sigma_2 - r)$. Choosing $s + \alpha \in \Omega_f$, we have that $x^{s-1} \in \mathscr{M}(\sigma_1 - r, \sigma_2 - r)$, and we may write

$$\langle x^\alpha f(x), x^{s-1} \rangle = \langle f(x), x^{s+\alpha-1} \rangle$$

This may be rewritten as the operation-transform formula:

$$\mathfrak{M}x^\alpha f = F(s + \alpha) \qquad s + \alpha \in \Omega_f \tag{3}$$

Another way to obtain these results is to use the change of variable $x = e^{-t}$, $t = -\log x$ and to invoke Theorem 4.2-2. Thus, for $\theta(x) \in \mathscr{M}(\sigma_1, \sigma_2)$ and $\phi(t) \triangleq e^{-t}\,\theta(e^{-t}) \in \mathscr{L}(\sigma_1, \sigma_2)$,

$$\langle x^\alpha f(x), \theta(x)\rangle = \langle e^{-\alpha t} f(e^{-t}), \phi(t)\rangle$$

Since $f(e^{-t}) \mapsto e^{-\alpha t} f(e^{-t})$ is an isomorphism from $\mathscr{L}'(\sigma_1, \sigma_2)$ onto $\mathscr{L}'(\sigma_1 - r, \sigma_2 - r)$ (see Sec. 3.4), Theorem 4.2-2 implies that $f(x) \mapsto x^\alpha f(x)$ is an isomorphism from $\mathscr{M}'(\sigma_1, \sigma_2)$ onto $\mathscr{M}'(\sigma_1 - r, \sigma_2 - r)$. Moreover, by substituting x^{s-1} for $\theta(x)$, we get from Sec. 3.4, Eq. (7)

$$\langle x^\alpha f(x), x^{s-1}\rangle = \langle e^{-\alpha t} f(e^{-t}), e^{-st}\rangle = F(s + \alpha) \qquad s + \alpha \in \Omega_f$$

which agrees with (3).

Differentiation: It is very easy to see that the operator $\theta \mapsto (-D)^k\,\theta$ is a continuous linear mapping of $\mathscr{M}_{a+k,\,b+k}$ into $\mathscr{M}_{a,\,b}$ and therefore a continuous linear mapping of $\mathscr{M}(w + k, z + k)$ into $\mathscr{M}(w, z)$. Consequently, the adjoint operator $f \mapsto D^k f$, which is defined by

$$(4) \qquad \langle D^k f, \theta\rangle \triangleq \langle f, (-D)^k\,\theta\rangle \qquad \theta \in \mathscr{M}(w + k, z + k)$$

is a continuous linear mapping of $\mathscr{M}'(w, z)$ into $\mathscr{M}'(w + k, z + k)$. Therefore, for $s - k \in \Omega_f = \{s: \sigma_1 < \operatorname{Re} s < \sigma_2\}$ and $f \in \mathscr{M}'(\sigma_1, \sigma_2)$, we may write

$$\begin{aligned}\langle D^k f(x), x^{s-1}\rangle &= \langle f(x), (-D)^k\, x^{s-1}\rangle \\ &= \langle f(x), (-1)^k\,(s - k)_k\, x^{s-k-1}\rangle\end{aligned}$$

where

$$(a)_k \triangleq a(a + 1)\cdots(a + k - 1) \qquad k = 1, 2, 3, \ldots$$

This yields the following operation-transform formula:

$$(5) \qquad \mathfrak{M} D^k f = (-1)^k\,(s - k)_k\, F(s - k) \qquad s - k \in \Omega_f$$

Other differential operators: By combining the preceding two operators, we see that $f \mapsto x^k D^k f$ is defined on $\mathscr{M}'(w, z)$ by

$$\langle x^k D^k f, \theta\rangle \triangleq \langle f, (-D)^k\, x^k\,\theta\rangle \qquad \theta \in \mathscr{M}(w, z)$$

and is a continuous linear mapping of $\mathscr{M}'(w, z)$ into itself. As a result, for $s \in \Omega_f = \{s: \sigma_1 < \operatorname{Re} s < \sigma_2\}$, $\theta(x) = x^{s-1}$, and $f(x) \in \mathscr{M}'(\sigma_1, \sigma_2)$, we have

$$\begin{aligned}\langle x^k D^k f(x), x^{s-1}\rangle &= \langle f(x), (-D)^k\, x^{s+k-1}\rangle \\ &= (-1)^k (s)_k \langle f(x), x^{s-1}\rangle\end{aligned}$$

or

$$(6) \qquad \mathfrak{M} x^k D^k f = (-1)^k (s)_k\, F(s) \qquad s \in \Omega_f$$

As a special case, the operator $f \mapsto xDf$ is a continuous linear mapping of $\mathscr{M}'(w, z)$ into $\mathscr{M}'(w, z)$. Consequently, $f \mapsto (xD)^k f$ is also such a mapping and is defined by

$$\langle (xD)^k f, \theta \rangle \triangleq \langle f, (-Dx)^k \theta \rangle \qquad \theta \in \mathscr{M}(w, z)$$

(Here, the notation $(-Dx)^k \theta$ denotes $(-1)^k D(x \ldots D(xD(x\theta)) \ldots)$, where there are k pairs of parentheses. The notation $(xD)^k$ is interpreted similarly.) Thus, for $s \in \Omega_f = \{s: \sigma_1 < \operatorname{Re} s < \sigma_2\}$, $\theta(x) = x^{s-1}$, and $f(x) \in \mathscr{M}'(\sigma_1, \sigma_2)$, we have

$$\langle (xD)^k f(x), x^{s-1} \rangle = \langle f(x), (-1)^k s^k x^{s-1} \rangle$$

so that

$$\mathfrak{M}(xD)^k f = (-1)^k s^k F(s) \qquad s \in \Omega_f \tag{7}$$

Similar arguments may be used to establish that

$$\mathfrak{M} D^k x^k f = (-1)^k (s - k)_k F(s) \qquad s \in \Omega_f \tag{8}$$

$$\mathfrak{M}(Dx)^k f = (-1)^k (s - 1)^k F(s) \qquad s \in \Omega_f \tag{9}$$

Incidentally, upon comparing (7) with Sec. 3.4, Eq. (1) and invoking the uniqueness theorems for the Laplace and Mellin transformations, we see that our change of variable converts generalized differentiation D_t into the generalized differential operator $-xD_x$. By using this fact, we can convert Corollary 3.6-1a into the following theorem, which presents another inversion formula.

THEOREM 4.4-1. *Let $\mathfrak{M}f = F(s)$ for $s \in \Omega_f$. Choose three fixed real numbers a, σ, and b in Ω_f such that $a < \sigma < b$. Also, choose a polynomial $Q(s)$ that has no zeros for $a \le \operatorname{Re} s \le b$ and satisfies*

$$\left| \frac{F(s)}{Q(s)} \right| \le \frac{K}{|s|^2} \qquad a < \operatorname{Re} s < b$$

where K is a constant. Then, in the sense of equality in $\mathscr{M}'(a, b)$,

$$f(x) = Q(-xD_x) \frac{1}{2\pi i} \int_{\sigma - i\infty}^{\sigma + i\infty} \frac{F(s)}{Q(s)} x^{-s} \, ds \qquad a < \sigma < b$$

where xD_x is a generalized differential operator in $\mathscr{M}'(a, b)$ and the integral converges in the conventional sense to a continuous function that generates a regular member of $\mathscr{M}'(a, b)$.

PROBLEM 4.4-1. Prove directly the assertions concerning the operators $\theta \mapsto (\log x)^k \theta$ and $f \mapsto (\log x)^k f$ without invoking Theorems 4.2-1 and 4.2-2.

PROBLEM 4.4-2. Let α be a complex number and $r = \operatorname{Re} \alpha$. Show that $\theta \mapsto x^{\alpha} \theta$ is an isomorphism from $\mathscr{M}(w - r, z - r)$ onto $\mathscr{M}(w, z)$.

PROBLEM 4.4-3. Show that $\theta \mapsto (-D)^k \theta$ is a continuous linear mapping of $\mathscr{M}(w + k, z + k)$ into $\mathscr{M}(w, z)$.

PROBLEM 4.4-4. Derive (7) by applying Theorem 4.2-2 to Sec. 3.4, Eq. (1).

PROBLEM 4.4-5. Derive formulas (8) and (9).

PROBLEM 4.4-6. Derive the following formulas wherein r is a fixed real number.

(a) $\mathfrak{M}f(rx) = r^{-s} F(s)$ $\qquad r > 0,\ s \in \Omega_f$

(b) $\mathfrak{M}f(x^{-1}) = F(-s)$ $\qquad -s \in \Omega_f$

(c) $\mathfrak{M}f(x^r) = |r|^{-1} F(r^{-1}s)$ $\qquad r \neq 0,\ r^{-1} s \in \Omega_f$

Hint: First discuss the following definitions:

(a′) $\langle f(rx), \theta(x) \rangle \triangleq \langle f(x), r^{-1} \theta(r^{-1}x) \rangle$

(b′) $\langle f(x^{-1}), \theta(x) \rangle \triangleq \langle f(x), x^{-2} \theta(x^{-1}) \rangle$

(c′) $\langle f(x^r), \theta(x) \rangle \triangleq \langle f(x), |r|^{-1} x^{(1-r)/r} \theta(x^{1/r}) \rangle$

4.5. An Operational Calculus for Euler Differential Equations

Consider the linear differential equation:

$$(1) \qquad Lu(x) \triangleq (a_n x^n D^n + a_{n-1} x^{n-1} D^{n-1} + \cdots + a_0) u(x) = g(x)$$

where the a_ν are constants and $a_n \neq 0$. Such equations are called at times *Euler differential equations*. The Mellin transformation generates an operational calculus by means of which (1) may be solved for the unknown $u(x)$ when $g(x)$ is a known Mellin-transformable generalized function.

Let $\mathfrak{M}g = G(s)$, where the strip of definition Ω_g is $\{s: \sigma_{g1} < \operatorname{Re} s < \sigma_{g2}\}$, and denote $\mathfrak{M}u$ by $U(s)$. By applying the Mellin transformation to (1) and employing Sec. 4.4, Eq. (6), we obtain $B(s)U(s) = G(s)$, where $B(s)$ is the polynomial

$$B(s) = a_n(-1)^n (s)_n + a_{n-1}(-1)^{n-1}(s)_{n-1} + \cdots + a_0$$

and

$$(s)_\nu \triangleq s(s+1) \cdots (s + \nu - 1)$$

If $B(s)$ possesses no zeros in the strip Ω_g, then by Theorem 4.3-5, there exists a Mellin-transformable generalized function $u(x)$ whose transform is equal to $G(s)/B(s)$ on Ω_g. By Theorem 4.3-4, $u(x)$ is unique as a member of $\mathscr{M}'(\sigma_{g1}, \sigma_{g2})$. Moreover, $u(x)$ is a solution of (1) in the sense of equality in $\mathscr{M}'(\sigma_{g1}, \sigma_{g2})$, according to Sec. 4.4, Eq. (6) and Theorem 4.3-4 again. As a convenient symbolism, we write

$$u(x) = \mathfrak{M}^{-1} \frac{G(s)}{B(s)} \qquad s \in \Omega_g$$

We can determine u (at least its restriction to some subspace of $\mathscr{M}(\sigma_{g1}, \sigma_{g2})$) either by using Theorems 4.3-3 or 4.4-1, or by referring to a table of Mellin transforms (Laughlin [1]).

On the other hand, assume that $B(s)$ does possess zeros in Ω_g. These will be finite in number, and we will have a set of m adjacent substrips:

(2) $\sigma_{g1} = \sigma_0 < \operatorname{Re} s < \sigma_1, \sigma_1 < \operatorname{Re} s < \sigma_2, \ldots, \sigma_{m-1} < \operatorname{Re} s < \sigma_m = \sigma_{g2}$

on which $G(s)/B(s)$ is analytic and satisfies the polynomial growth condition of Theorem 4.3-5. By the same reasons as above, we can conclude that for each such substrip, say, $\sigma_\nu < \operatorname{Re} s < \sigma_{\nu+1}$, there exists a unique member $u(x)$ of $\mathscr{M}'(\sigma_\nu, \sigma_{\nu+1})$ which satisfies (1) in the space $\mathscr{M}'(\sigma_\nu, \sigma_{\nu+1})$ and whose Mellin transform is equal to $G(s)/B(s)$ on $\sigma_\nu < \operatorname{Re} s < \sigma_{\nu+1}$. We denote this solution by

$$u(x) = \mathfrak{M}^{-1} \frac{G(s)}{B(s)} \qquad \sigma_\nu < \operatorname{Re} s < \sigma_{\nu+1}$$

For any other choice of substrip we will obtain a different solution. (It can be shown that the difference between any two such solutions is a smooth function on $0 < x < \infty$ that satisfies the homogeneous equation $Lu = 0$ in the conventional sense.)

EXAMPLE 4.5-1. Here is a simple example of a physical situation that leads to an Euler differential equation. Another example will be given by Sec. 4.7.

Consider the electrical circuit of Fig. 4.5-1 consisting of a series connection of a time-varying inductance $L(t) = t$ henrys and a fixed resistance $R = 1$ ohm. Here, t represents time. We assume that the circuit has no initial excitation in it and that the driving voltage $v(t)$ is zero until some positive instant of time $t = T > 0$. We wish to determine the responding current $j(t)$.

Kirchhoff's laws lead to the following differential equation:

$$v = D(Lj) + Rj = D(tj) + j = tDj + 2j$$

Assuming that $v(t)$ is a Mellin-transformable generalized function concentrated on $T \leq t < \infty$, we may apply the generalized Mellin transformation to get

$$V(s) = -sJ(s) + 2J(s)$$

where

$$V = \mathfrak{M}v, \qquad J = \mathfrak{M}j$$

The region of definition for $\mathfrak{M}v = V(s)$ must be a half-plane extending infinitely to the left because v is concentrated on $T \leq t < \infty$. (Indeed, v will be a member of $\mathscr{M}'_{a,b}$ for every value of a and some b, whence our assertion.) Denote this half-plane by $\Omega_v = \{s: \text{Re } s < \sigma_v\}$. Physical considerations dictate that $j(t)$ will also be concentrated on $T \leq t < \infty$. Hence, the solution $j(t)$ is that unique Mellin-transformable generalized function whose Mellin transform is equal to $V(s)/(2 - s)$ and has as its region of definition $\{s: \text{Re } s < \min (\sigma_v, 2)\}$.

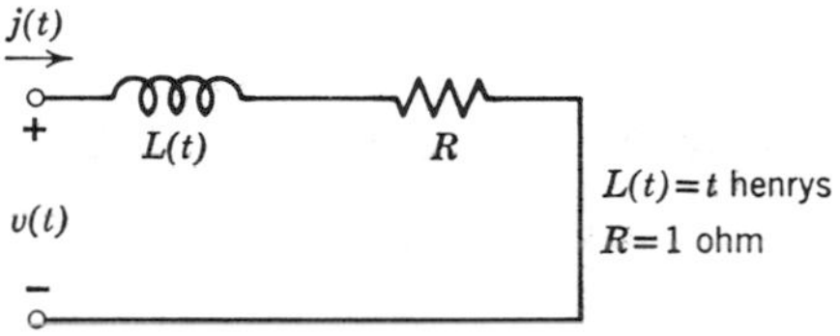

FIGURE 4.5-1

PROBLEM 4.5-1. Using the notation introduced in this section, assume that $B(s)$ has roots in Ω_g. Show that the difference u_d between two Mellin-transformable solutions of $Lu = g$, corresponding to different choices of the substrips (2) in Ω_g, is a smooth function on $0 < x < \infty$ that satisfies $Lu_d = 0$ in the conventional sense. *Hint*: Use Theorem 4.4-1.

PROBLEM 4.5-2. Let X be a fixed positive number and $1_+(x - X)$ the shifted unit step function:

$$1_+(x - X) = \begin{cases} 0 & 0 < x < X \\ \frac{1}{2} & x = X \\ 1 & X < x < \infty \end{cases}$$

Also, let L be the differential operator defined in (1) and $u(x)$ a solution to the homogeneous equation $Lu(x) = 0$, $0 < x < \infty$. It is a fact that $u(x)$ is smooth on $0 < x < \infty$. Show that

$$L[u(x)1_+(x - X)] = \sum_{\nu=0}^{n-1} b_\nu D_x{}^\nu \delta(x - X) \tag{3}$$

where

$$b_\nu = a_{\nu+1} x^{\nu+1} u_0 + a_{\nu+2} x^{\nu+2} u_1 + \cdots + a_n x^n u_{n-\nu-1}$$

$$u_\nu = u^{(\nu)}(X) = D^\nu u(x)|_{x=X}$$

(Actually, the right-hand side of (3) represents a convenient means of introducing initial conditions onto Euler differential equations in a form suitable for analysis by our generalized Mellin transformation. That is, if we solve (3) by using the Mellin transformation and taking for Ω_u a half-plane extending infinitely to the left, we will obtain the regular generalized function $u(x)1_+(x-X)$ where $u(x)$ is the conventional solution of the following initial-value problem: $Lu = 0$, $u^{(\nu)}(X) = u_\nu$.)

PROBLEM 4.5-3. In Example 4.5-1, let $v(t) = t^{-1}1_+(t-1)$. Determine $j(t)$. *Hint*: First determine the Mellin transform of $t^{-\alpha}1_+(t-1)$ where α is a constant.

PROBLEM 4.5-4. Find the Mellin transform of the current $j(t)$ in the electrical circuit of Fig. 4.5-2. The resistances are indicated in ohms and

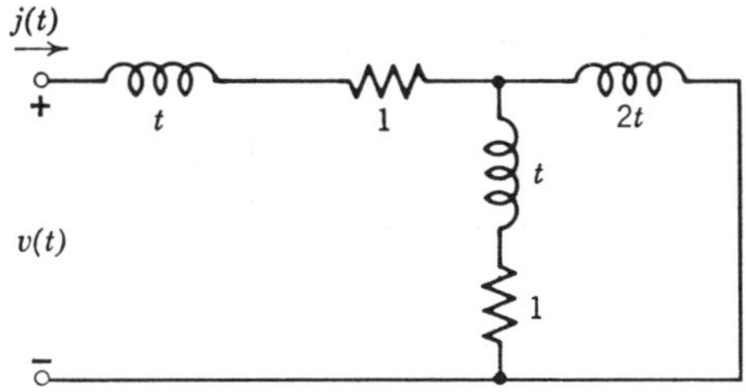

FIGURE 4.5-2

the inductances, which vary linearly with time, are indicated in henrys. Assume that the voltage $v(t)$ is a Mellin-transformable generalized function concentrated on the interval $T \le t < \infty$, where $T > 0$. Also, assume that the circuit has no initial excitation in it.

PROBLEM 4.5-5. Find a general class of electrical networks which under mesh analyses generate simultaneous systems of Euler differential equations. Do the same thing for nodal analyses.

4.6. Mellin-Type Convolution

There is a certain type of convolution for Mellin-transformable generalized functions. It can be derived by applying the customary change of variable to the convolution of Laplace-transformable generalized functions.

Let a and b be two real numbers with $a \le b$. *Mellin-type convolution* is an operation that assigns to each arbitrary choice of the pair $f \in \mathscr{M}'_{a,b}$ and $g \in \mathscr{M}'_{a,b}$ the *Mellin-type convolution product* $f \vee g$ defined by

$$\text{(1)} \qquad \langle f \vee g, \theta \rangle \triangleq \langle f(x), \langle g(y), \theta(xy) \rangle \rangle \qquad \theta \in \mathscr{M}_{a,b}$$

This product $f \vee g$ is also a member of $\mathcal{M}'_{a,b}$.

To prove this assertion, we make the change of variables $x = e^{-t}$ and $y = e^{-\tau}$ in accordance with Theorem 4.2-1. Thus, we set

$$\phi(t+\tau) = e^{-t-\tau}\,\theta(e^{-t-\tau})$$

or equivalently $\theta(xy) = (xy)^{-1}\phi(-\log xy)$. By Lemma 3.7-2 and Theorem 4.2-2,

$$\langle g(e^{-\tau}),\, \phi(t+\tau)\rangle \tag{2}$$

is a member of $\mathcal{L}_{a,b}$ whenever $\phi \in \mathcal{L}_{a,b}$. Therefore, by Theorems 4.2-1 and 4.2-2, we can convert (2) into the following member of $\mathcal{M}_{a,b}$:

$$x^{-1}\langle g(y),\, y^{-1}\phi(-\log x - \log y)\rangle = \langle g(y),\, \theta(xy)\rangle$$

This shows that the right-hand side of (1) has a sense.

To complete the proof that $f \vee g \in \mathcal{M}'_{a,b}$, we invoke Theorem 4.2-2 to write $f(e^{-t}) \in \mathcal{L}'_{a,b}$ and $g(e^{-t}) \in \mathcal{L}'_{a,b}$. Therefore, according to Theorem 3.7-1, $f(e^{-t}) * g(e^{-t}) \in \mathcal{L}'_{a,b}$. Theorem 4.2-2 now shows that $f \vee g \in \mathcal{M}'_{a,b}$ because

$$\begin{aligned}\langle f(e^{-t}) * g(e^{-t}),\, \phi(t)\rangle &= \langle f(e^{-t}),\, \langle g(e^{-\tau}),\, \phi(t+\tau)\rangle\rangle \\ &= \langle f(x),\, \langle g(y),\, \theta(xy)\rangle\rangle = \langle f \vee g,\, \theta\rangle\end{aligned}$$

This completely verifies our assertion, which we repeat:

THEOREM 4.6-1. *If f and g are members of $\mathcal{M}'_{a,b}$ where $a \le b$, then $f \vee g$, as defined by* (1), *is also a member of $\mathcal{M}'_{a,b}$.*

Equation (1) also defines Mellin-type convolution in any space $\mathcal{M}'(w, z)$ where $w < z$, but now θ is an arbitrary member of $\mathcal{M}(w, z)$. In other words, for any f and g in $\mathcal{M}'(w, z)$, the Mellin-type convolution product $f \vee g$ is that member of $\mathcal{M}'(w, z)$ whose restriction to each space $\mathcal{M}_{a,b}$, for which $w < a \le b < z$, coincides with the convolution product of the restrictions of f and g to $\mathcal{M}_{a,b}$.

The usual change of variables and Theorem 4.2-2 allow us to carry various properties of the convolution described in Secs. 3.7 and 3.8 over to Mellin-type convolution. For example, by substituting $t = -\log x$, $\tau = -\log y$, $h(\tau) = f(y)$, and $j(t-\tau) = g(x/y)$ into the conventional convolution product:

$$\int_{-\infty}^{\infty} h(\tau) j(t-\tau)\, d\tau$$

we get

$$\int_0^{\infty} f(y) g\left(\frac{x}{y}\right) \frac{1}{y}\, dy \qquad 0 < x < \infty \tag{3}$$

Therefore, by converting the conditions under which conventional convolution is a special case of generalized convolution (see Sec. 3.7,) we can conclude the following: If f and g are locally integrable functions on $0 < x < \infty$ and if $f/\zeta_{a,b}$ and $g/\zeta_{a,b}$ are absolutely integrable on $0 < x < \infty$, then $f \vee g$ is a regular member of $\mathcal{M}'_{a,b}$ corresponding to the conventional function (3).

A special case, which we shall make use of in the next section, arises when both g and θ are testing functions in $\mathcal{D}(I)$, where I is the interval $0 < x < \infty$. Indeed, $\langle g(y), \theta(xy)\rangle$ is now an integral which through a change of variable can be written as

$$\left\langle \theta(y), \frac{1}{x} g\left(\frac{y}{x}\right)\right\rangle$$

The supports of $\theta(y)$ and $x^{-1}g(y/x)$ intersect in a compact subset of the open first quadrant: $0 < x < \infty$, $0 < y < \infty$. Consequently, (1) is equal to

$$\left\langle f(x), \left\langle \theta(y), \frac{1}{x} g\left(\frac{y}{x}\right)\right\rangle\right\rangle = \left\langle f(x) \times \theta(y), \frac{1}{x} g\left(\frac{y}{x}\right)\right\rangle$$

where $f(x) \times \theta(y)$ represents the direct product of $f(x)$ and $\theta(y)$ (see Zemanian [1], Sec. 5.2). By the commutativity of the direct product (Zemanian [1], Theorem 5.3-2), we obtain

$$\langle f \vee g, \theta\rangle = \left\langle \theta(y), \left\langle f(x), \frac{1}{x} g\left(\frac{y}{x}\right)\right\rangle\right\rangle \tag{4}$$

Some direct computation shows that

$$\left\langle f(x), \frac{1}{x} g\left(\frac{y}{x}\right)\right\rangle \tag{5}$$

is a smooth function on $0 < y < \infty$. (See Problem 4.6-1.) However, a simpler way to establish the smoothness of (5) is to show that, through the usual change of variables, (5) is equal to

$$\langle f(e^{-t}), e^{t} g(e^{t-\tau})\rangle \tag{6}$$

and then to note that (6) is a regularization and hence a smooth function of τ (Zemanian [1], Theorem 5.5-1).

Referring back to (4), we see that we have established

THEOREM 4.6-2. *If $f \in \mathcal{M}'_{a,b}$ and $g \in \mathcal{D}(I)$, then, in the sense of equality in $\mathcal{D}'(I)$, $f \vee g$ is equal to a smooth function on I, namely,*

$$(f \vee g)(y) = \left\langle f(x), \frac{1}{x} g\left(\frac{y}{x}\right)\right\rangle \qquad y \in I \tag{7}$$

We shall call (7) the *Mellin-type regularization of f by g.*

The next theorem states the *exchange formula* for the Mellin transformation and two properties of Mellin-type convolution that can be derived from the exchange formula.

THEOREM 4.6-3. *If* $\mathfrak{M}f = F(s)$ *for* $s \in \Omega_f$, $\mathfrak{M}g = G(s)$ *for* $s \in \Omega_g$, *and* $\Omega_f \cap \Omega_g$ *is not empty, then* $f \vee g$ *exists in the sense of Mellin-type convolution in* $\mathscr{M}'(w, z)$ *where the interval* $w < \sigma < z$ *is the intersection of* $\Omega_f \cap \Omega_g$ with *the real axis. Moreover,*

$$\mathfrak{M}(f \vee g) = F(s)G(s) \qquad s \in \Omega_f \cap \Omega_g \tag{8}$$

Furthermore, $f \vee g = g \vee f$ *(commutativity) in the sense of equality in* $\mathscr{M}'(w, z)$. *If, in addition,* $\mathfrak{M}h = H(s)$ *for* $s \in \Omega_h$ *and* $\Omega_f \cap \Omega_g \cap \Omega_h$ *is not empty, then* $f \vee (g \vee h) = (f \vee g) \vee h$ *(associativity) in the sense of equality in* $\mathscr{M}'(w', z')$ *where the interval* $w' < \sigma < z'$ *is the intersection of* $\Omega_f \cap \Omega_g \cap \Omega_h$ *with the real axis.*

PROOF. The first sentence is established by Theorem 4.6-1 and the fact that both f and g are members of $\mathscr{M}'_{a,b}$ whenever $w < a \leq b < z$. Next, for any fixed $s \in \Omega_f \cap \Omega_g$, our definition (1) yields

$$\begin{aligned}\mathfrak{M}(f \vee g) &= \langle f(x), \langle g(y), (xy)^{s-1}\rangle\rangle = \langle f(x), x^{s-1}\rangle\langle g(y), y^{s-1}\rangle \\ &= F(s)G(s) = \langle g(y), \langle f(x), (xy)^{s-1}\rangle\rangle = \mathfrak{M}(g \vee f)\end{aligned}$$

The uniqueness theorem (Theorem 4.3-4) now implies that $f \vee g = g \vee f$ in $\mathscr{M}'(w, z)$. Finally, for the last sentence of the theorem, choose $s \in \Omega_f \cap \Omega_g \cap \Omega_h$. Then, as above.

$$\mathfrak{M}[f \vee (g \vee h)] = F(s)G(s)H(s) = \mathfrak{M}[(f \vee g) \vee h]$$

The uniqueness theorem implies again that $f \vee (g \vee h) = (f \vee g) \vee h$ in $\mathscr{M}'(w', z')$. Q.E.D.

PROBLEM 4.6-1. Show directly that (5) is a smooth function of y. *Hint*: For any fixed $y \in I$ consider $\langle f(x), \psi_{\Delta y}(x)\rangle$, where

$$\psi_{\Delta y}(x) = \frac{g\left(\dfrac{y + \Delta y}{x}\right) - g\left(\dfrac{y}{x}\right)}{x\Delta y} - \frac{1}{x} D_y g\left(\frac{y}{x}\right) \qquad \Delta y \neq 0$$

and show that $\psi_{\Delta y}$ converges in $\mathscr{D}(I)$ to zero as $\Delta y \to 0$. This will prove that

$$D_y\left\langle f(x), \frac{1}{x} g\left(\frac{y}{x}\right)\right\rangle = \left\langle f(x), \frac{1}{x} D_y g\left(\frac{y}{x}\right)\right\rangle$$

A similar equation for higher order derivatives follows by induction since $D^k g \in \mathscr{D}(I)$ whenever $g \in \mathscr{D}(I)$.

PROBLEM 4.6-2. Show that $\mathscr{M}'(w, z)$, where $w < z$, is a commutative Mellin-type convolution algebra. (See Zemanian [1], pp. 149–150 for the definition of such an algebra.) Does it have a unit element? Prove that it contains no divisors of zero (i.e., if $f \vee g = 0$, then either $f = 0$ or $g = 0$.)

PROBLEM 4.6-3. Let $a < b$, $f \in \mathscr{M}'_{1-b, 1-a}$, and $g \in \mathscr{M}'_{a, b}$. We define a second Mellin-type convolution product $f \wedge g$ by

$$(f \wedge g)(x) \triangleq \left[\frac{1}{x} f\left(\frac{1}{x}\right)\right] \vee g(x)$$

That is, for any $\theta \in \mathscr{M}_{a, b}$

$$\langle f \wedge g, \theta \rangle = \left\langle \frac{1}{x} f\left(\frac{1}{x}\right), \langle g(y), \theta(xy) \rangle \right\rangle$$

$$= \left\langle f(x), \left\langle g(y), \frac{1}{x} \theta\left(\frac{y}{x}\right)\right\rangle\right\rangle$$

Show that $f \wedge g$ is a member of $\mathscr{M}'_{a, b}$ and that

$$\mathfrak{M}(f \wedge g) = F(1-s)G(s)$$

for at least $a < \operatorname{Re} s < b$ where $\mathfrak{M}f = F(s)$ for at least $1 - b < \operatorname{Re} s < 1 - a$ and $\mathfrak{M}g = G(s)$ for at least $a < \operatorname{Re} s < b$. *Hint*: Use the results of Sec. 4.4 including those of Problem 4.4-6.

PROBLEM 4.6-4. Show that Mellin-type convolution is a continuous operation in the following sense: If $w < z$, if $\{f_\nu\}_{\nu=1}^{\infty}$ converges in $\mathscr{M}'(w, z)$ to f, and if $g \in \mathscr{M}'(w, z)$, then

$$f_\nu \vee g = g \vee f_\nu \to f \vee g = g \vee f$$

in $\mathscr{M}'(w, z)$ as $\nu \to \infty$.

PROBLEM 4.6-5. Let f and g be members of $\mathscr{M}'(w, z)$ where $w < z$. Show that for any positive integer k

$$(xD)^k (f \vee g) = [(xD)^k f] \vee g = f \vee [(xD)^k g]$$

On the other hand, is the formula:

$$D^k (f \vee g) = (D^k f) \vee g = f \vee (D^k g)$$

a valid one? Finally, for any complex number α, show that

$$(\log x)(f \vee g) = [(\log x) f] \vee g + f \vee [(\log x) g]$$

and

$$x^{\alpha}(f \vee g) = (x^{\alpha} f) \vee (x^{\alpha} g)$$

4.7. Dirichlet's Problem for a Wedge with a Generalized-Function Boundary Condition

Consider an infinite two-dimensional wedge as indicated in Fig. 4.7-1. We choose a polar coordinate system (r, θ) with the origin at the apex of the

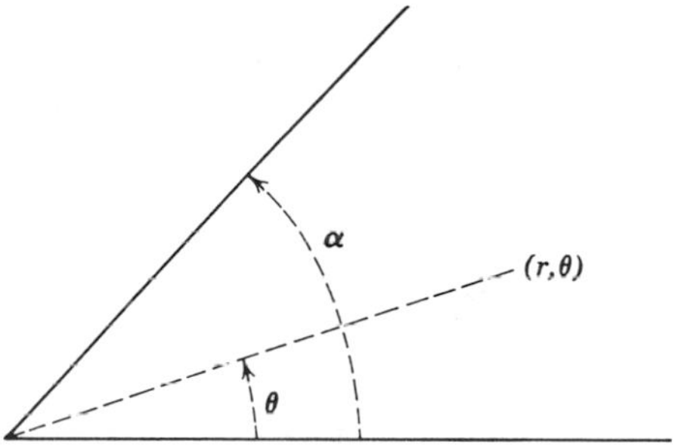

FIGURE 4.7-1

wedge and the sides of the wedge along the radial lines $\theta = 0$ and $\theta = \alpha$ $(0 < \alpha < 2\pi)$. Dirichlet's problem for the interior of this wedge is to find a function $u(r, \theta)$ that satisfies

$$(1) \qquad r^2 \frac{\partial^2 u}{\partial r^2} + r \frac{\partial u}{\partial r} + \frac{\partial^2 u}{\partial \theta^2} = 0 \qquad 0 < r < \infty, \qquad 0 < \theta < \alpha$$

and certain boundary conditions on $u(r, \theta)$. Eq. (1) is Laplace's equation in polar coordinates multiplied by r^2.

The boundary conditions we shall impose are:

1. As $\theta \to 0+$, $u(r, \theta)$ converges in $\mathscr{D}'(I)$ to some Mellin-transformable generalized function $f(r)$. (Here, I is the interval $0 < r < \infty$.)

2. As $\theta \to \alpha-$, $u(r, \theta)$ converges to zero uniformly on every compact subset of $0 < r < \infty$.

The Mellin transformation provides a means for solving this problem through an operational technique. Here again, we first derive the solution formally and subsequently prove that we truly have a solution.

When applying the Mellin transformation, we shall treat r as the independent variable and θ as a fixed parameter:

$$\mathfrak{M}u(r, \theta) = \langle u(r, \theta), r^{s-1} \rangle \triangleq U(s, \theta)$$

Because of the operation-transform formula Sec. 4.4, Eq. (6), $\mathfrak{M}$ transforms (1) into

$$s(s+1)U(s, \theta) - sU(s, \theta) + \frac{\partial^2}{\partial\theta^2} U(s, \theta) = 0$$

if we assume that $\partial^2/\partial\theta^2$ can be interchanged with $\mathfrak{M}$. Therefore,

$$U(s, \theta) = A(s)e^{is\theta} + B(s)e^{-is\theta} \tag{2}$$

where the unknown functions $A(s)$ and $B(s)$ do not depend on θ. To determine A and B, we first transform the boundary conditions. Then assuming that the limiting processes for $\theta \to 0+$ and $\theta \to \alpha-$ can be interchanged with $\mathfrak{M}$, we substitute the transformed boundary conditions into (2) and solve the resulting simultaneous equations. Thus, if $\mathfrak{M}f = F(s)$ for $s \in \Omega_f = \{s: \sigma_{f1} < \text{Re } s < \sigma_{f2}\}$, we obtain

$$A(s) + B(s) = F(s)$$

$$A(s)e^{is\alpha} + B(s)e^{-is\alpha} = 0$$

so that

$$A(s) = \frac{F(s)}{1 - e^{i2\alpha s}}$$

$$B(s) = \frac{F(s)}{1 - e^{-i2\alpha s}}$$

Consequently,

$$U(s, \theta) = \frac{\sin(\alpha s - \theta s)}{\sin \alpha s} F(s) \tag{3}$$

We shall now show that on certain strips in the s plane (3) satisfies the hypothesis of Theorem 4.3-6, which implies that (3) is the conventional Mellin transform of a continuous function. For $0 \leq \theta \leq \alpha$ and $s = \sigma + i\omega$,

$$\frac{\sin(\alpha s - \theta s)}{\sin \alpha s} = O(e^{-|\omega|\theta}) \qquad |\omega| \to \infty \tag{4}$$

uniformly on $-\infty < \sigma < \infty$. Moreover, for $0 < \theta \leq \alpha$, the singularities of $\sin(\alpha s - \theta s)/\sin \alpha s$ are simple poles and can occur only at the points $s = n\pi/\alpha$, n being any integer not equal to 0. (The point $s = 0$ is not a pole because both $\sin(\alpha s - \theta s)$ and $\sin \alpha s$ possess simple zeros at $s = 0$ which cancel. For certain values of θ there may be such cancellations of other poles; but, this cannot occur for all values of θ, for example, when θ/α is an irrational number). In view of Theorem 4.3-5 and the estimate (4), for $0 < \theta \leq \alpha$ and as $|\omega| \to \infty$, (3) tends to zero at some exponential rate uniformly on any closed substrip of Ω_f. Consequently, if $0<\theta\leq\alpha$, (3) satisfies the hypothesis of Theorem 4.3-6 on any strip $a\leq \operatorname{Re} s\leq b$ $(\sigma_{f1}<a<b<\sigma_{f2})$ which does not contain any point $n\pi/\alpha (n=\pm 1, \pm 2, \ldots)$.

Thus, upon invoking Theorem 4.3-6, we obtain as our possible solution

$$(5)\quad u(r, \theta)=\frac{1}{2\pi i}\int_{\sigma-i\infty}^{\sigma+i\infty} F(s)\frac{\sin(\alpha s-\theta s)}{\sin \alpha s} r^{-s}\,ds \qquad 0<\theta<\alpha, \qquad 0<r<\infty$$

where σ is any real number in Ω_f that does not equal $n\pi/\alpha$ $(n=\pm 1, \pm 2, \ldots)$. If Ω_f contains some of the points $n\pi/\alpha$, (5) is not unique; it will differ for two different choices of σ, say, σ_1 and σ_2, if any points $n\pi/\alpha$ occur between σ_1 and σ_2. We will discuss this later on.

Since our development was merely formal, we should now prove that (5) is a solution. That (5) satisfies the differential equation (1) in the sense of conventional differentiation follows from the facts that the differentiations $\partial/\partial r$, $\partial^2/\partial r^2$, $\partial/\partial\theta$, and $\partial^2/\partial\theta^2$ may be interchanged with the integration in (5) and that

$$\frac{\sin(\alpha s-\theta s)}{\sin \alpha s} r^{-s}$$

satisfies (1). We omit the details, which are straightforward.

Turning to the boundary conditions, let's verify the second one first. That one is easier. By (5) and the fact that $|r^{-i\omega}|=1$, we have

$$(6)\qquad |u(r, \theta)|\leq\frac{r^{-\alpha}}{2\pi}\int_{-\infty}^{\infty}\left|F(\sigma+i\omega)\frac{\sin(\alpha-\theta)(\sigma+i\omega)}{\sin\alpha(\sigma+i\omega)}\right|d\omega$$

where σ is fixed as before. Theorem 4.3-5 asserts that $F(\sigma+i\omega)$ is of slow growth as $|\omega|\to\infty$. Consequently, the estimate (4) implies that the integral in (6) converges uniformly on every interval $\beta\leq\theta\leq\alpha$, where $\beta>0$. Therefore, we may take the limit as $\theta\to\alpha-$ under the integral sign in (6), thereby verifying the second boundary condition.

To verify the first boundary condition, we have to show that for any $\phi\in\mathscr{D}(I)$

$$\lim_{\theta\to 0+}\langle u(r, \theta), \phi(r)\rangle=\langle f(r), \phi(r)\rangle$$

Since $u(r, \theta)$ is a continuous function,

$$\langle u(r, \theta), \phi(r)\rangle$$

$$=\frac{1}{2\pi}\int_0^\infty dr\int_{-\infty}^{\infty} F(\sigma+i\omega)\frac{\sin(\alpha-\theta)(\sigma+i\omega)}{\sin\alpha(\sigma+i\omega)} r^{-\sigma-i\omega}\phi(r)\,d\omega$$

where $0 < \theta < \alpha$ and σ is fixed as before. With θ fixed, the integrand is a smooth function of (r, ω) whose support is contained in some strip:

$$\{(r, \omega): r_1 \le r \le r_2,\ 0 < r_1 < r_2 < \infty,\ -\infty < \omega < \infty\}$$

On this strip it is bounded by $K/(1 + \omega^2)$, where K is a sufficiently large constant, in view of (4). Consequently, by Fubini's theorem we may change the order of integration.

$$(7) \quad \langle u(r, \theta), \phi(r)\rangle = \frac{1}{2\pi}\int_{-\infty}^{\infty} F(\sigma + i\omega)\Phi(\sigma + i\omega)\frac{\sin(\alpha - \theta)(\sigma + i\omega)}{\sin \alpha(\sigma + i\omega)}\, d\omega$$

where

$$(8) \qquad \Phi(s) = \int_0^\infty r^{-s}\phi(r)dr = \mathfrak{M}[r^{-1}\phi(r^{-1})] \qquad -\infty < \operatorname{Re} s < \infty$$

This shows that $\Phi(s)$ is an entire function (Theorem 4.3-2). Successively integrating by parts, we get

$$\Phi(s) = \frac{1}{(s-1)(s-2)\dots(s-n)}\int_0^\infty r^{-s+n}D^n\phi(r)dr \qquad n = 0, 1, 2, \dots$$

Since $|r^{i\omega}| = 1$,

$$(9) \quad |\Phi(\sigma + i\omega)| \le \frac{1}{|\sigma + i\omega - 1|\dots|\sigma + i\omega - n|}\int_0^\infty r^{-\sigma+n}|D^n\phi(r)|dr$$

for every n. This implies that $\Phi(\sigma + i\omega)$ is of rapid descent as $|\omega| \to \infty$. Moreover,

$$(10) \qquad \frac{\sin(\alpha - \theta)(\sigma + i\omega)}{\sin \alpha(\sigma + i\omega)}$$

is bounded for $-\infty < \omega < \infty$ and $0 \le \theta \le \alpha$, when σ is fixed as before ($\sigma \ne n\pi/\alpha, n = \pm 1, \pm 2, \dots$). The fact that $F(\sigma + i\omega)$ is of slow growth as $|\omega| \to \infty$ now implies that (7) converges uniformly for $0 \le \theta \le \alpha$, so that

$$(11) \qquad \lim_{\theta \to 0+} \langle u(r, \theta), \phi(r)\rangle = \frac{1}{2\pi}\int_{-\infty}^{\infty} F(\sigma + i\omega)\Phi(\sigma + i\omega)\, d\omega$$

By (9) and Theorem 4.3-5 once again, $F(\sigma + i\omega)\Phi(\sigma + i\omega)$ also satisfies the hypothesis of Theorem 4.3-6 on every closed substrip of Ω_f, and therefore $F(\sigma + i\omega)\Phi(\sigma + i\omega)$ is the Mellin transform of a continuous function $h(r)$ on $0 < r < \infty$. A comparison of (11) with Sec. 4.3, Eq. (3) shows that (11) is the value $h(1)$. Moreover, by (8) and Theorem 4.6-3, $F(\sigma + i\omega)$

$\Phi(\sigma + i\omega)$ is the transform of $h(r) = f(r) \vee [r^{-1} \phi(r^{-1})]$. Since $r^{-1} \phi(r^{-1}) \in \mathscr{D}(I)$, Theorem 4.6-2 indicates that

$$h(r) = \left\langle f(x), \frac{1}{x}\left[\frac{x}{r} \phi\left(\frac{x}{r}\right)\right]\right\rangle$$

By combining these results, we finally obtain

$$\lim_{\theta \to 0+} \langle u(r, \theta), \phi(r)\rangle = h(1) = \langle f(x), \phi(x)\rangle$$

which shows that the first boundary condition is also satisfied. This completes the argument showing that (5) is truly a solution to our Dirichlet problem for an infinite wedge.

As was mentioned before, (5) will not be a unique solution if Ω_f contains poles of the integrand of (5). In this case, two different choices of σ, say, σ_1 and σ_2, will yield solutions differing by the sum of the residues of the poles occurring between σ_1 and σ_2. These residues are of the form:

$$Ar^{-k\pi/\alpha} \sin \left(1 - \frac{\theta}{\alpha}\right) k\pi \tag{12}$$

where A is a constant and k an integer. It is easy to show that (12) satisfies the differential equation (1) and converges uniformly to zero on compact subsets of $0 < r < \infty$ as $\theta \to 0+$ or $\theta \to \alpha-$. This implies that (12) tends to zero in $\mathscr{D}'(I)$ as $\theta \to 0+$. Moreover, these comments remain true even when the point $k\pi/\alpha$ is not in Ω_f. Thus, there is even more arbitrariness: We may add linear combinations of such terms to (5) and still have a solution to our problem, whether or not the $k\pi/\alpha$ are in Ω_f. In terms of electrostatic potential, (12) is the potential due to an electric multipole either concentrated at the origin $r = 0$ or "smeared out over $r = \infty$." Since our boundary conditions did not impose any restrictions at these points, these arbitrary components in the potential $u(r, \theta)$ are to be expected.

PROBLEM 4.7-1. Prove that (5) satisfies the differential equation (1).

PROBLEM 4.7-2. Sketch the equipotential lines of the potential function (12) for various values of the integer k.

PROBLEM 4.7-3. Consider the finite wedge described by $0 < r < a$, $0 < \theta < \alpha$, where a is finite and $0 < \alpha < 2\pi$. Also, consider the Dirichlet problem for this wedge under the following boundary conditions on the potential function $u(r, \theta)$:

1. As $\theta \to 0+$, $u(r, \theta)$ tends to $f(r)$ in $\mathscr{D}'(J)$ where J is the open interval $0 < r < a$ and $f \in \mathscr{E}'(J)$.

2. As $\theta \to \alpha-$, $u(r, \theta)$ tends to zero uniformly on every compact subset of $0 < r < a$.

3. As $r \to a-$, $u(r, \theta)$ tends to zero pointwise.

Show that a solution to this problem is

$$u(r, \theta) = \frac{1}{2\pi i} \int_{-i\infty}^{i\infty} G(s) \frac{\sin(\alpha s - \theta s)}{\sin \alpha s} r^{-s} \, ds$$

where

$$G(s) = \left\langle f(r), r^{s-1} - \frac{a^{2s}}{r^{s+1}} \right\rangle$$

PROBLEM 4.7-4. The solution of the preceding problem also satisfies a Dirichlet problem for the region $a < r < \infty$, $0 < \theta < \alpha$. What are the boundary conditions corresponding to those stated above?

CHAPTER V

The Hankel Transformation

5.1. Introduction

Consider now the *conventional Hankel transformation*, which is defined by

$$F(y) = \mathfrak{H}_\mu f = \int_0^\infty f(x)\sqrt{xy}\, J_\mu(xy)\, dx \tag{1}$$

where $0 < y < \infty$, μ is a real number, and J_μ is the Bessel function of first kind and order μ. Our objective in this chapter is to describe how this transformation can be extended to certain generalized functions.

A standard result concerning (1) is the following inversion theorem (Watson [1]; p. 456). In the following $L_1(0, \infty)$ denotes the space of all (equivalence classes of) functions $f(x)$ that are Lebesgue integrable on $0 < x < \infty$.

THEOREM 5.1-1. *If $f(x) \in L_1(0, \infty)$, if $f(x)$ is of bounded variation in a neighborhood of the point $x = x_0 > 0$, if $\mu \geq -\frac{1}{2}$, and if $F(y)$ is defined by* (1), *then*

$$\tfrac{1}{2}[f(x_0 + 0) + f(x_0 - 0)] = \mathfrak{H}_\mu^{-1} F = \int_0^\infty F(y)\sqrt{x_0 y}\, J_\mu(x_0 y)\, dy \tag{2}$$

Note that, when $\mu \geq -\frac{1}{2}$, this conventional inverse Hankel transformation $\mathfrak{H}_\mu^{-1}$ is defined by precisely the same formula as is the direct Hankel transformation $\mathfrak{H}_\mu$; in symbols, $\mathfrak{H}_\mu = \mathfrak{H}_\mu^{-1}$.

Another result, which we shall need, is *Parseval's equation*:

THEOREM 5.1-2. If *$f(x)$ and $G(y)$ are in $L_1(0, \infty)$, if $\mu \geq -\frac{1}{2}$, if $F(y) = \mathfrak{H}_\mu[f(x)]$, and if $g(x) = \mathfrak{H}_\mu^{-1}[G(y)]$, then*

$$\int_0^\infty f(x)g(x)\, dx = \int_0^\infty F(y)G(y)\, dy \tag{3}$$

This theorem is readily proven by substituting $g(x) = \mathfrak{H}_\mu^{-1}[G(y)]$ into the left-hand side of (3) and then invoking Fubini's theorem to change the order of integration. Other conditions under which (3) holds were established by Macauley-Owen [1].

Apparently, the first one to extend the Hankel transformation to generalized functions in such a way that an inversion formula could be stated for it was J. L. Lions [1]. His results were obtained as a particular case in a study of certain general types of operators, called transmutation operators, acting on some testing-function spaces of Schwartz. In contrast to this, we present in this chapter an alternative theory (Zemanian [4–8]); it is designed specifically for the Hankel transformation, and the properties of the generalized Hankel transformation are investigated at some length. Still another theory that has recently appeared is due to Fenyo [1]; it is devised for the case where the order of the Hankel transformation is a nonnegative integer and is defined for all distributions on the open interval $(0, \infty)$.

Our procedure is reminiscent of Schwartz's method for extending the Fourier transformation to distributions of slow growth (Schwartz [1], Vol. 2, Chapter VII). We construct a testing-function space $\mathscr{H}_\mu$ on $0 < x < \infty$, on which the μth order Hankel transformation $\mathfrak{H}_\mu$ is an automorphism whenever $\mu \geq -\frac{1}{2}$. The generalized functions in the dual $\mathscr{H}_\mu'$ of $\mathscr{H}_\mu$ act like distributions of slow growth as $x \to \infty$ (Zemanian [1], Chapter 4). Moreover, $\mathscr{H}_\mu'$ is the domain of our generalized Hankel transformation $\mathfrak{H}_\mu'$, which is defined through a generalization of the Parseval equation (3); in particular, the generalized Hankel transformation $\mathfrak{H}_\mu'$ is defined as the adjoint of $\mathfrak{H}_\mu$ through the equation:

$$\langle \mathfrak{H}_\mu' f, \Phi \rangle \triangleq \langle f, \mathfrak{H}_\mu \Phi \rangle \tag{4}$$

where $\mu \geq -\frac{1}{2}$, $f \in \mathscr{H}_\mu'$, and $\Phi \in \mathscr{H}_\mu$. It follows that $\mathfrak{H}_\mu'$, is an automorphism on $\mathscr{H}_\mu'$. Moreover, it turns out that $\mathfrak{H}_\mu' f$ agrees with $\mathfrak{H}_\mu f$ when f is a regular generalized function corresponding to a testing function in $\mathscr{H}_\mu$. In other words, under certain conditions $\mathfrak{H}_\mu'$ contains $\mathfrak{H}_\mu$ as a special case, and for this reason we shall drop the prime on the notation for the generalized Hankel transformation.

Note that this procedure for generalizing the Hankel transformation is different from that used for the Laplace and Mellin transformations. Here, our definition is an indirect one based upon Parseval's equation. On the other hand, in each of the previous cases we constructed a testing-function space, which contained the kernel function, and we defined a transform directly as the application of a generalized function to the kernel function. This latter approach doesn't always work now because the kernel function $\sqrt{xy}\, J_\mu(xy)$ as a function of x is not a member of $\mathscr{H}_\mu$, and therefore the equation

$$(\mathfrak{H}_\mu f)(y) = \langle f(x), \sqrt{xy}\, J_\mu(xy) \rangle \tag{5}$$

does not possess a sense for every $f \in \mathscr{H}_\mu'$. However, under certain restriction on f, (5) will possess a sense and will agree with the definition (4); this is discussed in Sec. 5.6.

In Secs. 5.8 and 5.9 we present two examples of the use of our generalized Hankel transformation in solving boundary-value problems having generalized-function boundary conditions. The first is a Dirichlet problem and the second a Cauchy problem for the wave equation, both of which are in a cylindrical-coordinate system.

We end the chapter by briefly describing two extensions to our results. In Sec. 5.10 the restriction $\mu \geq -\frac{1}{2}$ is removed, and we obtain a generalized Hankel transformation which is defined for all real values of the order μ. In Sec. 5.11 our Hankel transformation is extended to certain generalized functions having no restriction on their growth as $x \to \infty$.

Let us take note here of two differentiation formulas (Jahnke, Emde, and Losch [1], p. 154) that we shall use quite a few times in this chapter.

$$D_x x^\mu J_\mu(xy) = yx^\mu J_{\mu-1}(xy) \tag{6}$$

$$D_x x^{-\mu} J_\mu(xy) = -yx^{-\mu} J_{\mu+1}(xy) \tag{7}$$

When μ is not an integer, the function $J_\mu(z)$ is multivalued and analytic everywhere on the complex z plane except for branch points at $z = 0$ and $z = \infty$ (Jahnke, Emde, and Losch [1], p. 134). When μ is an integer, it is single-valued. We shall always restrict it to its principal branch by requiring that $|\arg z| < \pi$, except when the opposite is specifically indicated. Thus, $J_\mu(z)$ takes on real values when z is real and positive. The same convention holds for $\sqrt{z}$ or z^α where α is any complex number; thus, for real α, $\sqrt{z}$ and z^α take on real positive values when z is real and positive.

A series expansion for the Bessel function $J_\mu(z)$ of any order μ is

$$J_\mu(z) = \sum_{k=0}^{\infty} \frac{(-1)^k (z/2)^{\mu+2k}}{k!\,\Gamma(\mu+k+1)}$$

(Jahnke, Emde, and Losch [1], p. 134).

5.2. The Testing-Function Space $\mathscr{H}_\mu$ and Its Dual

Throughout this Chapter, I again denotes the open interval $(0, \infty)$, and x is a real variable restricted to I. For each real number μ we define a countably multinormed space $\mathscr{H}_\mu$ as follows. A function $\phi(x)$ is in $\mathscr{H}_\mu$ if and only if it is defined on $0 < x < \infty$, it is complex-valued and smooth, and, for each pair of nonnegative integers m and k,

$$\gamma_{m,k}^\mu(\phi) \triangleq \sup_{0<x<\infty} |x^m (x^{-1}D)^k [x^{-\mu-1/2}\phi(x)]| \tag{1}$$

exists (i.e., is finite). $\mathscr{H}_\mu$ is a linear space. Also, each $\gamma^\mu_{m,k}$ is a seminorm on $\mathscr{H}_\mu$, and the collection $\{\gamma^\mu_{m,k}\}^\infty_{m,k=0}$ is a multinorm because the $\gamma^\mu_{m,0}$ are norms. The topology of $\mathscr{H}_\mu$ is that generated by $\{\gamma^\mu_{m,k}\}^\infty_{m,k=0}$. We shall see shortly that $\mathscr{H}_\mu$ is a testing-function space as defined in Sec. 2.4.

LEMMA 5.2-1. *$\phi(x)$ is a member of $\mathscr{H}_\mu$ if and only if it satisfies the following three conditions:*

(i) *$\phi(x)$ is a smooth complex-valued function on $0 < x < \infty$.*

(ii) *For each nonnegative integer k,*

$$\phi(x) = x^{\mu+1/2}[a_0 + a_2 x^2 + \cdots + a_{2k} x^{2k} + R_{2k}(x)] \tag{2}$$

where the a's are constants given by

$$a_{2k} = \frac{1}{k!2^k} \lim_{x\to 0+} (x^{-1}D)^k x^{-\mu-1/2}\phi(x) \tag{3}$$

and the remainder term $R_{2k}(x)$ satisfies

$$(x^{-1}D)^k R_{2k}(x) = o(1) \qquad x \to 0+ \tag{4}$$

(iii) *For each nonnegative integer k, $D^k\phi(x)$ is of rapid descent as $x \to \infty$ [i.e., $D^k\phi(x)$ tends to zero faster than any power of $1/x$ as $x \to \infty$].*

PROOF. Assume that $\phi(x) \in \mathscr{H}_\mu$. Condition (*i*) is satisfied by definition. Next, set

$$\psi(x) = (x^{-1}D)^k x^{-\mu-1/2}\phi(x) \tag{5}$$

$\psi(x)$ is a smooth function on $0 < x < \infty$. By the seminorm $\gamma^\mu_{0,k+1}$, $D\psi(x) = O(x)$ as $x \to 0+$. Therefore, the quantity:

$$\int_1^x D_t\psi(t)\,dt = \psi(x) - \psi(1)$$

tends to a finite limit as $x \to 0+$, which implies that $\psi(x)$ does the same. By substituting (2) into (5), we obtain

$$\psi(x) = k!2^k a_{2k} + (x^{-1}D)^k R_{2k}(x)$$

Upon choosing a_{2k} according to (3), we obtain (4). Conversely, if conditions (*i*) and (*ii*) hold, then $\psi(x)$ and therefore $x^m\psi(x)$ $(m = 0, 1, 2, \ldots)$ must be bounded on $0 < x \leq 1$.

Next, by a direct computation we get

$$\begin{aligned}\psi(x) &= (x^{-1}D)^k x^{-\mu-1/2}\phi(x) \\ &= x^{-\mu-1/2}\left(b_{k0}\frac{\phi}{x^{2k}} + b_{k1}\frac{D\phi}{x^{2k-1}} + \cdots + b_{kk}\frac{D^k\phi}{x^k}\right)\end{aligned} \tag{6}$$

where the b's denote constants ($b_{kk} = 1$). If $\phi \in \mathscr{H}_\mu$, then, by the seminorm $\gamma^{\mu}_{m,k}$, $x^m\phi(x)$ is bounded on $1 \le x < \infty$ for every positive integer m; it follows by induction on k that $D^k\phi$ is of rapid descent as $x \to \infty$, so that condition (iii) is fulfilled. Conversely, conditions (i) and (iii) imply that $x^m\phi(x)$ is bounded on $1 < x < \infty$ for every m; this statement and the last sentence of the preceding paragraph show that conditions (i), (ii), and (iii) imply that ϕ is in $\mathscr{H}_\mu$. Q.E.D.

Note that, for any fixed $y \in I$, $\sqrt{xy}\, J_\mu(xy)$ as a function of x satisfies conditions (i) and (ii) of Lemma 5.2-1. However, it does not satisfy condition (iii) since

$$\sqrt{xy}\, J_\mu(xy) \sim \sqrt{\frac{2}{\pi}} \cos\left(xy - \frac{\mu\pi}{2} - \frac{\pi}{4}\right) \qquad x \to \infty$$

(See Jahnke, Emde, and Losch [1], p. 134 and p. 147.) Hence, $\sqrt{xy}\, J_\mu(xy)$ is not a member of $\mathscr{H}_\mu$.

LEMMA 5.2-2 *$\mathscr{H}_\mu$ is complete and therefore a Fréchet space.*

PROOF. Let $\{\phi_\nu\}_{\nu=1}^{\infty}$ be a Cauchy sequence in $\mathscr{H}_\mu$. In view of (6) and the seminorms $\gamma^{\mu}_{0,k}$, it follows by induction on k that, for each k, $\{D^k\phi_\nu\}_{\nu=1}^{\infty}$ converges uniformly on every compact subset of I. Therefore, there exists a smooth function $\phi(x)$ defined on I such that, for each k and x, $D^k\phi_\nu(x) \to D^k\phi(x)$ as $\nu \to \infty$.

Moreover, again by the fact that $\{\phi_\nu\}_{\nu=1}^{\infty}$ is a Cauchy sequence, for each m and k and a given $\varepsilon > 0$ there exists a real number $N_{m,k}$ such that, for every $\nu, \eta > N_{m,k}$, $\gamma^{\mu}_{m,k}(\phi_\nu - \phi_\eta) < \varepsilon$. Taking the limit as $\eta \to \infty$, we obtain

$$\gamma^{\mu}_{m,k}(\phi_\nu \quad \phi) \le \varepsilon \tag{7}$$

for all $\nu > N_{m,k}$, which shows that $\gamma^{\mu}_{m,k}(\phi_\nu - \phi) \to 0$ as $\nu \to \infty$.

Finally, there exists a constant $C_{m,k}$ not depending on ν such that $\gamma^{\mu}_{m,k}(\phi_\nu) < C_{m,k}$. Therefore, from (7) and Sec. 1.5, Eq. (1) we obtain

$$\gamma^{\mu}_{m,k}(\phi) < C_{m,k} + \varepsilon$$

We have shown therefore that $\phi \in \mathscr{H}_\mu$ and that $\{\phi_\nu\}$ converges in $\mathscr{H}_\mu$ to ϕ, as was required.

Incidentally, our discussion up to this point has shown that $\mathscr{H}_\mu$ is a testing-function space since the three conditions stated at the beginning of Sec. 2.4 are satisfied.

$\mathscr{H}_\mu'$ denotes the dual of $\mathscr{H}_\mu$. According to Theorem 1.8-3, $\mathscr{H}_\mu'$ is also complete. The members of $\mathscr{H}_\mu'$ are the generalized functions on which our Hankel transformation will be defined. We now list some other properties of $\mathscr{H}_\mu$ and $\mathscr{H}_\mu'$ that we shall have need of subsequently.

I. In view of (6), it is clear that $\mathscr{D}(I)$ is a subspace of $\mathscr{H}_\mu$ for every choice of μ, and that convergence in $\mathscr{D}(I)$ implies convergence in $\mathscr{H}_\mu$. Consequently, the restriction of any $f \in \mathscr{H}_\mu'$ to $\mathscr{D}(I)$ is a member of $\mathscr{D}'(I)$.

However, $\mathscr{D}(I)$ is not dense in $\mathscr{H}_\mu$. Indeed, if $\phi \in \mathscr{H}_\mu$ and if the constant a_0 in (2) is not equal to zero, then $\gamma^\mu_{0,0}(\phi - \psi) \geq a_0$ for every $\psi \in \mathscr{D}(I)$, and the balloon $\{\theta: \theta \in \mathscr{H}_\mu, \gamma^\mu_{0,0}(\phi - \theta) \leq a_0/2\}$ does not contain any member of $\mathscr{D}(I)$, whence our assertion.

II. If q is an even positive integer, then $\mathscr{H}_{\mu+q} \subset \mathscr{H}_\mu$, and the topology of $\mathscr{H}_{\mu+q}$ is stronger than that induced on it by $\mathscr{H}_\mu$. To see this, note that, for every m, $\gamma^\mu_{m,0}(\phi) = \gamma^{\mu+2}_{m+2,0}(\phi)$; on the other hand, for every m and every positive k,

$$\begin{aligned}(x^{-1}D)^k x^{-\mu-1/2}\phi &= (x^{-1}D)^k[x^2 x^{-\mu-5/2}\phi]\\ &= 2(x^{-1}D)^{k-1}x^{-\mu-5/2}\phi + (x^{-1}D)^{k-1}x^2(x^{-1}D)x^{-\mu-5/2}\phi\\ &= \cdots = 2k(x^{-1}D)^{k-1}x^{-\mu-5/2}\phi + x^2(x^{-1}D)^k x^{-\mu-5/2}\phi\end{aligned}$$

so that

$$\gamma^\mu_{m,k}(\phi) \leq 2k\gamma^{\mu+2}_{m,k-1}(\phi) + \gamma^{\mu+2}_{m+2,k}(\phi)$$

Hence, our assertion is true for $q = 2$. The general case follows by induction on q.

We can now conclude that the restriction of $f \in \mathscr{H}_\mu'$ to $\mathscr{H}_{\mu+q}$ is in $\mathscr{H}'_{\mu+q}$. However, $\mathscr{H}_{\mu+q}$ is not dense in $\mathscr{H}_\mu$, and two different members of $\mathscr{H}_\mu'$ may have the same restriction to $\mathscr{H}_{\mu+q}$. (Show this.)

III. For each μ, $\mathscr{H}_\mu$ is clearly a subspace of $\mathscr{E}(I)$. Moreover, it is dense in $\mathscr{E}(I)$ because $\mathscr{H}_\mu$ contains $\mathscr{D}(I)$ and $\mathscr{D}(I)$ is dense in $\mathscr{E}(I)$.

Moreover, the topology of $\mathscr{H}_\mu$ is stronger than that induced on it by $\mathscr{E}(I)$. To see this, first recall that the seminorms for $\mathscr{E}(I)$ are

$$\gamma_{K,k}(\psi) = \sup_{x\in K} |D^k\psi(x)| \qquad \psi \in \mathscr{E}(I)$$

where $k = 0, 1, 2, \ldots$ and K ranges through all compact subsets of I. For any $\phi \in \mathscr{H}_\mu$, consider $\phi(x) = x^{\mu+1/2}x^{-\mu-1/2}\phi(x)$. If $C_{0,0}$ is the supremum of $x^{\mu+1/2}$ on K, then $\gamma_{K,0}(\phi) \leq C_{0,0}\gamma^\mu_{0,0}(\phi)$. Moreover, for any positive integer k, we may rearrange (6) into

$$D^k\phi = x^k\left[x^{\mu+1/2}(x^{-1}D)^k x^{-\mu-1/2}\phi - \frac{b_{k,0}}{x^{2k}}\phi - \cdots - \frac{b_{k,k-1}}{x^{k+1}}D^{k-1}\phi\right]$$

By induction on k, this shows that, for any fixed compact subset K of I, there exist constants $C_{k,q}$ such that

$$\gamma_{K,k}(\phi) \leq C_{k,0}\gamma^\mu_{0,0}(\phi) + C_{k,1}\gamma^\mu_{0,1}(\phi) + \ldots + C_{k,k}\gamma^\mu_{0,k}(\phi)$$

Lemma 1.6-3 completes the proof of our assertion.

Theorem 1.8-2 now shows that $\mathscr{E}'(I)$ is a subspace of $\mathscr{H}_\mu'$ whatever be the choice of μ.

IV. For each choice of the nonnegative integer r, set

$$\rho_r^\mu(\phi) = \max_{\substack{0\leq m\leq r\\0\leq k\leq r}} \gamma^\mu_{m,k}(\phi) \tag{8}$$

By Theorem 1.8-1, for each $f \in \mathscr{H}_\mu'$ there exist a positive constant C and a nonnegative integer r such that

$$|\langle f, \phi\rangle| \leq C\rho_r^\mu(\phi)$$

for every $\phi \in \mathscr{H}_\mu$. Here, C and r depend on f but not on ϕ.

V. Let $f(x)$ be a locally integrable function on $0 < x < \infty$ such that $f(x)$ is of slow growth as $x \to \infty$ and $x^{\mu+1/2}f(x)$ is absolutely integrable on $0 < x < 1$. Then, $f(x)$ generates a regular generalized function f in $\mathscr{H}_\mu'$ by

$$\langle f, \phi\rangle \triangleq \int_0^\infty f(x)\phi(x)\,dx \tag{9}$$

Indeed,

$$\langle f, \phi\rangle = \int_0^1 x^{\mu+1/2}f(x)x^{-\mu-1/2}\phi(x)\,dx + \int_1^\infty x^{-m+\mu+1/2}f(x)x^{m-\mu-1/2}\phi(x)\,dx$$

Since $f(x)$ is of slow growth, we can choose m so large that $x^{-m+\mu+1/2}f(x)$ is absolutely integrable on $1 < x < \infty$. Then,

$$|\langle f, \phi\rangle| \leq \gamma^\mu_{0,0}(\phi)\int_0^1 |x^{\mu+1/2}f(x)|\,dx + \gamma^\mu_{m,0}(\phi)\int_1^\infty |x^{-m+\mu+1/2}f(x)|\,dx$$

which implies our assertion.

This argument also shows that f generates through (9) a regular generalized function in $\mathscr{H}_\mu'$ if $\mu \geq -\frac{1}{2}$ and $f \in L_1(0, \infty)$.

Furthermore, $\mathscr{H}_\mu$ can be identified with a subspace of $\mathscr{H}_\mu'$ when $\mu \geq -\frac{1}{2}$. Indeed, our assumptions on f given at the beginning of this note are certainly satisfied if $\mu \geq -\frac{1}{2}$ and $f \in \mathscr{H}_\mu$. Thus, each $f \in \mathscr{H}_\mu$ truly generates through (9) a unique member of $\mathscr{H}_\mu'$. On the other hand, two members, say, f and g of $\mathscr{H}_\mu$ that generate the same member of $\mathscr{H}_\mu'$ must be identical. For, if f and g differed somewhere, they would have to differ everywhere on some open nonvoid interval $J \subset I$ since they are continuous. But then, we could choose a testing function $\phi \in \mathscr{D}(I)$ whose support is contained in J such that $\langle f, \phi\rangle \neq \langle g, \phi\rangle$; this would imply that f and g differ as members of $\mathscr{H}_\mu'$. Thus, we have a one-to-one correspondence between $\mathscr{H}_\mu$ and a subspace of $\mathscr{H}_\mu'$, and we are justified in writing $\mathscr{H}_\mu \subset \mathscr{H}_\mu'$.

PROBLEM 5.2-1. We have shown that, for every positive even integer q, $\mathscr{H}_{\mu+q} \subset \mathscr{H}_\mu$. Now, show that $\mathscr{H}_{\mu+q}$ is not dense in $\mathscr{H}_\mu$ and that two different members of $\mathscr{H}_\mu'$ may have the same restriction to $\mathscr{H}_{\mu+q}$.

PROBLEM 5.2-2. It was shown in note V that $\mathscr{H}_\mu \subset \mathscr{H}_\mu'$. Now, prove that the topology of $\mathscr{H}_\mu$ is stronger than that induced on it by $\mathscr{H}_\mu'$.

PROBLEM 5.2-3. Let f be a distribution whose support is contained in $X \leq x < \infty (X > 0)$. Show that $f \in \mathscr{H}_\mu'$ if and only if $f \in \mathscr{S}'$. Then, show that every $f \in \mathscr{H}_\mu'$ coincides with some distribution of slow growth on the domain consisting of those members of $\mathscr{S}$ whose supports are contained in $X \leq x < \infty$.

5.3. Some Operations on $\mathscr{H}_\mu$ and $\mathscr{H}_\mu'$

Multipliers in $\mathscr{H}_\mu$: Let $\mathcal{O}$ denote the linear space of all smooth functions $\theta(x)$ defined on $0 < x < \infty$ such that, for each nonnegative integer ν, there is an integer n_ν for which

$$\frac{(x^{-1}D)^\nu \theta(x)}{1 + x^{n_\nu}}$$

is bounded on $0 < x < \infty$. The product of any two members of $\mathcal{O}$ is also in $\mathcal{O}$. (Show this.)

Any $\theta \in \mathcal{O}$ is a multiplier for $\mathscr{H}_\mu$ for every μ. Indeed, for $\phi \in \mathscr{H}_\mu$

$$(x^{-1}D)^k x^{-\mu-1/2}\theta\phi = \sum_{\nu=0}^{k} \binom{k}{\nu} \frac{(x^{-1}D)^\nu \theta}{1 + x^{n_\nu}} (1 + x^{n_\nu})(x^{-1}D)^{k-\nu} x^{-\mu-1/2}\phi$$

so that

$$\gamma_{m,k}^\mu(\theta\phi) \leq \sum_{\nu=0}^{k} \binom{k}{\nu} B_\nu[\gamma_{m,k-\nu}^\mu(\phi) + \gamma_{m+n_\nu,k-\nu}^\mu(\phi)]$$

where the B_ν are constants. This shows that the linear operator $\phi \mapsto \theta\phi$ is a continuous mapping of $\mathscr{H}_\mu$ into itself, which is what we had to show.

By Sec. 2.5, the adjoint operator $f \mapsto \theta f$, which is defined on $\mathscr{H}_\mu'$ by

$$(1) \qquad \langle \theta f, \phi \rangle \triangleq \langle f, \theta\phi \rangle \qquad f \in \mathscr{H}_\mu', \quad \phi \in \mathscr{H}_\mu, \quad \theta \in \mathcal{O}$$

is a continuous linear mapping of $\mathscr{H}_\mu'$ into itself. We emphasize that the space $\mathcal{O}$ does not depend upon μ; it is a space of multipliers for $\mathscr{H}_\mu$ no matter what real value μ assumes.

LEMMA 5.3-1. *If $P(x)$ and $Q(x)$ are polynomials and $Q(x)$ has no zeros on $0 \leq x < \infty$, then $P(x^2)/Q(x^2)$ is a member of $\mathcal{O}$.*

PROOF. Clearly, $P(x^2)$ is in $\mathcal{O}$. Since the product of two members of $\mathcal{O}$ is also in $\mathcal{O}$, we need merely show that $1/Q(x^2)$ is also in $\mathcal{O}$. This can be done by applying the operator $(x^{-1}D)^\nu$ to $1/Q(x^2)$. This yields a rational function of the form:

$$\frac{N_\nu(x^2)}{[Q(x^2)]^{\nu+1}}$$

where $N_\nu(x^2)$ is a polynomial whose degree is less than the degree of $[Q(x^2)]^{\nu+1}$. Hence,

$$(x^{-1}D)^\nu \frac{1}{Q(x^2)}$$

is bounded on $0 \leq x < \infty$, which proves the lemma.

Multiplication by an integral power of x:

LEMMA 5.3-2. *For any positive or negative integer n and for any μ, the mapping $\phi(x) \mapsto x^n\phi(x)$ is an isomorphism from $\mathscr{H}_\mu$ onto $\mathscr{H}_{\mu+n}$. Consequently, the operator $f(x) \mapsto x^n f(x)$, which is defined by*

(2) $$\langle x^n f(x), \phi(x)\rangle \triangleq \langle f(x), x^n\phi(x)\rangle$$

is an isomorphism from $\mathscr{H}'_{\mu+n}$ onto $\mathscr{H}_\mu'$.

Proof. If $\phi \in \mathscr{H}_\mu$, then $\gamma_{m,k}^{\mu+n}(x^n\phi) = \gamma_{m,k}^{\mu}(\phi)$. The first assertion is an immediate consequence of this. The second assertion follows from Theorem 1.10-2.

Some differential and integral operators: We define two linear differential operators N_μ and M_μ and a linear integral operator N_μ^{-1} by

(3) $$N_\mu\phi(x) \triangleq x^{\mu+1/2}Dx^{-\mu-1/2}\phi(x)$$

(4) $$M_\mu\phi(x) \triangleq x^{-\mu-1/2}Dx^{\mu+1/2}\phi(x)$$

(5) $$N_\mu^{-1}\phi(x) \triangleq x^{\mu+1/2}\int_\infty^x t^{-\mu-1/2}\phi(t)\,dt$$

For the moment, we interpret the differentiations in N_μ and M_μ in the conventional sense. Also, N_μ^{-1} is certainly defined on every locally integrable function of rapid descent and therefore on every $\phi \in \mathscr{H}_{\mu+1}$. Moreover, N_μ and N_μ^{-1} are inverses of one another whenever ϕ and its derivative are continuous on $0 < x < \infty$ and of rapid descent as $x \to \infty$; thus, our notation is consistent.

N_μ is a continuous linear mapping of $\mathscr{H}_\mu$ into $\mathscr{H}_{\mu+1}$ because clearly $\gamma_{m,k}^{\mu+1}(N_\mu\phi) = \gamma_{m,k+1}^{\mu}(\phi)$ for every $\phi \in \mathscr{H}_\mu$ and every choice of m and k. On the other hand, N_μ^{-1} is a continuous linear mapping of $\mathscr{H}_{\mu+1}$ into $\mathscr{H}_\mu$; we prove this in two steps:

Assume that $\phi(x) \in \mathscr{H}_{\mu+1}$ and that k is a fixed positive integer. Then,

$$(x^{-1}D)^k x^{-\mu-1/2} N_\mu^{-1}\phi(x) = (x^{-1}D)^k x^{-\mu-1/2} x^{\mu+1/2} \int_\infty^x t^{-\mu-1/2}\phi(t)\,dt$$
$$= (x^{-1}D)^{k-1} x^{-\mu-3/2}\phi(x)$$

Hence,

(6) $$\gamma_{m,k}^{\mu}(N_\mu^{-1}\phi) = \gamma_{m,k-1}^{\mu+1}(\phi) \qquad k = 1, 2, 3, \ldots; \quad m = 0, 1, 2, \ldots$$

A similar result for the case $k = 0$ can be derived as follows:

$$|x^m x^{-\mu-1/2} N_\mu^{-1}\phi(x)| \le x^m \int_x^\infty |t^{-\mu-1/2}\phi(t)|\,dt$$
$$\le \int_x^\infty |t^m t^{-\mu-1/2}\phi(t)|\,dt$$
$$\le \int_0^\infty \left|\frac{1}{1+t^2}(t^{m+1} + t^{m+3}) t^{-\mu-3/2}\phi(t)\right| dt$$
$$\le \int_0^\infty \frac{dt}{1+t^2} \sup_{0<t<\infty} |(t^{m+1} + t^{m+3}) t^{-\mu-3/2}\phi(t)|$$

Therefore,

(7) $$\gamma_{m,0}^{\mu}(N_\mu^{-1}\phi) \le \frac{\pi}{2}\,[\gamma_{m+1,0}^{\mu+1}(\phi) + \gamma_{m+3,0}^{\mu+1}(\phi)] \qquad m = 0, 1, 2, \ldots$$

The results (6) and (7) prove that $\phi \mapsto N_\mu^{-1}\phi$ is a continuous linear mapping of $\mathscr{H}_{\mu+1}$ into $\mathscr{H}_\mu$.

Since N_μ and N_μ^{-1} are inverses of each other, we can now conclude that N_μ is a one-to-one mapping of $\mathscr{H}_\mu$ onto $\mathscr{H}_{\mu+1}$; the same may be said of N_μ^{-1} as a mapping of $\mathscr{H}_{\mu+1}$ onto $\mathscr{H}_\mu$. Altogether then, we have proven that N_μ is an isomorphism from $\mathscr{H}_\mu$ onto $\mathscr{H}_{\mu+1}$ with N_μ^{-1} being the inverse mapping.

Turning to the linear operator M_μ, we prove that $\phi \mapsto M_\mu\phi$ is a continuous linear mapping of $\mathscr{H}_{\mu+1}$ into $\mathscr{H}_\mu$. Indeed, for $\phi \in \mathscr{H}_{\mu+1}$ and any choice of m and k,

$$\gamma_{m,k}^{\mu}(M_\mu\phi) = \sup_{0<x<\infty} |x^m(x^{-1}D)^k\, x^{-2\mu-1} D x^{2\mu+2}\, x^{-\mu-3/2}\,\phi(x)|$$
$$= \sup_{0<x<\infty} |(2\mu+2)x^m(x^{-1}D)^k\, x^{-\mu-3/2}\,\phi(x)$$
$$+ x^m(x^{-1}D)^k\, x^2(x^{-1}D)x^{-\mu-3/2}\,\phi(x)|$$
$$= \cdots = \sup_{0<x<\infty} |2(\mu+k+1)x^m(x^{-1}D)^k\, x^{-\mu-3/2}\,\phi(x)$$
$$+ x^{m+2}(x^{-1}D)^{k+1}\, x^{-\mu-3/2}\,\phi(x)|$$
$$\le 2(\mu+k+1)\gamma_{m,k}^{\mu+1}(\phi) + \gamma_{m+2,k+1}^{\mu+1}(\phi)$$

This implies our assertion.

In the dual spaces we follow the notational convention described in Sec. 2.5 and define N_μ, when it is acting on certain generalized functions, as the adjoint of $-M_\mu$, and M_μ, in similar circumstances, as the adjoint of $-N_\mu$. More specifically, N_μ is defined as a generalized differential operator on $\mathscr{H}_\mu'$ by

(8) $$\langle N_\mu f, \phi\rangle \triangleq \langle f, -M_\mu \phi\rangle \qquad f \in \mathscr{H}_\mu', \qquad \phi \in \mathscr{H}_{\mu+1}$$

Consequently, $f \mapsto N_\mu f$ is a continuous linear mapping of $\mathscr{H}_\mu'$ into $\mathscr{H}'_{\mu+1}$. On the other hand, M_μ is defined as a generalized differential operator on $\mathscr{H}'_{\mu+1}$ by

(9) $$\langle M_\mu f, \phi\rangle \triangleq \langle f, -N_\mu \phi\rangle \qquad f \in \mathscr{H}'_{\mu+1}, \qquad \phi \in \mathscr{H}_\mu$$

Therefore, $f \mapsto M_\mu f$ is an isomorphism from $\mathscr{H}'_{\mu+1}$ onto $\mathscr{H}_\mu'$.

As was mentioned in Sec. 2.5, the reader must interpret the symbols N_μ and M_μ according to whether the operators are being applied to generalized functions or testing functions. (We do not use them in any other way.) In the first case, they are generalized differential operators defined by (8) or (9); in the second case, they are conventional differential operators defined by (3) or (4). Actually, when $\mu \geq -\frac{1}{2}$, we know from Sec. 5.2, note V that $\mathscr{H}_\mu \subset \mathscr{H}_\mu'$; in this case, the generalized operator N_μ, when acting on $\mathscr{H}_\mu$, can be identified with the conventional operator, and similarly for the generalized operator M_μ when it is acting on $\mathscr{H}_{\mu+1}$. (See Problem 5.3-2.)

We summarize these results by

LEMMA 5.3-3.

(i) *The conventional differential operator N_μ, defined by (3), is an isomorphism from $\mathscr{H}_\mu$ onto $\mathscr{H}_{\mu+1}$, the inverse mapping being N_μ^{-1}.*

(ii) *The conventional differential operator M_μ, defined by (4), is a continuous linear mapping of $\mathscr{H}_{\mu+1}$ into $\mathscr{H}_\mu$.*

(iii) *The generalized differential operator N_μ, defined by (8) is a continuous linear mapping of $\mathscr{H}_\mu'$ into $\mathscr{H}'_{\mu+1}$.*

(iv) *The generalized differential operator M_μ, defined by (9), is an isomorphism from $\mathscr{H}'_{\mu+1}$ onto $\mathscr{H}_\mu'$.*

Assertions (i) and (ii) imply that the conventional differential operator:

$$M_\mu N_\mu = D^2 - \frac{4\mu^2 - 1}{4x^2}$$

is a continuous linear mapping of $\mathscr{H}_\mu$ into $\mathscr{H}_\mu$. In agreement with (8) and (9) we define the generalized differential operator $M_\mu N_\mu$ by

(10) $$\langle M_\mu N_\mu f, \phi\rangle \triangleq \langle f, M_\mu N_\mu \phi\rangle \qquad f \in \mathscr{H}_\mu', \qquad \phi \in \mathscr{H}_\mu$$

and obtain thereby a continuous linear mapping of $\mathscr{H}_\mu'$ into $\mathscr{H}_\mu'$.

Note that, if $g(x)$ is a twice differentiable function, then the equation

$$\frac{1}{\sqrt{x}} M_\mu N_\mu \sqrt{x}\, g + g = 0$$

can be rewritten as

$$D^2 g + x^{-1} Dg + (1 - x^{-2}\mu^2) g = 0$$

which is Bessel's differential equation multiplied by x^{-2}.

We illustrate some of the results of this section in Fig. 5.3-1.

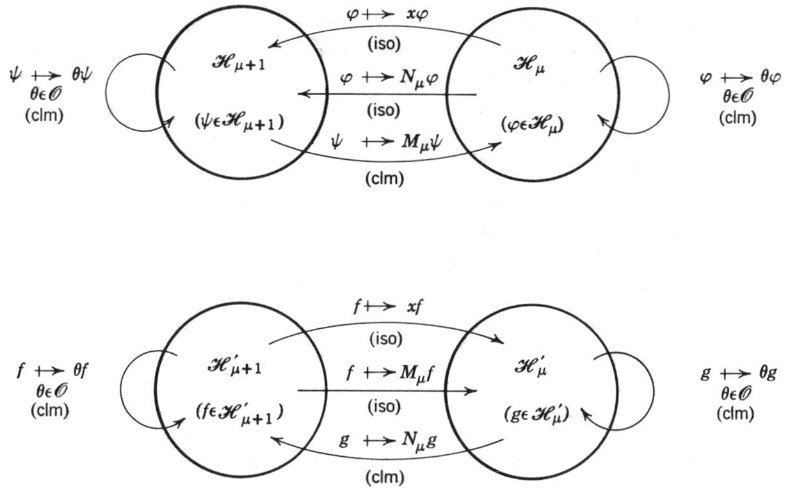

FIGURE 5.3–1. An illustration of the behavior of certain operators. Here "clm" means "continuous linear mapping," and "iso" means "isomorphism."

PROBLEM 5.3-1. Show that the product of two members of $\mathscr{O}$ is also a member of $\mathscr{O}$.

PROBLEM 5.3-2. Let f be a testing function in $\mathscr{H}_\mu$, and let $\mu \geq -\frac{1}{2}$. Show that (8) has a meaning as an integration by parts. This indicates that, under the stated conditions, the generalized operator N_μ can be identified with the conventional operator N_μ. In view of (9), this also indicates that the generalized operator M_μ can be identified with the conventional operator M_μ when $\mu \geq -\frac{1}{2}$ and $f \in \mathscr{H}_{\mu+1}$.

5.4. The Conventional Hankel Transformation on $\mathscr{H}_\mu$

The fundamental theorem in our theory of a generalized Hankel transformation asserts that the conventional Hankel transformation is an automorphism on $\mathscr{H}_\mu$. The proof of this fact is the objective of this section. But first, we shall establish some operation-transform formulas for the space $\mathscr{H}_\mu$. Note that, if $\mu \geq -\frac{1}{2}$, the conventional Hankel transform:

$$(1) \qquad \Phi(y) = (\mathfrak{H}_\mu \phi)(y) = \int_0^\infty \phi(x) \sqrt{xy}\, J_\mu(xy)\, dx \qquad 0 < y < \infty$$

exists for every $\phi \in \mathscr{H}_\mu$ by virtue of Lemma 5.2-1 and the facts that $\sqrt{xy}\, J_\mu(xy) = O(x^{\mu+1/2})$ as $x \to 0+$ and $\sqrt{xy}\, J_\mu(xy) = O(1)$ as $x \to \infty$ (Jahnke, Emde, and Losch [1], p. 134 and p. 147). In the following, it is always understood that the independent variable of $\phi \in \mathscr{H}_\mu$ is x, whereas the independent variable of its Hankel transform $\Phi = \mathfrak{H}_\mu \phi$ is y.

LEMMA 5.4-1. *Let $\mu \geq -\frac{1}{2}$. If $\phi \in \mathscr{H}_\mu$, then*

$$(2) \qquad \mathfrak{H}_{\mu+1}(-x\phi) = N_\mu \mathfrak{H}_\mu \phi$$

$$(3) \qquad \mathfrak{H}_{\mu+1}(N_\mu \phi) = -y \mathfrak{H}_\mu \phi$$

$$(4) \qquad \mathfrak{H}_\mu(-x^2 \phi) = M_\mu N_\mu \mathfrak{H}_\mu \phi$$

$$(5) \qquad \mathfrak{H}_\mu(M_\mu N_\mu \phi) = -y^2 \mathfrak{H}_\mu \phi$$

If $\phi \in \mathscr{H}_{\mu+1}$, then

$$(6) \qquad \mathfrak{H}_\mu(x\phi) = M_\mu \mathfrak{H}_{\mu+1}\phi$$

$$(7) \qquad \mathfrak{H}_\mu(M_\mu \phi) = y \mathfrak{H}_{\mu+1}\phi$$

PROOF. Equation (2) is established by using the formula (Jahnke, Emde, and Losch [1], p. 154):

$$(8) \qquad D_y y^{-\mu} J_\mu(xy) = -x y^{-\mu} J_{\mu+1}(xy)$$

and differentiating under an integral sign as follows.

$$(9) \qquad D_y y^{-\mu-1/2} \Phi(y) = \int_0^\infty \phi(x) \sqrt{x}\, D_y[y^{-\mu} J_\mu(xy)]\, dx$$
$$= -\int_0^\infty \phi(x) x^{3/2} y^{-\mu} J_{\mu+1}(xy)\, dx$$

This procedure is permissible. Indeed, for $\mu \geq -\frac{1}{2}$, $J_{\mu+1}(xy)$ is a smooth bounded function on $0 < xy < \infty$, and $\phi(x)x^{3/2}$ is a member of $L_1(0, \infty)$ when $\phi \in \mathscr{H}_\mu$ according to Lemma 5.2-1. Consequently, the right-hand side of (9) is a uniformly convergent integral on any compact subset of $0 < y < \infty$. Upon multiplying (9) by $y^{\mu+1/2}$, we obtain (2).

To obtain (3), first note that $N_\mu \phi$ is in $\mathscr{H}_{\mu+1}$ according to Lemma 5.3-3(i), so that we may take its $(\mu+1)$th-order Hankel transform. An integration by parts and the use of the formula (Jahnke, Emde, and Losch [1], p. 154):

$$D_x x^{\mu+1} J_{\mu+1}(xy) = yx^{\mu+1} J_\mu(xy) \tag{10}$$

yields

$$\begin{aligned}\mathfrak{H}_{\mu+1}(N_\mu \phi) &= \sqrt{y}\int_0^\infty [D_x x^{-\mu-1/2}\phi(x)]x^{\mu+1} J_{\mu+1}(xy)\,dx \\ &= \phi(x)\sqrt{xy}\,J_{\mu+1}(xy)\Bigg|_{x\to 0+}^{x\to\infty} - y\int_0^\infty \phi(x)\sqrt{xy}\,J_\mu(xy)\,dx\end{aligned}$$

The limit terms are equal to zero since $\phi(x)$ is of rapid descent as $x \to \infty$ and, as $x \to 0+$, $\sqrt{xy}\,J_{\mu+1}(xy) = O(x)$ and $\phi(x) = O(1)$ when $\mu \geq -\frac{1}{2}$. This verifies (3).

Now, assume that $\phi \in \mathscr{H}_{\mu+1}$ and set $\Phi(y) = \mathfrak{H}_{\mu+1}[\phi(x)]$. Formula (6) is derived by differentiating under the integral sign and using formula (10) with x and y interchanged.

$$\begin{aligned}D_y y^{\mu+1/2}\Phi(y) &= \int_0^\infty \phi(x)\sqrt{x}\,D_y[y^{\mu+1}J_{\mu+1}(xy)]\,dx \\ &= \int_0^\infty \phi(x)x^{3/2}\,y^{\mu+1}J_\mu(xy)\,dx\end{aligned} \tag{11}$$

This procedure is valid because, for $\mu \geq -\frac{1}{2}$, $\sqrt{xy}\,J_\mu(xy)$ is smooth and bounded for $0 < xy < \infty$ and $x\phi(x) \in L_1(0, \infty)$ by virtue of Lemma 5.2-1; consequently, the right-hand side of (11) converges uniformly on every compact subset of $0 < y < \infty$. The multiplication of (11) by $y^{-\mu-1/2}$ yields (6).

To get (7), invoke Lemma 5.3-3(ii) and apply $\mathfrak{H}_\mu$. An integration by parts and the use of (8) with x and y interchanged then yields

$$\begin{aligned}\mathfrak{H}_\mu(M_\mu \phi) &= \sqrt{y}\int_0^\infty [D_x x^{\mu+1/2}\phi(x)]x^{-\mu} J_\mu(xy)\,dx \\ &= \phi(x)\sqrt{xy}\,J_\mu(xy)\Bigg|_{x\to 0+}^{x\to\infty} + y\int_0^\infty \phi(x)\sqrt{xy}\,J_{\mu+1}(xy)\,dx\end{aligned}$$

By Lemma 5.2-1 and the fact that $\mu \geq -\frac{1}{2}$, $\phi(x) = O(x)$ as $x \to 0+$, and $\phi(x)$ is of rapid descent as $x \to \infty$. Moreover, $\sqrt{xy}\,J_\mu(xy)$ is bounded for $0 < xy < \infty$. Hence, the limit terms are equal to zero, and we have obtained (7).

Formula (4) is obtained by combining (2) and (6), and formula (5) by combining (3) and (7).

We now turn to our fundamental theorem.

THEOREM 5.4-1. *For* $\mu \geq -\frac{1}{2}$, *the conventional Hankel transformation* $\mathfrak{H}_\mu$ *is an automorphism on* $\mathscr{H}_\mu$.

PROOF. As before, let $\Phi(y) = \mathfrak{H}_\mu[\phi(x)]$, where $\phi \in \mathscr{H}_\mu$, and let m and k be any pair of nonnegative integers. By applying (2) k times and (3) m times and noting that

$$N_{\mu+k+m-1} \cdots N_{\mu+k+1} N_{\mu+k} x^k \phi(x) = x^k N_{\mu+m-1} \cdots N_{\mu+1} N_\mu \phi(x)$$

we obtain

$$(-y)^m N_{\mu+k-1} \cdots N_\mu \Phi(y) = \int_0^\infty (-x)^k [N_{\mu+m-1} \cdots N_\mu \phi(x)] \sqrt{xy}\, J_{\mu+k+m}(xy)\, dx$$

This is the same as

$$(12) \quad (-1)^{m+k} y^m (y^{-1} D_y)^k y^{-\mu-1/2} \Phi(y) = \int_0^\infty x^{2\mu+2k+m+1} [(x^{-1}D_x)^m x^{-\mu-1/2} \phi(x)] \frac{J_{\mu+k+m}(xy)}{(xy)^{\mu+k}}\, dx$$

Furthermore, since $\mu \geq -\frac{1}{2}$, $J_{\mu+k+m}(xy)/(xy)^{\mu+k}$ is bounded on $0 < xy < \infty$ by, say, the constant $B_{k,m}$. It follows that the last integral converges uniformly on $0 < y < \infty$ so that the left-hand side of (12) is a continuous function for every m and k. By an inductive argument based upon Sec. 5.2, Eq. (6), we can conclude that $\Phi(y)$ is a smooth function on $0 < y < \infty$.

Now, assume that n is an integer no less than $\mu + k + \frac{1}{2}(m+1)$. Then, $x^{2\mu+2k+m+1} < (1+x^2)^n$ for $x > 0$. Hence, (12) yields

$$\gamma^\mu_{m,k}(\Phi) \leq \int_0^\infty (1+x^2)^{n+1} |(x^{-1}D_x)^m x^{-\mu-1/2} \phi(x)| \frac{B_{k,m}}{1+x^2}\, dx$$
$$\leq \frac{\pi}{2} B_{k,m} \sum_{\nu=0}^{n+1} \binom{n+1}{\nu} \gamma^\mu_{2\nu,m}(\phi)$$

This inequality proves that Φ is also in $\mathscr{H}_\mu$ and that the linear mapping $\mathfrak{H}_\mu$ is also continuous from $\mathscr{H}_\mu$ into $\mathscr{H}_\mu$.

Furthermore, the classical inversion theorem (Theorem 5.1-1) applies in this case since $\mathscr{H}_\mu \subset L_1(0, \infty)$ when $\mu \geq -\frac{1}{2}$. Also, $\mathfrak{H}_\mu = \mathfrak{H}_\mu^{-1}$. These facts imply that $\mathfrak{H}_\mu$ is a one-to-one mapping of $\mathscr{H}_\mu$ onto $\mathscr{H}_\mu$ and indeed an automorphism on $\mathscr{H}_\mu$.

5.5. The Generalized Hankel Transformation

In this section μ is always restricted to the interval $-\frac{1}{2} \leq \mu < \infty$. We define the generalized Hankel transformation $\mathfrak{H}_\mu'$ on $\mathscr{H}_\mu'$ as the adjoint of

$\mathfrak{H}_\mu$ on $\mathscr{H}_\mu$. More specifically, for arbitrary $\Phi \in \mathscr{H}_\mu$ and $\phi = \mathfrak{H}_\mu \Phi$ and for arbitrary $f \in \mathscr{H}_\mu'$, we define $F = \mathfrak{H}_\mu' f$ by

$$\langle F, \Phi \rangle \triangleq \langle f, \phi \rangle \tag{1}$$

or, using different symbols, we write

$$\langle \mathfrak{H}_\mu' f, \Phi \rangle \triangleq \langle f, \mathfrak{H}_\mu \Phi \rangle \tag{2}$$

From Theorems 1.10-2 and 5.4-1, we immediately obtain

THEOREM 5.5-1. *For $\mu \geq -\frac{1}{2}$, the generalized Hankel transformation $\mathfrak{H}_\mu'$ is an automorphism on $\mathscr{H}_\mu'$.*

Since $\mathfrak{H}_\mu = \mathfrak{H}_\mu^{-1}$ on $\mathscr{H}_\mu$, the last statement of Theorem 1.10-2 indicates that $(\mathfrak{H}_\mu')^{-1} = \mathfrak{H}_\mu'$; in words, the generalized Hankel transformation is its own inverse when it is acting on $\mathscr{H}_\mu'$. The fact that $\mathfrak{H}_\mu = \mathfrak{H}_\mu^{-1}$ on $\mathscr{H}_\mu$ also permits us to rewrite (1) as

$$\langle \mathfrak{H}_\mu' f, \mathfrak{H}_\mu \phi \rangle = \langle f, \phi \rangle$$

and in this form it appears as a generalization of the Parseval equation (Sec. 5.1, Eq. (3)).

The conventional Hankel transformation $\mathfrak{H}_\mu$ of order $\mu \geq -\frac{1}{2}$, when acting on a function $f \in L_1(0, \infty)$, is a special case of our generalized Hankel transformation. Indeed, set

$$F_c(y) = \mathfrak{H}_\mu f = \int_0^\infty f(x) \sqrt{xy}\, J_\mu(xy)\, dx \qquad \mu \geq -\tfrac{1}{2},\ f \in L_1(0, \infty) \tag{3}$$

Since $\sqrt{xy}\, J_\mu(xy)$ is bounded on $0 < xy < \infty$, the integral in (3) converges uniformly on $0 < y < \infty$. Consequently, $F_c(y)$ is both continuous and bounded on $0 < y < \infty$. According to Sec. 5.2, note V, $F_c(y)$ generates a regular generalized function in $\mathscr{H}_\mu'$. Furthermore, by our definition (1) of $F = \mathfrak{H}_\mu' f$ and by the second paragraph of Sec. 5.2, note V, we have for any $\phi \in \mathscr{H}_\mu$ and $\Phi = \mathfrak{H}_\mu^{-1} \phi = \mathfrak{H}_\mu \phi$ that

$$\langle F, \Phi \rangle = \langle f, \phi \rangle = \int_0^\infty f(x)\phi(x)\, dx$$

Since $\Phi \in \mathscr{H}_\mu \subset L_1(0, \infty)$ when $\mu \geq -\frac{1}{2}$, we may invoke Parseval's equation (Sec. 5.1, Eq. (3)) to write

$$\int_0^\infty f(x)\phi(x)\, dx = \int_0^\infty F_c(y)\Phi(y)\, dy$$

Therefore,

$$\langle F, \Phi \rangle = \int_0^\infty F_c(y)\Phi(y)\, dy$$

Thus, our generalized Hankel transform $F = \mathfrak{H}_\mu' f$ is the regular generalized function corresponding to the conventional Hankel transform $F_c = \mathfrak{H}_\mu f$.

Because we can identify $\mathfrak{H}_\mu'$ with $\mathfrak{H}_\mu$ under suitable conditions on f, we shall henceforth drop the prime on $\mathfrak{H}_\mu'$ when $\mu \geq -\frac{1}{2}$. Thus, our definition (2) of the generalized Hankel transform $\mathfrak{H}_\mu f$ of any $f \in \mathscr{H}_\mu'$ now becomes

$$(4) \qquad \langle \mathfrak{H}_\mu f, \Phi \rangle \triangleq \langle f, \mathfrak{H}_\mu \Phi \rangle \qquad f \in \mathscr{H}_\mu', \quad \Phi \in \mathscr{H}_\mu, \quad \mu \geq -\tfrac{1}{2}$$

(In Sec. 5.10, we shall extend our generalized Hankel transformation to the case where $\mu < -\frac{1}{2}$. In that case the prime on $\mathfrak{H}_\mu'$ will not be discarded because an identification between the conventional and generalized Hankel transformation will not be made.)

We end this section by proving a number of operation-transform formulas for the generalized Hankel transformation. These formulas read just like those of Lemma 5.4-1, but it must now be borne in mind that the present formulas deal with generalized operations (i.e., the adjoints of conventional operations). In the following we let x be the independent variable corresponding to ϕ and f, and y the independent variable corresponding to Φ and F.

THEOREM 5.5-2. Let $\mu \geq -\frac{1}{2}$. *If* $f \in \mathscr{H}_\mu'$, *then*

$$(5) \qquad \mathfrak{H}_{\mu+1}(-xf) = N_\mu \mathfrak{H}_\mu f$$

$$(6) \qquad \mathfrak{H}_{\mu+1}(N_\mu f) = -y \mathfrak{H}_\mu f$$

$$(7) \qquad \mathfrak{H}_\mu(-x^2 f) = M_\mu N_\mu \mathfrak{H}_\mu f$$

$$(8) \qquad \mathfrak{H}_\mu(M_\mu N_\mu f) = -y^2 \mathfrak{H}_\mu f$$

If $f \in \mathscr{H}'_{\mu+1}$, *then*

$$(9) \qquad \mathfrak{H}_\mu(xf) = M_\mu \mathfrak{H}_{\mu+1} f$$

$$(10) \qquad \mathfrak{H}_\mu(M_\mu f) = y \mathfrak{H}_{\mu+1} f$$

Note: Here, (5) and (6) are understood in the sense of equality in $\mathscr{H}'_{\mu+1}$, and (7) through (10) in the sense of equality in $\mathscr{H}_\mu'$.

PROOF. We first establish (6). Let $\Phi \in \mathscr{H}_{\mu+1}$ and $\phi = \mathfrak{H}_{\mu+1}\Phi$. We invoke our definition (4) and Lemmas 5.3-2, 5.3-3, and 5.4-1 to write

$$(11) \quad \langle \mathfrak{H}_{\mu+1} N_\mu f, \Phi \rangle = \langle N_\mu f, \phi \rangle = \langle f, -M_\mu \phi \rangle$$
$$= \langle \mathfrak{H}_\mu f, -y\Phi \rangle = \langle -y \mathfrak{H}_\mu f, \Phi \rangle$$

The right-hand side of (11) has a sense because $-y\mathfrak{H}_\mu f$ is in $\mathscr{H}'_{\mu-1}$ so that its restriction to $\mathscr{H}_{\mu+1}$ is in $\mathscr{H}'_{\mu+1}$, according to Sec. 5.2, note *II*. We thus have the desired result.

Equation (5) can be obtained from (6) as follows. We have already noted that the generalized Hankel transformation is its own inverse when it is acting on $\mathscr{H}_\mu'$. In other words, $\mathfrak{H}_\mu \mathfrak{H}_\mu$ is the identity operator on $\mathscr{H}_\mu'$. Let $F = \mathfrak{H}_\mu f$. Then, by (6),

$$N_\mu \mathfrak{H}_\mu F = \mathfrak{H}_{\mu+1} \mathfrak{H}_{\mu+1} N_\mu f = \mathfrak{H}_{\mu+1}(-y\mathfrak{H}_\mu f) = \mathfrak{H}_{\mu+1}(-yF)$$

This is the same as (5) with f and x replaced by F and y respectively.

Similarly, for $\Phi \in \mathscr{H}_\mu$ and $\phi = \mathfrak{H}_\mu \Phi$, (10) is established by using the definition (4) and Lemmas 5.3-2, 5.3-3, and 5.4-1 as follows:

$$(12) \quad \langle \mathfrak{H}_\mu M_\mu f, \Phi \rangle = \langle M_\mu f, \phi \rangle = \langle f, -N_\mu \phi \rangle$$
$$= \langle \mathfrak{H}_{\mu+1} f, y\Phi \rangle = \langle y\mathfrak{H}_{\mu+1} f, \Phi \rangle$$

Then, by setting $F = \mathfrak{H}_{\mu+1} f$ and noting that $\mathfrak{H}_\mu \mathfrak{H}_\mu$ is the identity operator on $\mathscr{H}_\mu'$, we derive (9) from (10) by

$$M_\mu \mathfrak{H}_{\mu+1} F = \mathfrak{H}_\mu \mathfrak{H}_\mu M_\mu f = \mathfrak{H}_\mu y \mathfrak{H}_{\mu+1} f = \mathfrak{H}_\mu y F$$

Formulas (7) and (8) follow directly from these results.

PROBLEM 5.5-1. Let a be a real positive number and ν a nonnegative integer. The generalized function $\delta^{(\nu)}(x - a)$ is defined on $\mathscr{H}_\mu$ by

$$\langle \delta^{(\nu)}(x-a), \phi(x) \rangle \triangleq (-1)^\nu \phi^{(\nu)}(a) \qquad \phi \in \mathscr{H}_\mu$$

Clearly, $\delta^{(\nu)}(x-a) \in \mathscr{H}_\mu'$. Show that, for $\mu \geq -\frac{1}{2}$,

$$\mathfrak{H}_\mu \delta^{(\nu)}(x-a) = (-1)^\nu D_a{}^\nu[\sqrt{ay}\, J_\mu(ay)]$$
$$\mathfrak{H}_\mu D_a{}^\nu[\sqrt{ax}\, J_\mu(ax)] = (-1)^\nu \delta^{(\nu)}(y-a)$$

Here, $D_a{}^\nu$ is a conventional differentiation of νth order, but $D_a{}^\nu[\sqrt{ax}$ $J_\mu(ax)]$ denotes a regular generalized function in $\mathscr{H}_\mu'$. (Actually, Theorem 5.6-3 in the next section provides a very simple proof of these formulas. But, don't use it.)

PROBLEM 5.5-2. Define f as a functional on $\mathscr{H}_\mu$ by

$$\langle f, \phi \rangle \triangleq \sum_{\nu=1}^{\infty} \int_0^\infty \phi(x) \sqrt{a\nu x}\, J_\mu(a\nu x)\, dx$$

where $\mu \geq -\frac{1}{2}$, $\phi \in \mathscr{H}_\mu$, and a is a real positive number. Show that f is a member of $\mathscr{H}_\mu'$, and find its Hankel transform.

PROBLEM 5.5-3. Show that

$$\mathfrak{H}_\mu[x^{\mu+1/2} 1_+(a-x)] = a^{\mu+1} y^{-1/2} J_{\mu+1}(ay)$$

and

$$\mathfrak{H}_\mu[x^{\mu+5/2} 1_+(a-x)] = a^{\mu+2} y^{-1/2}[aJ_{\mu+1}(ay) - 2y^{-1} J_{\mu+2}(ay)]$$

where $\mu \geq -\frac{1}{2}$ and $0 < a < \infty$.

PROBLEM 5.5-4. Let n be a positive integer. Establish the following formulas.

(*a*) In the sense of equality in $\mathscr{H}'_{\mu+n}$ and for $\mu \geq -\frac{1}{2}$,

$$\mathfrak{H}_{\mu+n}\, x^{n-1/2} = (\mu+1)(\mu+3)\ldots(\mu+2n-1)y^{-n-1/2}$$

(*b*) In the sense of equality in $\mathscr{H}_\mu'$ and for $\mu \geq 2n-\frac{1}{2}$,

$$\mathfrak{H}_\mu\, x^{2n-1/2} = (\mu^2-1)(\mu^2-9)\ldots[\mu^2-(2n-1)^2]y^{-2n-1/2}$$

(*c*) In the sense of equality in $\mathscr{H}'_{\mu+1}$ and for $\mu \geq 2n-\frac{1}{2}$,

$$\mathfrak{H}_{\mu+1}\, x^{2n+1/2} = (\mu^2-1)(\mu^2-9)\ldots[\mu^2-(2n-1)^2](\mu+2n-1)y^{-2n-3/2}$$

Hint: Start from the fact that the formula in part (*a*) remains valid when n is set equal to zero.

5.6. The Hankel Transformation on $\mathscr{E}'(I)$

As was mentioned in the introduction to this chapter, the Hankel transform of certain (but not all) members of $\mathscr{H}_\mu'$ takes on the form,

$$F(y) = \langle f(x), \sqrt{xy}\, J_\mu(xy)\rangle \tag{1}$$

This happens, for example, when $f \in \mathscr{E}'(I)$ and $\mu \geq -\frac{1}{2}$, as we shall now prove. More general conditions, under which (1) holds, are given in Koh and Zemanian [1].

We shall first establish that, when $f \in \mathscr{E}'(I)$, $F(y)$ is a smooth function on $0 < y < \infty$. Indeed, it can be extended into an analytic function on the complex plane whose only singularities are branch points at the origin and at infinity. To do this, let η be a complex variable, and set

$$F(\eta) = \langle f(x), \sqrt{x\eta}\, J_\mu(x\eta)\rangle \tag{2}$$

For any fixed $x > 0$, $\sqrt{x\eta}\, J_\mu(x\eta)$ is a multivalued function of η except when $1+2\mu$ is an even integer. Moreover, it satisfies

$$(x\eta e^{i2n\pi})^{1/2} J_\mu(x\eta e^{i2n\pi}) = e^{in\pi(1+2\mu)}(x\eta)^{1/2} J_\mu(x\eta)$$

where $\eta \neq 0$ and n is any integer. (See Jahnke, Emde, and Losch [1], p. 134.) Consequently, $F(\eta)$ also satisfies

$$F(\eta e^{i2n\pi}) = e^{in\pi(1+2\mu)} F(\eta) \qquad \eta \neq 0$$

thus being multivalued when $1+2_\mu$ is not an even integer.

THEOREM 5.6-1. *Let f be a member of $\mathscr{E}'(I)$, μ any real number, and η a complex variable. Then, $F(\eta)$, as defined by (2), is an analytic function on the finite η plane except for a branch point at $\eta = 0$. Moreover,*

$$D_\eta F(\eta) = \langle f(x), D_\eta \sqrt{x\eta}\, J_\mu(x\eta)\rangle \qquad \eta \neq 0 \tag{3}$$

PROOF. For η fixed and $\eta \neq 0$, choose two concentric circles C and C_1 centered at η and of radii r and r_1, respectively, such that $0 < r < r_1 < |\eta|$. Let $\Delta\eta$ be a complex increment satisfying $0 < |\Delta\eta| < r$. Consider

$$\frac{F(\eta + \Delta\eta) - F(\eta)}{\Delta\eta} - \langle f(x), D_\eta \sqrt{x\eta}\, J_\mu(x\eta)\rangle = \langle f(x), \psi_{\Delta\eta}(x)\rangle \tag{4}$$

where

$$\psi_{\Delta\eta}(x) = \frac{1}{\Delta\eta}[\sqrt{x\eta + x\Delta\eta}\, J_\mu(x\eta + x\Delta\eta) - \sqrt{x\eta}\, J_\mu(x\eta)] - D_\eta \sqrt{x\eta}\, J_\mu(x\eta)$$

Note that, for any fixed $x \in I$ and any fixed nonnegative integer k, $D_x^k \sqrt{x\eta}\, J_\mu(x\eta)$ is analytic inside and on C_1. Let's restrict it to a particular branch. By Cauchy's integral formulas (Churchill [1], p. 120).

$$D_x^k \psi_{\Delta\eta}(x) =$$

$$\frac{1}{2\pi i}\int_{C_1} [D_x^k \sqrt{xz}\, J_\mu(xz)]\left[\frac{1}{\Delta\eta}\left(\frac{1}{z - \eta - \Delta\eta} - \frac{1}{z - \eta}\right) - \frac{1}{(z-\eta)^2}\right] dz$$

$$= \frac{\Delta\eta}{2\pi i}\int_{C_1} \frac{D_x^k \sqrt{xz}\, J_\mu(xz)}{(z - \eta - \Delta\eta)(z - \eta)^2}\, dz$$

But, for all $z \in C_1$ and x restricted to a compact subset of $0 < x < \infty$, $D_x^k \sqrt{xz}\, J_\mu(xz)$ is bounded by, say, the constant B. Since $|z - \eta| = r_1$ and $|z - \eta - \Delta\eta| > r_1 - r$, we therefore have

$$|D_x^k \psi_{\Delta\eta}(x)| \leq \frac{|\Delta\eta| B}{(r_1 - r)r_1}$$

This shows that, as $|\Delta\eta| \to 0$, $D_x^k \psi_{\Delta\eta}(x)$ tends to zero uniformly on compact subsets of $0 < x < \infty$; that is, $\psi_{\Delta\eta}(x)$ converges in $\mathscr{E}(I)$ to zero. Since $f \in \mathscr{E}'(I)$, (4) tends to zero. Thus, for η restricted as stated, we have obtained (3) and have proven the analyticity of $F(\eta)$.

Our next objective is to show that $F(y)$ satisfies certain order conditions as $y \to 0+$ and $y \to \infty$.

THEOREM 5.6-2. *Let f be a member of $\mathscr{E}'(I)$, let $\mu \geq -\frac{1}{2}$, and let $F(y)$ be defined by (1). Then, $F(y)$ satisfies the inequality:*

$$|F(y)| \leq \begin{cases} Ky^{\mu+1/2} & 0 < y < 1 \\ Ky^p & 1 < y < \infty \end{cases} \tag{5}$$

where K and p are sufficiently large real numbers.

PROOF. By Theorem 2.3-1, the support of f is a compact subset of I. Choose $\lambda \in \mathscr{D}(I)$ such that $\lambda(x) \equiv 1$ on a neighborhood of the support of f. Thus,

$$F(y) = \langle f(x), \lambda(x)\sqrt{xy}\, J_\mu(xy)\rangle$$

By Sec. 5.2, note III, f is a member of $\mathscr{H}_\mu'$, and, by Sec. 5.2, note I, $\lambda(x)\sqrt{xy}\, J_\mu(xy)$ is a member of $\mathscr{H}_\mu$. Consequently, by Sec. 5.2, note IV, there exist a positive constant C and a nonnegative integer r such that

$$|F(y)| \le C \max_{\substack{0\le m\le r\\ 0\le k\le r}} \sup_{0<x<\infty} |x^m (x^{-1}D_x)^k[\lambda(x)x^{-\mu-1/2}\sqrt{xy}\, J_\mu(xy)]|$$

$$\le C \max_{m,\,k} \sup_{x} \left| \sum_{\nu=0}^{k} \binom{k}{\nu} [x^m(x^{-1}D_x)^{k-\nu}\lambda(x)][(x^{-1}D_x)^\nu x^{-\mu-1/2}\sqrt{xy}\, J_\mu(xy)]\right|$$

Let the constant $C_{m,\,k-\nu}$ be a bound on $x^m(x^{-1}D_x)^{k-\nu}\lambda(x)$. Also, for every nonnegative integer ν, we have the identity

$$(x^{-1}D_x)^\nu[x^{-\mu-1/2}\sqrt{xy}\, J_\mu(xy)] = (-1)^\nu y^{\mu+1/2+2\nu}\frac{J_{\mu+\nu}(xy)}{(xy)^{\mu+\nu}} \tag{6}$$

which can be derived from Sec. 5.1, Eq. (7). But, since $\mu \ge -\frac{1}{2}$, $J_{\mu+\nu}(z)/z^{\mu+\nu}$ is bounded on $0 < z < \infty$ by a constant, say, B_ν. Therefore,

$$|F(y)| \le C \max_{\substack{0\le m\le r\\ 0\le k\le r}} \sum_{\nu=0}^{k}\binom{k}{\nu} C_{m,\,k-\nu}B_\nu y^{\mu+1/2+2\nu}$$

The inequality (5) follows directly from this.

The paramount theorem of this section is

THEOREM 5.6-3. *If $f \in \mathscr{E}'(I)$ and $\mu \ge -\frac{1}{2}$, then the generalized Hankel transform of f, as defined by Sec. 5.5, Eq. (4), is the regular generalized function in $\mathscr{H}_\mu'$ that is generated by the smooth function $F(y)$ defined by* (1).

Note that $F(y)$ truly does generate a regular member of $\mathscr{H}_\mu'$ in view of Sec. 5.2, note V, Theorems 5.6-1 and 5.6-2, and the condition $\mu \ge -\frac{1}{2}$. The proof of Theorem 5.6-3 proceeds by the way of two lemmas.

LEMMA 5.6-1. *If $\mu \ge -\frac{1}{2}$ and Φ is a member of $\mathscr{H}_\mu$, then*

$$\psi_Y(x) \triangleq \int_Y^\infty \Phi(y)\sqrt{xy}\, J_\mu(xy)\, dy$$

converges in $\mathscr{E}(I)$ to zero as $Y \to \infty$.

PROOF. By repeatedly differentiating under the integral sign and using (6), we obtain

$$(7) \qquad (x^{-1}D)^k x^{-\mu-1/2} \psi_Y(x) = (-1)^k \int_Y^\infty \Phi(y) y^{\mu+1/2+2k} \frac{J_{\mu+k}(xy)}{(xy)^{\mu+k}} \, dy$$

These differentiations under the integral sign are justified by the boundedness of $J_{\mu+k}(z)/z^{\mu+k}$ on $0 < z < \infty$ and the consequent uniform convergence on $0 < x < \infty$ of the integral in (7) for each k; thus, (7) tends to zero uniformly on $0 < x < \infty$ as $Y \to \infty$. But,

$$(8) \quad (x^{-1}D)^k x^{-\mu-1/2} \psi_Y(x) = x^{-\mu-1/2}\left[b_{k0} \frac{\psi_Y(x)}{x^{2k}} + b_{k1} \frac{D\psi_Y(x)}{x^{2k-1}} + \cdots + \frac{D^k \psi_Y(x)}{x^k} \right]$$

where the b's are constants. Our conclusion follows from this by induction on k; that is, for every k, $D^k\psi_Y(x)$ converges uniformly to zero on every compact subset of $0 < x < \infty$ as $Y \to \infty$.

LEMMA 5.6-2. *If $\mu \geq -\frac{1}{2}$, $\Phi \in \mathscr{H}_\mu$, $f \in \mathscr{E}'(I)$, and Y is a fixed (finite) positive number, then*

$$(9) \quad \int_0^Y \Phi(y) \langle f(x), \sqrt{xy}\, J_\mu(xy) \rangle \, dy = \langle f(x), \int_0^Y \Phi(y) \sqrt{xy}\, J_\mu(xy) \, dy \rangle$$

PROOF. To prove this, we use "Riemann sums" to approximate the integrals on y. Partition the interval $0 < y < Y$ into m subintervals whose lengths are all equal to $(\Delta y)_m$, and let $y_{m\nu}$ be a point in the νth subinterval. Since $\Phi(y)$ and $\langle f(x), \sqrt{xy}\, J_\mu(xy) \rangle$ are smooth bounded functions on $0 < y < Y$ (see Lemma 5.2-1 and Theorem 5.6-2), we can choose M_1 so large, having been given any $\varepsilon > 0$, that for all $m > M_1$

$$(10) \qquad \sum_{\nu=1}^m \Phi(y_{m\nu}) \langle f(x), \sqrt{xy_{m\nu}}\, J_\mu(xy_{m\nu}) \rangle (\Delta y)_m$$

differs from the left-hand side of (9) by a quantity whose magnitude is less than ε. Moreover, (10) is equal to

$$(11) \qquad \langle f(x), \sum_{\nu=0}^m \Phi(y_{m\nu}) \sqrt{xy_{m\nu}}\, J_\mu(xy_{m\nu}) (\Delta y)_m \rangle$$

Now, as $m \to \infty$, the testing function in (11) converges in $\mathscr{E}(I)$ to the testing function in the right-hand side of (9). Indeed, $D_x^k \sqrt{xy}\, J_\mu(xy)$ is a continuous function of the two-dimensional variable (x, y) on the domain $\Omega = \{(x, y) \colon x \in K, 0 \leq y \leq Y\}$ where K is any compact subset of $0 < x < \infty$. Therefore, it is permissible to differentiate under the integral sign according to

$$(12) \quad D_x^k \int_0^Y \Phi(y) \sqrt{xy}\, J_\mu(xy)\, dy = \int_0^Y \Phi(y)\, D_x^k \sqrt{xy}\, J_\mu(xy)\, dy$$

Moreover,

$$\sum_{\nu=1}^{m} \Phi(y_{m\nu})[D_x^k \sqrt{xy_{m\nu}}\, J_\mu(xy_{m\nu})](\Delta y)_m$$

converges to (12) uniformly on K as $m \to \infty$ because the integrand in the right-hand side of (12) is continuous, and therefore uniformly continuous, on the closed bounded domain Ω. This verifies the assertion at the beginning of this paragraph.

Since $f \in \mathscr{E}'(I)$, given the same $\varepsilon > 0$ as before, we can choose an M_2 such that, for all $m > M_2$, (11) differs from the right-hand side of (9) by a quantity whose magnitude is less than ε. Thus, we have shown that the difference between the two sides of (9) is bounded by 2ε, and, since $\varepsilon > 0$ is arbitrary, our proof is complete.

PROOF OF THEOREM 5.6-3. We have to show that, for any $\Phi \in \mathscr{H}_\mu$,

$$\langle\langle f(x), \sqrt{xy}\, J_\mu(xy)\rangle, \Phi(y)\rangle = \langle f, \mathfrak{H}_\mu \Phi\rangle \tag{13}$$

In view of Lemma 5.2-1 and Theorems 5.6-1 and 5.6-2, the left-hand side of (13) is equal to

$$\int_0^Y \langle f(x), \sqrt{xy}\, J_\mu(xy)\rangle\, \Phi(y)\, dy + \int_Y^\infty \langle f(x), \sqrt{xy}\, J_\mu(xy)\rangle\, \Phi(y)\, dy \tag{14}$$

Moreover, given any $\varepsilon > 0$, there exists a Y_1 such that, for all $Y > Y_1$, the second term in (14) is bounded in magnitude by ε. On the other hand, the right-hand side of (13) can be written as

$$\langle f(x), \int_0^Y \Phi(y) \sqrt{xy}\, J_\mu(xy)\, dy\rangle + \langle f(x), \int_Y^\infty \Phi(y) \sqrt{xy}\, J_\mu(xy)\, dy\rangle \tag{15}$$

By virtue of Lemma 5.6-1, there is a Y_2 such that the second term in (15) is bounded in magnitude by ε whenever $Y > Y_2$. Finally, the first terms in (14) and (15) are equal to each other according to Lemma 5.6-2. Thus, the difference between the two side of (13) is less than 2ε. Since $\varepsilon > 0$ can be chosen arbitrarily small, the proof of Theorem 5.6-3 is complete.

PROBLEM 5.6-1. Let $f \in \mathscr{E}'(I)$ and $\mu \geq -\frac{1}{2}$. Verify that $\mathfrak{H}_{\mu+1}(N_\mu f) = -y\mathfrak{H}_\mu f$ and $\mathfrak{H}_\mu(M_\mu f) = y\mathfrak{H}_{\mu+1} f$ by using Theorem 5.6-3 to compute directly the left-hand sides of these formulas.

PROBLEM 5.6-2. Assume that $\mu \geq -\frac{1}{2}$, $f \in \mathscr{E}'(I)$, and $F = \mathfrak{H}_\mu f$. Show that, in the sense of convergence in $\mathscr{H}_\mu'$,

$$f(x) = \lim_{Y \to \infty} \int_0^Y F(y) \sqrt{xy}\, J_\mu(xy)\, dy$$

This is an inversion formula for (1).

PROBLEM 5.6-3. Assume that $\mu \geq -\frac{1}{2}$, $f \in \mathscr{E}'(I)$, and $F = \mathfrak{H}_\mu f$. Also, let q be a positive integer such that $2q > p + 2$, where p is the number indicated in Theorem 5.6-2. Show that, in the sense of equality in $\mathscr{H}_\mu'$.

$$f(x) = [1 + (-M_\mu N_\mu)^q] \int_0^\infty \frac{F(y)}{1 + y^{2q}} \sqrt{xy}\, J_\mu(xy)\, dy$$

where the differential operator $M_\mu N_\mu$ is understood in the generalized sense and the integral is understood in the Riemann sense. This is another inversion formula for (1).

PROBLEM 5.6-4. For $\mu \geq -\frac{1}{2}$ and $0 < a < b < \infty$, show that

$$\mathfrak{H}_\mu\, Pf \frac{1_+(x-a)1_+(b-x)}{x-a} =$$

$$\int_a^{a+1} \frac{\sqrt{xy}\, J_\mu(xy) - \sqrt{ay}\, J_\mu(ay)}{x-a}\, dx + \int_{a+1}^b \frac{\sqrt{xy}\, J_\mu(xy)}{x-a}\, dx$$

Here, Pf denotes a pseudofunction; see Zemanian [1], Secs. 1.4 and 2.5.

5.7. An Operational Calculus

Our Hankel transformation generates an operational calculus by means of which certain differential equations involving generalized functions can be solved. Let $P(x)$ be a polynomial having no roots on $-\infty < x \leq 0$, and let $\mu \geq -\frac{1}{2}$. The differential equations that can be solved by using the generalized Hankel transformation are of the form:

$$P(M_\mu N_\mu)u = g \qquad 0 < x < \infty \tag{1}$$

where g is a given member of $\mathscr{H}_\mu'$ and u is unknown but also required to be in $\mathscr{H}_\mu'$.

By applying $\mathfrak{H}_\mu$ to (1) and invoking Sec. 5.5, Eq. (8), we obtain

$$P(-y^2)U(y) = G(y) \tag{2}$$

where U and G are the Hankel transforms of u and g respectively. According to Lemma 5.3-1, $1/P(-y^2)$ is a multiplier for $\mathscr{H}_\mu$. Consequently, we may multiply (2) by $1/P(-y^2)$ to get the transform $U(y)$ of our solution. Taking the inverse Hankel transform, we finally get as our solution

$$u(x) = \mathfrak{H}_\mu \frac{G(y)}{P(-y^2)} \tag{3}$$

(Recall that $\mathfrak{H}_\mu = \mathfrak{H}_\mu^{-1}$ on $\mathscr{H}_\mu'$.) Thus, u is that generalized function in $\mathscr{H}_\mu'$ which assigns to each $\phi \in \mathscr{H}_\mu$ the number:

$$\langle u, \phi \rangle = \left\langle \frac{G(y)}{P(-y^2)}, \quad \Phi(y) \right\rangle$$

$$= \left\langle g(x), \int_0^\infty \frac{\Phi(y)}{P(-y^2)} \sqrt{xy}\, J_\mu(xy)\, dy \right\rangle$$

where $\Phi = \mathfrak{H}_\mu \phi$. This solution is unique in $\mathscr{H}_\mu'$. For, every solution to (1) must have $G(y)/P(-y^2)$ as its transform (prove this), and $\mathfrak{H}_\mu$ is a one-to-one mapping of $\mathscr{H}_\mu'$ onto itself.

If $P(x)$ has a zero of multiplicity n at $x = 0$ but has no zeros on $-\infty < x < 0$, this operational technique for solving (1) still works if g is restricted in still another way. In particular, assume that g is such that there exists a $\tilde{G} \in \mathscr{H}'_{\mu-2n}$ whose restriction to $\mathscr{H}_\mu$ is equal to $G = \mathfrak{H}_\mu g$. (If one such $\tilde{G}$ exists, there will be an infinity of such $\tilde{G}$.) Under this assumption, (1) has a solution, which is no longer unique in $\mathscr{H}_\mu'$, as we shall see.

Let us factor $P(x)$ into $P(x) = x^n Q(x)$ where $Q(x)$ has no roots on $-\infty < x \leq 0$. Then, (1) can be rewritten as

$$(M_\mu N_\mu)^n Q(M_\mu N_\mu) u = g \tag{4}$$

An application of $\mathfrak{H}_\mu$ to (4) yields

$$(-y^2)^n Q(-y^2) U(y) = G(y) \tag{5}$$

By Lemma 5.3-2, the generalized function:

$$Q(-y^2) U(y) = (-y^2)^{-n} \tilde{G}(y) \tag{6}$$

is a member of $\mathscr{H}_\mu'$; also, it satisfies (5) in the sense of equality in $\mathscr{H}_\mu'$ in view of Sec. 5.2, note *II*.

However, the result (6) is not unique; there are other members of $\mathscr{H}_\mu'$ that satisfy (5) in the sense of equality in $\mathscr{H}_\mu'$. In particular, we may add to (6) any solution $H(y)$ in $\mathscr{H}_\mu'$ of the homogeneous equation:

$$(-y^2)^n H(y) = 0$$

Such a solution is

$$H(y) = \sum_{\nu=0}^{n-1} \alpha_\nu F_\nu(y) \tag{7}$$

where the α_ν are arbitrary constants and $F_\nu(y) \in \mathscr{H}_\mu'$ is defined by

$$\langle F_\nu(y), \Phi(y) \rangle \triangleq \lim_{y \to 0+} (y^{-1} D_y)^\nu\, y^{-\mu-1/2}\, \Phi(y) \qquad \Phi \in \mathscr{H}_\mu$$

That this truly defines $F_\nu(y)$ as a member of $\mathscr{H}_\mu'$ follows directly from Lemma 5.2-1, condition (ii). To see that (7) satisfies the homogeneous equation in $\mathscr{H}_\mu'$, observe that for any $\Phi \in \mathscr{H}_\mu$

$$\begin{aligned}\langle y^{2n} F_\nu(y), \Phi(y)\rangle &= \langle F_\nu(y), y^{2n}\Phi(y)\rangle \\ &= \lim_{y\to 0+} (y^{-1}D)^\nu[y^{2n} y^{-\mu-1/2}\Phi(y)] \\ &= 0\end{aligned}$$

since $0 \leq \nu \leq n-1$.

Thus, we may add (7) to the right-hand side of (6) to obtain an equation for a whole class of solutions to (5). Our operational calculus now shows that

$$u = \mathfrak{H}_\mu \left\{\frac{1}{Q(-y^2)}\left[\frac{\tilde{G}(y)}{(-y^2)^n} + \sum_{\nu=0}^{n-1} \alpha_\nu F_\nu(y)\right]\right\} \tag{8}$$

is a solution to (4); bear in mind that the α_ν are arbitrary constants.

A still more complicated case arises when $P(x)$ has roots on $-\infty < x < 0$. This leads to a division problem; namely, find a generalized function $U(y)$ that satisfies (2). (See Zemanian [8].)

Finally, we note that differential equations of the type:

$$P(B_\mu)v = f \tag{9}$$

where

$$B_\mu = D^2 + x^{-1}D - x^{-2}\mu^2$$

can be solved if we first multiply v and f by $\sqrt{x}$. This is because

$$P(B_\mu)v = \frac{1}{\sqrt{x}} P(M_\mu N_\mu)\sqrt{x}\, v \tag{10}$$

so that, with $u = \sqrt{x}v$ and $g = \sqrt{x}f$, (9) is the same as (1). In other words, we define a new space of generalized functions, say, $\hat{\mathscr{H}}_\mu'$ for which the mapping $f \leftrightarrow \sqrt{x}f$ is an isomorphism from $\hat{\mathscr{H}}_\mu'$ onto $\mathscr{H}_\mu'$. Then, our present technique allows us to find a solution $v \in \hat{\mathscr{H}}_\mu'$ for (9).

Before leaving this section, it is worth mentioning that Ditkin and Prudnikov [1], pp. 131–146, have developed a Mikusinski-type operational calculus for the differential operator DxD, which can be converted into $M_0 N_0$ by multiplying the generalized functions under considerations by $\sqrt{x}$. Moreover, Meller [1] has extended this result to differential operators of the form $x^{-\mu}Dx^{\mu+1}D$ where $-1 < \mu < 1$, and Dimovski [1] has done the same for more general operators of this type and of higher order.

PROBLEM 5.7-1. Assume that the polynomial $P(x)$ has no roots on $-\infty < x \leq 0$. Prove that any solution in $\mathscr{H}_\mu'$ to (1) must have $G(y)/P(-y^2)$ as its Hankel transform.

PROBLEM 5.7-2. Find the solutions in $\mathscr{H}_\mu'$ of the following two differential equations wherein $\mu \geq -\frac{1}{2}$.

$$u - M_\mu N_\mu u = \sqrt{x}\, J_\mu(x)$$

$$u - M_\mu N_\mu u = \delta(x-1)$$

Hint: See Problem 5.5-1 for the transforms of the right-hand members.

PROBLEM 5.7-3. Consider the simultaneous differential equations:

$$P_{11}(M_\mu N_\mu)u_1 + P_{12}(M_\mu N_\mu)u_2 = g_1$$

$$P_{21}(M_\mu N_\mu)u_1 + P_{22}(M_\mu N_\mu)u_2 = g_2$$

where $\mu \geq -\frac{1}{2}$, g_1 and g_2 are given members of $\mathscr{H}_\mu'$, the P's are polynomials, and the determinant:

$$\begin{vmatrix} P_{11}(x) & P_{12}(x) \\ P_{21}(x) & P_{22}(x) \end{vmatrix}$$

has no roots on $-\infty < x \leq 0$. Find a pair of solutions u_1 and u_2 in $\mathscr{H}_\mu'$. Are these solutions unique in $\mathscr{H}_\mu'$?

PROBLEM 5.7-4. Assume that the polynomial $P(x)$ has some roots on $-\infty < x < 0$ at, say, $x = -\xi_1^2, \ldots, -\xi_q^2\ (\xi_\nu > 0)$ with multiplicities $k_1, \ldots, k_q$ respectively, That is,

$$P(x) = Q(x)(x+\xi_1^2)^{k_1} \ldots (x + \xi_q^2)^{k_q}$$

where $Q(x)$ is never equal to zero anywhere on $-\infty < x < 0$. Show that a solution in $\mathscr{H}_\mu'$ of the homogeneous equation $P(M_\mu N_\mu)u_c = 0$, where $\mu \geq -\frac{1}{2}$, is

$$u_c(x) = \sum_{p=1}^{q} \sum_{\nu=0}^{k_p-1} c_{p\nu} D_y^\nu [\sqrt{xy}\, J_\mu(xy)] \bigg|_{y=\xi_p}$$

where the $c_{p\nu}$ are arbitrary constants. Here, D_y^ν denotes conventional νth-order differentiation, whereas $D_y^\nu[\sqrt{xy}\, J_\mu(xy)]$ with y fixed denotes a regular generalized function in $\mathscr{H}_\mu'$.

PROBLEM 5.7-5. Find a solution in $\mathscr{H}_\mu'$ for the differential equation:

$$(M_\mu N_\mu)^2 u - u = D^2 \delta(x-1) + \left(1 - \frac{4\mu^2 - 1}{4x^2}\right)\delta(x-1)$$

where again $\mu \geq -\frac{1}{2}$. Can you find other solutions in $\mathscr{H}_\mu'$?

PROBLEM 5.7-6. Verify the validity of (10) in the sense of conventional differentiation.

5.8. A Dirichlet Problem in Cylindrical Coordinates

This section is devoted to an application of the preceding theory to a Dirichlet problem in cylindrical coordinates having a generalized-function boundary condition. The problem can be stated as follows: Find a conventional function $v(r, z)$ on the domain $\{(r, z): 0 < r < \infty, 0 < z < \infty\}$ that satisfies Laplace's equation in cylindrical coordinates (assuming no θ variation):

$$\frac{\partial^2 v}{\partial r^2} + \frac{1}{r}\frac{\partial v}{\partial r} + \frac{\partial^2 v}{\partial z^2} = 0 \tag{1}$$

and the following boundary conditions:

(a) As $z \to 0+$, $v(r, z)$ converges in some generalized sense to the distribution $f(r) \in \mathscr{E}'(I)$. Here, I denotes the interval $0 < r < \infty$.

(b) As $z \to \infty$, $v(r, z)$ converges uniformly to zero on $0 < r < \infty$.

(c) As $r \to \infty$, $v(r, z)$ converges to zero for every $z > 0$.

(d) As $r \to 0+$, $v(r, z)$ remains finite for every $z > 0$.

We adhere to our customary technique of first deriving the solution formally and leaving as a subsequent step the proof that the result we have obtained truly satisfies the differential equation (1) and the boundary conditions. The differential equation (1) can be converted into a form that can be analyzed by our zero-order Hankel transformation by using the change of variables:

$$u(r, z) = \sqrt{r}\, v(r, z), \qquad g(r) = \sqrt{r}\, f(r)$$

Here, $g(r)$ is a member of $\mathscr{E}'(I)$. According to the last paragraph of the preceding section, (1) becomes

$$M_0 N_0 u + \frac{\partial^2 u}{\partial z^2} = 0 \tag{2}$$

where

$$M_0 N_0 u = r^{-1/2} \frac{\partial}{\partial r} r \frac{\partial}{\partial r} r^{-1/2} u(r, z)$$

By applying $\mathfrak{H}_0$ to (2), formally interchanging $\mathfrak{H}_0$ with $\partial^2/\partial z^2$, and setting $U(\rho, z) = \mathfrak{H}_0[u(r, z)]$, we convert (2) into

$$-\rho^2 U(\rho, z) + \frac{\partial^2}{\partial z^2} U(\rho, z) = 0$$

Thus,

$$U(\rho, z) = A(\rho)e^{-\rho z} + B(\rho)e^{\rho z}$$

In view of boundary condition (*b*), we set $B(\rho) = 0$. Then, boundary condition (*a*) suggests that $A(\rho) = \mathfrak{H}_0[g(r)]$ so that by Theorem 5.6-3 we may write

$$A(\rho) = \langle g(x), \sqrt{x\rho}\, J_0(x\rho)\rangle \tag{3}$$

Furthermore, Theorems 5.6-1 and 5.6-2 state that, for each fixed $z > 0$, $A(\rho)e^{-\rho z}$ is a smooth function of ρ in $L_1(0, \infty)$. Therefore, we may apply the conventional inverse Hankel transformation to get as our formal solution:

$$u(r, z) = \int_0^\infty \langle g(x), \sqrt{x\rho}\, J_0(x\rho)\rangle\, e^{-\rho z} \sqrt{r\rho}\, J_0(r\rho)\, d\rho \tag{4}$$

$$v(r, z) = r^{-1/2} u(r, z) \qquad 0 < r < \infty, \qquad 0 < z < \infty$$

That (4) is truly the solution we seek can be shown as follows. First of all, $J_0(r\rho)$ and $J_1(r\rho)$ are bounded on $0 < r\rho < \infty$ and $e^{-\rho z} \leq e^{-\rho Z}$ for $Z \leq z < \infty, 0 < \rho < \infty$. These facts and Theorems 5.6-1 and 5.6-2 allow us to interchange the differentiations in (2) with the integration in (4) since at every step the resulting integral converges uniformly on every compact subset of the domain $\{(r, z)\colon 0 < r < \infty,\ 0 < z < \infty\}$. Since $e^{-\rho z}\sqrt{r\rho}\, J_0(r\rho)$ satisfies the differential equation (2) for each fixed ρ, we can conclude that $u(r, z)$ also satisfies (2). Hence, $v(r, z)$ satisfies (1).

We now prove that boundary condition (*a*) is satisfied. As a function of ρ,

$$\langle g(x), \sqrt{x\rho}\, J_0(x\rho)\rangle e^{-\rho z} \tag{5}$$

is smooth, and, for each fixed $z > 0$, it is a member of $L_1(0, \infty)$ by virtue of Theorems 5.6-1 and 5.6-2. (The same is true for the product of (5) and $\sqrt{\rho}$; we shall need this fact later on.) Thus, (5) satisfies the conditions under which the conventional Hankel transformation is a special case of our generalized Hankel transformation (see Sec. 5.5). According to (4), its Hankel transform is $u(r, z)$, so that, for any $\phi \in \mathscr{H}_0$ and $\Phi = \mathfrak{H}_0\phi$, our definition of the generalized Hankel transformation (Sec. 5.5, Eq. (4)) yields

$$\langle u(r, z), \phi(r)\rangle = \int_0^\infty \langle g(x), \sqrt{x\rho}\, J_0(x\rho)\rangle\, e^{-\rho z}\, \Phi(\rho)\, d\rho$$

The integral on the right-hand side converges uniformly on $0 \leq z < \infty$ because its integrand is bounded by

$$|\langle g(x), \sqrt{x\rho}\, J_0(x\rho)\rangle\, \Phi(\rho)| \in L_1(0, \infty)$$

Thus, we may interchange the limiting process $z \to 0+$ with the integration to get

$$\lim_{z \to 0+} \langle u(r, z), \phi(r) \rangle = \int_0^\infty \langle g(x), \sqrt{x\rho}\, J_0(x\rho) \rangle\, \Phi(\rho)\, d\rho$$

Again by Sec. 5.5, Eq. (4), the right-hand side is equal to $\langle g(r), \phi(r) \rangle$. Thus, we have shown that, in the sense of convergence in $\mathscr{H}_0'$, $u(r, z) \to g(r)$ as $z \to 0+$. In other words, $v(r, z) \to f(r)$ in a generalized sense (i.e., in the space $\hat{\mathscr{H}}_0'$ mentioned at the end of the preceding section).

To verify boundary condition (*b*), let B be a bound on $J_0(r\rho)$ for $0 < r\rho < \infty$. For $z > 1$, (4) yields

$$\begin{aligned} |v(r, z)| &= \left| \frac{u(r, z)}{\sqrt{r}} \right| \\ &\leq B \int_0^\infty |\langle g(x), \sqrt{x\rho}\, J_0(x\rho) \rangle|\, e^{-\rho} e^{-\rho(z-1)} \sqrt{\rho}\, d\rho \\ &\leq B \int_0^\delta |\langle g(x), \sqrt{x\rho}\, J_0(x\rho) \rangle|\, e^{-\rho} \sqrt{\rho}\, d\rho \qquad (6) \\ &\quad + Be^{-\delta(z-1)} \int_\delta^\infty |\langle g(x), \sqrt{x\rho}\, J_0(x\rho) \rangle|\, e^{-\rho} \sqrt{\rho}\, d\rho \end{aligned}$$

We now exploit the fact that, as a function of ρ and with $z = 1$, the product of (5) and $\sqrt{\rho}$ is a member of $L_1(0, \infty)$. Given any $\varepsilon > 0$, choose $\delta > 0$ so small that the first term on the right-hand side of (6), which is independent of z, is bounded by $\varepsilon/2$. Fix δ this way. Then, there exists a Z such that, for all $z > Z$, the second term is bounded by $\varepsilon/2$. This proves that, as $z \to \infty$, $v(r, z) \to 0$ uniformly on $0 < r < \infty$.

That boundary condition (*c*) is satisfied follows from the following Riemann-Lebesgue lemma for the Hankel transform (Watson [1], p. 457): If $h(\rho) \in L_1(0, \infty)$ and $\mu \geq -\frac{1}{2}$, then, as $r \to \infty$,

$$\int_0^\infty h(\rho) \sqrt{\rho}\, J_\mu(r\rho)\, d\rho = o(1/\sqrt{r})$$

We need merely identify $h(\rho)$ with (5) and set $\mu = 0$ in order to get the desired conclusion.

Finally, the fulfillment of boundary condition (*d*) follows immediately from our parenthetical comment about (5) and the fact that $J_0(r\rho)$ is bounded on $0 < r\rho < \infty$.

PROBLEM 5.8-1. Fill in the details of the proof of the fact that (4) satisfies the differential equation (2) for $0 < r < \infty$ and $0 < z < \infty$.

PROBLEM 5.8-2. Solve the following Dirichlet problem on the domain $\{(r, z)\colon 0 < r < \infty,\ 0 < z < 1\}$. The differential equation is (1), and the boundary conditions are as follows:

(*a*) As $z \to 0+$, $\sqrt{r}\, v(r, z)$ converges in $\mathscr{H}_0{}'$ to $f(r) \in \mathscr{E}'(I)$.

(*b*) As $z \to 1-$, $\sqrt{r}\, v(r, z)$ converges in $\mathscr{H}_0{}'$ to $g(r) \in \mathscr{E}'(I)$.

(*c*) As $r \to \infty$, $v(r, z)$ tends to zero pointwise on $0 < z < 1$.

(*d*) As $r \to 0+$, $v(r, z)$ remains finite at each point of the interval $0 < z < 1$.

PROBLEM 5.8-3. Let $v(r, \theta, z)$ be a potential function of the cylindrical coordinates (r, θ, z) on the domain:

$$\{(r, \theta, z)\colon 0 < r < \infty, 0 \le \theta < 2\pi, 0 < z < \infty\}.$$

Find a solution to Laplace's equation:

$$\frac{\partial^2 v}{\partial r^2} + r^{-1}\frac{\partial v}{\partial r} + r^{-2}\frac{\partial^2 v}{\partial \theta^2} + \frac{\partial^2 v}{\partial z^2} = 0$$

under the following boundary conditions

(*a*) As $z \to 0+$ and for each fixed θ, $\sqrt{r}\, v(r, \theta, z)$, as a function of r, converges in $\mathscr{H}_n{}'$ to $f(r) \cos n\theta$, where n is a fixed positive integer and $f(r) \in \mathscr{E}'(I)$.

(*b*) As $z \to \infty$, $v(r, \theta, z)$ converges uniformly to zero on $\{(r, \theta)\colon 0 < r < \infty, 0 \le \theta < 2\pi\}$.

(*c*) As $r \to \infty$, $v(r, \theta, z)$ converges to zero for every choice of the coordinates θ and z.

(*d*) As $r \to 0+$, $v(r, \theta, z)$ converges to zero for every choice of the coordinates θ and z.

Hint: First show that the change of variable $v(r, \theta, z) = r^{-1/2} u(r, z) \cos n\theta$ converts the differential equation into a form to which the nth-order Hankel transformation can be profitably applied. Also, for boundary condition (*d*), use the fact that $|J_n(r\rho)| < K(r\rho)^n$ for some constant K and $0 < r\rho < \infty$.

5.9. A Cauchy Problem for Cylindrical Waves

As another application of our generalized Hankel transformation, we shall solve a Cauchy problem for cylindrical waves. In particular, let $v = v_t(r)$ be a generalized function in a cylindrical-coordinate system (r, θ, z) that does not depend on the variables θ and z but does depend parametrically upon the time variable t. Moreover, assume that it satisfies the wave equation:

$$\frac{\partial^2 v}{\partial r^2} + r^{-1}\frac{\partial v}{\partial r} = \frac{\partial^2 v}{\partial t^2}$$

for $0 < r < \infty$, $-\infty < t < \infty$. Here, we have set the speed of the wave equal to 1. Also, the differentiations with respect to r are generalized differentiations whereas that with respect to t is parametric differentiation. Let us make the change of variable $u_t(r) = \sqrt{r}\, v_t(r)$ and state the Cauchy problem for $u = u_t(r)$: The wave equation becomes

$$(1) \qquad M_0 N_0 u = \frac{\partial^2 u}{\partial t^2} \qquad 0 < r < \infty, \qquad -\infty < t < \infty$$

where

$$M_0 N_0 u = r^{-1/2}\frac{\partial}{\partial r} r \frac{\partial}{\partial r} r^{-1/2} u$$

The initial conditions for our problem are:

(a) as $t \to 0$, $u_t(r)$ converges in $\mathscr{H}_0'$ to $g(r) \in \mathscr{H}_0'$,

(b) as t $\to 0$, $\partial u_t(r)/\partial t$ converges in $\mathscr{H}_0'$ to $h(r) \in \mathscr{H}_0'$.

The solution that we shall find will satisfy (1) in the sense of equality in $\mathscr{H}_0'$.

Proceeding formally to get a possible solution, we set $U_t(\rho) = \mathfrak{H}_0[u_t(r)]$, apply $\mathfrak{H}_0$ to the differential equation (1), and interchange $\mathfrak{H}_0$ with $\partial^2/\partial t^2$ to write

$$-\rho^2 U_t(\rho) = \frac{\partial^2}{\partial t^2} U_t(\rho)$$

Therefore,

$$(2) \qquad U_t(\rho) = A(\rho)e^{i\rho t} + B(\rho)e^{-i\rho t}$$

where $A(\rho)$ and $B(\rho)$ are unknown generalized functions which do not depend on t. To determine $A(\rho)$ and $B(\rho)$ we match the initial conditions at $t=0$. Set $G(\rho) = \mathfrak{H}_0[g(r)]$ and $H(\rho) = \mathfrak{H}_0[h(r)]$. Initial condition ($a$) suggests that

$$(3) \qquad U_0(\rho) = A(\rho) + B(\rho) = G(\rho)$$

Initial condition (b) suggests that, if $\mathfrak{H}_0$ and $\partial/\partial t$ are formally interchanged, then

$$(4) \qquad \frac{\partial}{\partial t} U_t(\rho)|_{t=0} = i\rho A(\rho) - i\rho B(\rho) = H(\rho)$$

Upon solving (3) and (4) for $A(\rho)$ and $B(\rho)$ and substituting the results into (2), we obtain

$$U_t(\rho) = G(\rho)\cos\rho t + H(\rho)\rho^{-1}\sin\rho t \tag{5}$$

Thus, our solution appears to be

$$u_t(r) = \mathfrak{H}_0^{-1}[U_t(\rho)] \tag{6}$$

To justify this, we first show that, for each fixed value of t, $U_t(\rho)$ is a member of $\mathscr{H}_0'$ on $0 < \rho < \infty$. Since both $g(r)$ and $h(r)$ are members of $\mathscr{H}_0'$, it follows that $G(\rho)$ and $H(\rho)$ are also members of $\mathscr{H}_0'$. Thus, we need merely show that, again for any fixed t, $\cos\rho t$ and $\rho^{-1}\sin\rho t$ are members of $\mathcal{O}$, the space of multipliers for $\mathscr{H}_\mu$. (See Sec. 5.3.) But, Maclaurin's series for $\cos\rho t$ has only even powers of ρt so that $(\rho^{-1}D_\rho)^\nu\cos\rho t$ is bounded on $0 < \rho < 1$ for every nonnegative integer ν and every fixed t. On the other hand, some computation shows that

$$(\rho^{-1}D_\rho)^\nu \cos\rho t = O(\rho^{-\nu})$$

as $\rho \to \infty$. Therefore, $(\rho^{-1}D_\rho)^\nu\cos\rho t$ is bounded on $0 < \rho < \infty$, which verifies that $\cos\rho t$ truly is a member of $\mathcal{O}$. A similar argument leads to the same conclusion for $\rho^{-1}\sin\rho t$.

Because of these results, (6) is a generalized function in $\mathscr{H}_0'$ on $0 < r < \infty$ depending parametrically on t. We specify it as a functional on $\mathscr{H}_0$ by

$$\langle u_t(r), \phi(r)\rangle = \langle G(\rho)\cos\rho t + H(\rho)\rho^{-1}\sin\rho t, \Phi(\rho)\rangle \tag{7}$$

where ϕ is any member of $\mathscr{H}_0$ and $\Phi = \mathfrak{H}_0\phi$.

We now set about proving that $u_t(r)$, as defined by (7), satisfies the differential equation (1) in the sense of equality in $\mathscr{H}_0'$. Let $\phi(r)$ be any member of $\mathscr{H}_0$ and $\Phi(\rho) = \mathfrak{H}_0[\phi(r)]$. By Sec. 5.5, Eq. (8),

$$\langle M_0 N_0 u_t(r), \phi(r)\rangle = \langle -\rho^2 U_t(\rho), \Phi(\rho)\rangle \tag{8}$$

On the other hand, by the definition of the parametric differentiation $D_t = \partial/\partial t$ (see Sec. 2.7),

$$\begin{aligned}\langle D_t^2 u_t(r), \phi(r)\rangle &= D_t^2\langle u_t(r), \phi(r)\rangle = D_t^2\langle U_t(\rho), \Phi(\rho)\rangle \\ &= D_t^2\langle G(\rho), \Phi(\rho)\cos\rho t\rangle + D_t^2\left\langle H(\rho), \Phi(\rho)\frac{\sin\rho t}{\rho}\right\rangle\end{aligned} \tag{9}$$

For the moment, let us take for granted

LEMMA 5.9-1. *For $F(\rho) \in \mathscr{H}_\mu'$, $\Phi(\rho) \in \mathscr{H}_\mu$, and $n = 1$ or 2,*

$$D_t^n\langle F(\rho), \Phi(\rho)\cos\rho t\rangle = \langle F(\rho), \Phi(\rho)D_t^n\cos\rho t\rangle \tag{10}$$

and

$$D_t^n\left\langle F(\rho), \Phi(\rho)\frac{\sin \rho t}{\rho}\right\rangle = \left\langle F(\rho), \Phi(\rho) D_t^n \frac{\sin \rho t}{\rho}\right\rangle \tag{11}$$

In accordance with this lemma, the right-hand side of (9) can be rewritten as

$$\langle G(\rho), \Phi(\rho)(-\rho^2)\cos \rho t\rangle + \left\langle H(\rho), \Phi(\rho)(-\rho^2)\frac{\sin \rho t}{\rho}\right\rangle$$
$$= \langle -\rho^2 U_t(\rho), \Phi(\rho)\rangle$$

By comparing this result with (8), we see that $u_t(r)$ truly satisfies (1) in the stated sense.

Thus, we are left with the

PROOF OF LEMMA 5.9-1. We shall only establish (10) with $n = 1$ since the proofs of the other conclusions are hardly different. Since $F(\rho)$ is a member of $\mathscr{H}_\mu'$, our desired conclusion will be established if we show that

$$\Phi(\rho)\left\{\frac{1}{\Delta t}[\cos(\rho t + \rho\Delta t) - \cos \rho t] - D_t \cos \rho t\right\} \qquad \Delta t \neq 0$$

converges in $\mathscr{H}_\mu$ to zero as $\Delta t \to 0$. In view of the second paragraph of Sec. 5.3, this will be accomplished when we show that, for each nonnegative integer ν,

$$(1+\rho^2)^{-1}(\rho^{-1}D_\rho)^\nu\left\{\frac{1}{\Delta t}[\cos(\rho t + \rho\Delta t) - \cos \rho t] - D_t \cos \rho t\right\} \tag{12}$$

converges uniformly to zero on $0 < \rho < \infty$ as $\Delta t \to 0$.

Now, for any fixed t,

$$\frac{1}{\Delta t}[\cos(\rho t + \rho\Delta t) - \cos \rho t] - D_t \cos \rho t = \frac{1}{\Delta t}\int_0^{\Delta t} d\tau \int_0^\tau D_\eta^2 \cos(\rho t + \rho\eta)\, d\eta$$
$$= -\frac{\rho^2}{\Delta t}\int_0^{\Delta t} d\tau \int_0^\tau \cos(\rho t + \rho\eta)\, d\eta$$

So, for $\nu = 0$, the magnitude of (12) is bounded by

$$\frac{\rho^2}{(1+\rho^2)|\Delta t|}\left|\int_0^{\Delta t} d\tau \int_0^\tau \cos(\rho t + \rho\eta)\, d\eta\right| \leq \frac{\rho^2|\Delta t|}{2(1+\rho^2)}$$

a quantity that tends uniformly to zero on $0 < \rho < \infty$ as $\Delta t \to 0$. On the other hand, for $\nu > 0$, we may repeatedly differentiate under the integral sign to write

$$(\rho^{-1} D_\rho)^\nu \left\{\frac{1}{\Delta t}[\cos(\rho t + \rho\Delta t) - \cos \rho t] - D_t \cos \rho t\right\}$$

$$= -\frac{1}{\Delta t}\int_0^{\Delta t} d\tau \int_0^{\tau} (\rho^{-1}D_\rho)^\nu[\rho^2 \cos(\rho t + \rho\eta)]\, d\eta$$

$$= -\frac{1}{\Delta t}\int_0^{\Delta t} d\tau \int_0^{\tau} [2\nu(\rho^{-1}D_\rho)^{\nu-1} \cos(\rho t + \rho\eta)$$

$$+ \rho^2 (\rho^{-1}D_\rho)^\nu \cos(\rho t + \rho\eta)]\, d\eta$$

Some computation shows that, for any nonnegative integer k, $(\rho^{-1}D_\rho)^k \cos(\rho t + \rho\eta)$ is bounded for $0 < \rho < \infty$ and $-1 < \eta < 1$ by, say, the constant B_k. Therefore, for $0 < |\tau| \le |\Delta t| \le 1$, the magnitude of (12) is now bounded by

$$\frac{(2\nu B_{\nu-1} + \rho^2 B_\nu)|\Delta t|}{2(1+\rho^2)}$$

which also tends uniformly to zero on $0 < \rho < \infty$ as $\Delta t \to 0$. This completes our proof of (10) for $n = 1$.

We now verify that our solution $u_t(r)$ satisfies initial condition (a). That is, we shall show that, for any $\phi \in \mathscr{H}_0$ and as $t \to 0$,

$$\langle u_t(r), \phi(r)\rangle \to \langle g(r), \phi(r)\rangle \tag{13}$$

With $\Phi(\rho) = \mathfrak{H}_0[\phi(r)]$ as before, the left-hand side of (13) is equal to

$$\langle G(\rho), \Phi(\rho)\cos \rho t\rangle + \left\langle H(\rho), \Phi(\rho)\frac{\sin \rho t}{\rho}\right\rangle \tag{14}$$

by virtue of (7). In the next paragraph we shall show that, as $t \to 0$, $\Phi(\rho)\cos \rho t$ converges in $\mathscr{H}_0$ to $\Phi(\rho)$; similarly, $\Phi(\rho)\rho^{-1}\sin \rho t$ converges in $\mathscr{H}_0$ to zero. The limit (13) follows directly from these results and the facts that $G(\rho) \in \mathscr{H}_0'$ and $H(\rho) \in \mathscr{H}_0'$.

The second paragraph of Sec. 5.3 again shows that our assertion concerning $\Phi(\rho)\cos \rho t$ will be established as soon as we prove that as $t \to 0$

$$\frac{\cos \rho t}{1+\rho} \to \frac{1}{1+\rho} \tag{15}$$

and, for each positive integer ν,

$$(\rho^{-1}D_\rho)^\nu \cos \rho t \to 0 \tag{16}$$

where in every case the convergence is uniform on $0 < \rho < \infty$. The result (16) follows from the facts that

$$(\rho^{-1}D_\rho)^\nu \cos \rho t = t^{2\nu}(z^{-1}D_z)^\nu \cos z \qquad z = \rho t$$

and that $(z^{-1}D_z)^\nu \cos z$ is bounded for $-\infty < z < \infty$. On the other hand, to show (15), we first note that, given any $\varepsilon > 0$, there exists an $R < \infty$ such that, for all $\rho > R$ and $-\infty < t < \infty$,

$$0 \leq \frac{1 - \cos \rho t}{1 + \rho} \leq \frac{2}{1 + R} < \varepsilon$$

Having fixed R this way, we restrict ρ to $0 < \rho \leq R$ and t to $|t| < \pi/R$ and then write

$$0 \leq \frac{1 - \cos \rho t}{1 + \rho} \leq 1 - \cos Rt \to 0 \qquad t \to 0$$

Therefore, there exists a $T > 0$ such that, for all $|t| < T$,

$$0 \leq \frac{1 - \cos \rho t}{1 + \rho} \leq \varepsilon \qquad 0 < \rho < \infty$$

Since $\varepsilon > 0$ is arbitrary, our assertion that $\Phi(\rho) \cos \rho t$ converges in $\mathscr{H}_0$ to $\Phi(\rho)$ is hereby established.

That $\Phi(\rho)\rho^{-1} \sin \rho t$ converges in $\mathscr{H}_0$ to zero as $t \to 0$ follows by similar reasoning from the equation:

$$(\rho^{-1}D_\rho)^\nu \rho^{-1} \sin \rho t = t^{2\nu+1}(z^{-1}D_z)^\nu z^{-1} \sin z \qquad \nu = 0, 1, 2, \ldots; \qquad z = \rho t$$

and the observation that $(z^{-1}D_z)^\nu z^{-1} \sin z$ is bounded on $-\infty < z < \infty$. This completes our argument showing that $u_t(r)$ satisfies initial condition (*a*).

That $u_t(r)$ satisfies initial condition (*b*) now follows readily from the results we have already established.

PROBLEM 5.9-1. Show that, for any nonnegative integer k and for any fixed t, $(\rho^{-1}D_\rho)^k \cos(\rho t + \rho\eta)$ is bounded on the domain $\{(\rho, \eta)\colon 0 < \rho < \infty, -1 < \eta < 1\}$.

PROBLEM 5.9-2. Prove (10) for $n = 2$ and (11) for $n = 1$ and 2.

PROBLEM 5.9-3. Show that $u_t(r)$, as specified by (7), satisfies boundary condition (*b*).

PROBLEM 5.9-4. State a set of conditions on the generalized functions $g(r)$ and $h(r)$ occurring in boundary conditions (*a*) and (*b*) and on the integers p and q under which the solution $u_t(r)$ to the Cauchy problem of this section can be written in the explicit form:

$$u_t(r) = [1 + (-M_0 N_0)^p] \int_0^\infty \frac{G(\rho) \cos \rho t}{1 + \rho^{2p}} \sqrt{r\rho}\, J_0(r\rho)\, d\rho$$

$$+ [1 + (-M_0 N_0)^q] \int_0^\infty \frac{H(\rho)\rho^{-1} \sin \rho t}{1 + \rho^{2q}} \sqrt{r\rho}\, J_0(r\rho)\, d\rho$$

where, as before, $G(\rho) = \mathfrak{H}_0[g(r)]$ and $H(\rho) = \mathfrak{H}_0[h(r)]$. Here, the integrals converge in the conventional sense, but $M_0 N_0$ denotes a generalized operation.

5.10. Hankel Transforms of Arbitrary Order

The theory presented in Secs. 5.3 and 5.4 provides a means of defining a generalized Hankel transformation for any real value of the order μ (including values of μ less than $-\frac{1}{2}$) in such a way that an inverse Hankel transform also exists (Zemanian [8]). It is the existence of the inverse transformation that makes this extension of the Hankel transformation significant; indeed, if no attempt is to be made to obtain an inverse transformation, it is quite easy to define a direct Hankel transformation of order $\mu < -\frac{1}{2}$ by restricting sufficiently the generalized functions on which it will act. The extension of the Hankel transformation that we shall present in this section possesses the following properties:

(i) The direct transformation possesses an inverse for every real value of the order μ.

(ii) The direct and inverse transformations of order μ are defined on the generalized-function space $\mathscr{H}_\mu'$.

(iii) If $\mu \geq -\frac{1}{2}$, the extended direct and inverse transformation coincide with the Hankel transformation discussed previously (i.e., in Sec. 5.5).

This is not the only way of extending the Hankel transformation and its inverse to larger ranges of μ. J. L. Lions [1] presents such a transformation that is valid for all real and complex values of μ excluding $\mu = -1, -2, -3, \ldots$.

Throughout this section, x is the independent variable corresponding to the (testing or generalized) functions ϕ, f, u, and g, whereas y is the independent variable corresponding to their Hankel transforms Φ, F, U, and G.

We first define two transformations on the testing-function space $\mathscr{H}_\mu$ which coincide with the conventional $\mathfrak{H}_\mu$ whenever $\mu \geq -\frac{1}{2}$. We recall that the latter transformation is defined by

$$(1) \qquad (\mathfrak{H}_\mu f)(y) \triangleq \int_0^\infty f(x)\sqrt{xy}\, J_\mu(xy)\, dx$$

Let μ be any fixed real number and k any positive integer such that $\mu + k \geq -\frac{1}{2}$. We define the transformation $\mathfrak{H}_{\mu,\,k}$ on any $\Phi \in \mathscr{H}_\mu$ by

$$(2) \qquad \phi(x) \triangleq \mathfrak{H}_{\mu,\,k}[\Phi(y)] \triangleq (-1)^k x^{-k}\, \mathfrak{H}_{\mu+k} N_{\mu+k-1} \cdots N_{\mu+1} N_\mu \Phi(y)$$

Similarly, $\mathfrak{H}_{\mu,\,k}^{-1}$ is defined on any $\phi \in \mathscr{H}_\mu$ by

$$(3) \qquad \Phi(y) \triangleq \mathfrak{H}_{\mu,\,k}^{-1}[\phi(x)] \triangleq (-1)^k N_\mu^{-1} N_{\mu+1}^{-1} \cdots N_{\mu+k-1}^{-1}\, \mathfrak{H}_{\mu+k}\, x^k \phi(x)$$

In (2) and (3), $\mathfrak{H}_{\mu+k}$ is defined by (1) wherein μ is replaced by $\mu + k$.

LEMMA 5.10-1. *The transformation* $\mathfrak{H}_{\mu,\,k}$, *as defined by* (2) *with* μ *and* k *restricted as stated, is an automorphism on* $\mathscr{H}_\mu$ *whatever be the real number* μ. *Its inverse is* $\mathfrak{H}_{\mu,\,k}^{-1}$ *as defined by* (3). *Finally,* $\mathfrak{H}_{\mu,\,k}$ *coincides with* $\mathfrak{H}_\mu$ *as defined by* (1) *whenever* $\mu \geq -\frac{1}{2}$ *and* $\mathfrak{H}_{\mu,\,k}$ *is acting on* $\mathscr{H}_\mu$.

PROOF. The first assertion follows from the facts that $\Phi \mapsto N_{\mu+k-1} \cdots N_{\mu+1} N_\mu \Phi$ is an isomorphism from $\mathscr{H}_\mu$ onto $\mathscr{H}_{\mu+k}$, $\Phi \mapsto \mathfrak{H}_{\mu+k}\Phi$ is an automorphism on $\mathscr{H}_{\mu+k}$, and $\phi \mapsto x^{-k}\phi$ is an isomorphism from $\mathscr{H}_{\mu+k}$ onto $\mathscr{H}_\mu$. See Lemmas 5.3-2 and 5.3-3 and Theorem 5.4-1.

By assumption, $\mu + k \geq -\frac{1}{2}$. Hence, the second assertion follows from Lemmas 5.3-2 and 5.3-3(i) and the fact (see Theorem 5.1-1) that $\mathfrak{H}_{\mu+k}$ is its own inverse when acting on $\mathscr{H}_{\mu+k}$.

For the third assertion, assume that $\Phi(y) \in \mathscr{H}_\mu$ and $\mu \geq -\frac{1}{2}$, and consider the case $k = 1$:

$$\begin{aligned}\mathfrak{H}_{\mu,\,1}\Phi &= -x^{-1}\,\mathfrak{H}_{\mu+1} N_\mu \Phi \\ &= -x^{-1}\int_0^\infty y^{\mu+1/2}[D_y\, y^{-\mu-1/2}\,\Phi(y)]\sqrt{xy}\, J_{\mu+1}(xy)\, dy\end{aligned}$$

An integration by parts and the formula:

$$D_y\, y^{\mu+1} J_{\mu+1}(xy) = x y^{\mu+1} J_\mu(xy) \tag{4}$$

yield

$$-x^{-1}\sqrt{xy}\, J_{\mu+1}(xy)\Phi(y)\Bigg|_{y=0}^{y=\infty} + \int_0^\infty \Phi(y)\sqrt{xy}\, J_\mu(xy)\, dy$$

In view of Lemma 5.2-1, the limit terms are zero; indeed, $\Phi(y)$ is of rapid descent and $\sqrt{xy}\, J_{\mu+1}(xy)$ remains bounded as $y \to \infty$, whereas, for $y \to 0+$, $\Phi(y) = O(y^{\mu+1/2})$ and $\sqrt{xy}\, J_{\mu+1}(xy) = O(y^{\mu+3/2})$ where $\mu \geq -\frac{1}{2}$. Thus,

$$\mathfrak{H}_{\mu,\,1}\Phi = -x^{-1}\,\mathfrak{H}_{\mu+1} N_\mu \Phi = \int_0^\infty \Phi(y)\sqrt{xy}\, J_\mu(xy)\, dy \tag{5}$$

The general statement for larger values of k follows by induction from this result. This ends the proof.

A consequence of the last lemma is that $\mathfrak{H}_{\mu,\,k} = \mathfrak{H}_{\mu,\,p}$ so long as the positive integers k and p are no less than $-\mu - \frac{1}{2}$. Indeed, assuming that $k > p$, we have that $\mathfrak{H}_{\mu+p,\,k-p} = \mathfrak{H}_{\mu+p}$ according to the last statement of Lemma 5.10-1, and therefore, for $\Phi \in \mathscr{H}_\mu$,

$$\mathfrak{H}_{\mu,\,k}\Phi = (-1)^p x^{-p}\,\mathfrak{H}_{\mu+p,\,k-p} N_{\mu+p-1} \cdots N_\mu \Phi = \mathfrak{H}_{\mu,\,p}\Phi$$

Lemma 5.10-1 also implies that $\mathfrak{H}_{\mu,\,k}^{-1}$ coincides with $\mathfrak{H}_\mu$ on $\mathscr{H}_\mu$ when $\mu \geq -\frac{1}{2}$. For, since $\mathfrak{H}_{\mu,\,k}$ and $\mathfrak{H}_\mu$ agree when they are acting on $\mathscr{H}_\mu$, they must have the same inverse; therefore, $\mathfrak{H}_{\mu,k}^{-1} = \mathfrak{H}_{\mu}^{-1} = \mathfrak{H}_{\mu}$. Moreover, $\mathfrak{H}_{\mu,k}^{-1}$ is independent of the choice of k so long as $\mu + k \geq -\frac{1}{2}$ and $\mathfrak{H}_{\mu,k}^{-1}$ is acting on $\mathscr{H}_{\mu}$. This is because, by the preceding paragraph, $\mathfrak{H}_{\mu,k} = \mathfrak{H}_{\mu,p}$ if neither k nor p are less than $-\mu-\frac{1}{2}$; consequently, their inverses must agree.

In view of these results, it is reasonable to consider the mapping $\mathfrak{H}_{\mu,k}$, where k is any positive integer no less than $-\mu-\frac{1}{2}$, as an extension to all real μ of the conventional direct Hankel transformation $\mathfrak{H}_{\mu}(\mu \geq -\frac{1}{2})$. The inverse mapping $\mathfrak{H}_{\mu,k}^{-1}$ is taken to be the extension to all real μ of the conventional inverse Hankel transformation $\mathfrak{H}_{\mu}^{-1}(\mu \geq -\frac{1}{2})$. As in the preceding sections, the direct and inverse transformations coincide (i.e., $\mathfrak{H}_{\mu,k} = \mathfrak{H}_{\mu,k}^{-1}$) when $\mu \geq -\frac{1}{2}$, but this is not true when $\mu < -\frac{1}{2}$.

Actually, we have a choice here as to which mapping, $\mathfrak{H}_{\mu,k}$ or $\mathfrak{H}_{\mu,k}^{-1}$, we are going to call the direct transformation and which the inverse. This is merely a matter of terminology, and we choose to call $\mathfrak{H}_{\mu,k}$ the direct transformation.

We turn now to the definition of the Hankel transformation of any real order μ when it is acting on $\mathscr{H}_{\mu}'$. As before, k is any positive integer $\geq -\mu - \frac{1}{2}$. Then, the generalized Hankel transformation $\mathfrak{H}_{\mu}'$ is defined on $\mathscr{H}_{\mu}'$ as the adjoint of $\mathfrak{H}_{\mu,k}$ on $\mathscr{H}_{\mu}$. That is, the Hankel transform $\mathfrak{H}_{\mu}'f$ of any $f \in \mathscr{H}_{\mu}'$ is defined as a functional on $\mathscr{H}_{\mu}$ by

$$\langle \mathfrak{H}_{\mu}'f, \Phi \rangle \triangleq \langle f, \mathfrak{H}_{\mu,k}\Phi \rangle \qquad \Phi \in \mathscr{H}_{\mu} \tag{6}$$

Because of Theorem 1.10-2 and Lemma 5.10-1, we immediately obtain

THEOREM 5.10-1. *The generalized Hankel transformation $\mathfrak{H}_{\mu}'$ is an automorphism on $\mathscr{H}_{\mu}'$, whatever be the real number μ.*

Equation (6) also defines the inverse of $\mathfrak{H}_{\mu}'$ as the adjoint of $\mathfrak{H}_{\mu,k}^{-1}$; in particular, if we set $F = \mathfrak{H}_{\mu}'f$ and $\phi = \mathfrak{H}_{\mu,k}\Phi$, we get

$$\langle F, \mathfrak{H}_{\mu,k}^{-1}\phi \rangle \triangleq \langle (\mathfrak{H}_{\mu}')^{-1}F, \phi \rangle \tag{7}$$

When $\mu \geq -\frac{1}{2}$, the definition (6) coincides with that used in the preceding sections (namely, Sec. 5.5, Eq. (4)) because in this case $\mathfrak{H}_{\mu,k}$ agrees with $\mathfrak{H}_{\mu}$ as defined by (1).

In this section we shall not discard the prime on the notation $\mathfrak{H}_{\mu}'$ for the generalized Hankel transformation. We did so previously because, when $\mu \geq -\frac{1}{2}$, $\mathfrak{H}_{\mu}'f$ could be identified with (1) under suitable restrictions on f (for example, $f \in \mathscr{H}_{\mu}$). But here, such an identification has not been made when $\mu < -\frac{1}{2}$.

The operation-transform formula:

$$\mathfrak{H}_{\mu}'(M_{\mu}N_{\mu}f) = -y^2\,\mathfrak{H}_{\mu}'f \qquad f \in \mathscr{H}_{\mu}' \tag{8}$$

which we previously established only for $\mu \geq -\frac{1}{2}$ (see Theorem 5.5-2), remains valid when μ takes on any real value and the equality is understood to be in $\mathscr{H}_\mu'$. As a consequence, our extended Hankel transformation $\mathfrak{H}_\mu'$ also generates an operational calculus by means of which certain differential equations involving generalized functions can be resolved. To establish (8), we need

LEMMA 5.10-2. *Let μ be any fixed real number and k a positive integer $\geq -\mu - \frac{1}{2}$. Then, for every $\Phi \in \mathscr{H}_\mu$,*

$$M_\mu N_\mu \mathfrak{H}_{\mu, k} \Phi = \mathfrak{H}_{\mu, k}(-y^2 \Phi) \tag{9}$$

The proof of this lemma is rather tedious but involves no more than a judicious use of integration by parts and differentiation under the integral sign. We leave the details to the reader as Problem 5.10-2.

THEOREM 5.10-2. *For arbitrary real μ,* (8) *is valid in the sense of equality in $\mathscr{H}_\mu'$.*

PROOF. Let $\Phi \in \mathscr{H}_\mu$, and let k be restricted as in Lemma 5.10-2. Then, by the definition of $\mathfrak{H}_\mu'$, the definitions of $M_\mu N_\mu$ and multiplication-by-y as generalized operators, and Lemma 5.10-2, we may write

$$\langle \mathfrak{H}_\mu' M_\mu N_\mu f, \Phi \rangle = \langle M_\mu N_\mu f, \mathfrak{H}_{\mu, k} \Phi \rangle = \langle f, M_\mu N_\mu \mathfrak{H}_{\mu, k} \Phi \rangle$$
$$= \langle f, \mathfrak{H}_{\mu, k} (-y^2 \Phi) \rangle = \langle \mathfrak{H}_\mu' f, -y^2 \Phi \rangle = \langle -y^2 \mathfrak{H}_\mu' f, \Phi \rangle \quad \text{Q.E.D.}$$

The operational calculus generated by (8) is applicable to differential equations of the form:

$$P(M_\mu N_\mu) u = g \tag{10}$$

where g is a given member of $\mathscr{H}_\mu'$, u is unknown but required to be in $\mathscr{H}_\mu'$, and $P(z)$ is a polynomial having no roots on the nonpositive real axis $-\infty < z \leq 0$. However, in contrast to the situation in Sec. 5.7, μ is now allowed to assume any real value. Proceeding exactly as in Sec. 5.7, we find the solution to (10) to be

$$u = (\mathfrak{H}_\mu')^{-1} \frac{G(y)}{P(-y^2)} \tag{11}$$

More specifically, if k is a positive integer $\geq -\mu - \frac{1}{2}$, $\phi \in \mathscr{H}_\mu$, and $\Phi = \mathfrak{H}_{\mu, k}^{-1} \phi$, then u is that member of $\mathscr{H}_\mu'$ that assigns to ϕ the number:

$$\langle u, \phi \rangle = \left\langle (\mathfrak{H}_\mu')^{-1} \frac{G(y)}{P(-y^2)}, \phi \right\rangle = \left\langle G(y), \frac{\Phi(y)}{P(-y^2)} \right\rangle$$
$$= \left\langle g, \mathfrak{H}_{\mu, k} \frac{\Phi(y)}{P(-y^2)} \right\rangle$$

Incidentally, if μ is a negative integer, then every $g \in \mathscr{H}_\mu{}'$ is also a member of $\mathscr{H}'_{-\mu}$ according to Sec. 5.2, note *II*. In view of the fact that

$$M_\mu N_\mu = D^2 - \frac{4\mu^2 - 1}{4x^2} = M_{-\mu} N_{-\mu}$$

we may apply $\mathfrak{H}'_{-\mu}$ to (10) to obtain a solution in $\mathscr{H}'_{-\mu}$. However, the use of $\mathfrak{H}_\mu{}'$ leads to a stronger result. This is because $\mathscr{H}_\mu$ contains $\mathscr{H}_{-\mu}$ as a nondense proper subspace so that equality in $\mathscr{H}_\mu{}'$ is stronger than equality in $\mathscr{H}'_{-\mu}$. Thus, a solution of (10) in $\mathscr{H}_\mu{}'$ is stronger than a solution in $\mathscr{H}'_{-\mu}$ In other words, if u satisfies (10) in the sense of equality in $\mathscr{H}_\mu{}'$, it certainly does so in $\mathscr{H}'_{-\mu}$, but the converse is not necessarily true. See the last statement of Sec. 5.2, note *II*.

In the case where μ is negative but not an integer, there are members of $\mathscr{H}_\mu{}'$ that are not members of $\mathscr{H}'_{-\mu}$. (For example, the functional g defined on $\mathscr{H}_\mu$ $(\mu < 0)$ by

$$\langle g, \phi \rangle \triangleq \lim_{x \to 0+} x^{-1} D x^{-\mu-1/2} \phi(x) \tag{12}$$

is a member of $\mathscr{H}_\mu{}'$ as is evident from Lemma 5.2-1. However, g is not a member of $\mathscr{H}'_{-\mu}$ if $-1 < \mu < 0$; indeed, (12) doesn't exist if $\phi(x) \in \mathscr{H}_{-\mu}$ is identical to $\sqrt{x}\, J_{-\mu}(x)$ on $0 < x < 1$, as is possible.) Thus, when $\mu < 0$ and $\mu \neq -1, -2, \ldots, \mathfrak{H}'_{-\mu}$ does not generate an operational calculus for (10) with arbitrary $g \in \mathscr{H}_\mu{}'$.

PROBLEM 5.10-1. Show that, for any $\mu < -\frac{1}{2}$ and any positive integer k, $\mathfrak{H}_{\mu,k}$ coincides with $\mathfrak{H}_\mu$ when $\mathfrak{H}_{\mu,k}$ is acting on a testing function $\Phi(y) \in \mathscr{H}_\mu$ such that $\Phi(y) \equiv 0$ on some interval $0 < y < \varepsilon$.

PROBLEM 5.10-2. Prove Lemma 5.10-2. *Hint*: By differentiating under the integral sign and using Sec. 5.1, Eqs. (6) and (7), we can convert the left-hand side of (9) into

$$(-1)^k 2k \int_0^\infty y^{\mu+k+2} [(y^{-1}D_y)^k y^{-\mu-1/2} \Phi(y)] x^{-k-1/2} J_{\mu+k+1}(xy)\, dy$$
$$+ (-1)^{k+1} \int_0^\infty y^{\mu+k+3} [(y^{-1}D_y)^k y^{-\mu-1/2} \Phi(y)] x^{-k+1/2} J_{\mu+k}(xy)\, dy$$

By integrating the first term by parts and noting that the limit terms are equal to zero, we find that

$$M_\mu N_\mu \mathfrak{H}_{\mu,k} \Phi = (-1)^{k+1} \int_0^\infty y^{\mu+k+1} [2k(y^{-1}D_y)^{k-1} y^{-\mu-1/2} \Phi(y)$$
$$+ y^2 (y^{-1}D_y)^k y^{-\mu-1/2} \Phi(y)] x^{1/2-k} J_{\mu+k}(xy)\, dy$$

A straightforward manipulation then shows that $\mathfrak{H}_{\mu,k}(-y^2\Phi)$ is also equal to the last expression. Justify all the steps of this argument.

PROBLEM 5.10-3. Establish the following operation-transform formulas wherein μ is an arbitrary real number, k is a positive integer $\geq -\mu - \frac{1}{2}$, $\Phi \in \mathscr{H}_\mu$, and $f \in \mathscr{H}'_{\mu+1}$.

$$N_\mu \mathfrak{H}_{\mu,k}\Phi = \mathfrak{H}_{\mu+1,k}(-y\Phi)$$

$$\mathfrak{H}_{\mu+1,k}(N_\mu \Phi) = -x\mathfrak{H}_{\mu,k}\Phi$$

$$\mathfrak{H}_\mu'(M_\mu f) = y\mathfrak{H}'_{\mu+1}f$$

$$M_\mu \mathfrak{H}'_{\mu+1}f = \mathfrak{H}_\mu'(xf)$$

5.11. Hankel Transforms of Certain Generalized Functions of Arbitrary Growth

The Hankel transformation discussed up till now is defined on the generalized functions $f(x)$ in $\mathscr{H}_\mu'$, and these have the property that they cannot grow too fast as $x \to \infty$. (In fact, they are distributions of slow growth on any interval of the form $X < x < \infty$ where $X > 0$. See Problem 5.2-3.) In this section we call attention to the fact that the Hankel transformation can be extended to certain generalized functions $f(x)$ having no restriction on their growth as $x \to \infty$. Here, we shall merely sketch the theory of this extension, and refer the reader to Zemanian [5], [7] for a more thorough discussion. Our procedure is similar to the methods used by Ehrenpreis [1] and Gelfand and Shilov [1], Vol. 1, to extend the Fourier transformation to all distributions.

Let μ and b be fixed real numbers with $b > 0$. $\mathscr{B}_{\mu,b}$ is defined as the linear space of all smooth complex-valued functions $\phi(x)$ on $0 < x < \infty$ such that $\phi(x) \equiv 0$ on $b < x < \infty$ and

$$\gamma_k{}^\mu(\phi) \triangleq \sup_{0<x<\infty} |(x^{-1}D)^k x^{-\mu-1/2}\phi(x)| < \infty \qquad k = 0, 1, 2, \ldots$$

We assign to $\mathscr{B}_{\mu,b}$ the topology generated by the countable multinorm $\{\gamma_k{}^\mu\}_{k=0}^\infty$. It can be shown that $\mathscr{B}_{\mu,b}$ is complete. Also, if $b < c$, then $\mathscr{B}_{\mu,b} \subset \mathscr{B}_{\mu,c}$, and the topology induced on $\mathscr{B}_{\mu,b}$ by $\mathscr{B}_{\mu,c}$ is identical to the topology of $\mathscr{B}_{\mu,b}$.

Next, we choose a sequence of real positive numbers $\{b_n\}_{n=1}^\infty$ that tends monotonically to infinity and define $\mathscr{B}_\mu$ as the strict countable-union space $\bigcup_{n=1}^\infty \mathscr{B}_{\mu,b_n}$. The space $\mathscr{B}_\mu$ does not depend on the choice of the sequence $\{b_n\}_{n=1}^\infty$. The dual $\mathscr{B}_\mu'$ of $\mathscr{B}_\mu$ is the space of generalized functions onto which the Hankel transformation will be extended.

We need still another testing-function space $\mathscr{Y}_{\mu,b}$. Here again, μ and b are fixed real numbers with $b > 0$. Let y and ω be real variables and set $\eta = y + i\omega$. Φ is a member of $\mathscr{Y}_{\mu,b}$ if and only if $\eta^{-\mu-1/2}\Phi(\eta)$ is an even entire function of η (defined by continuity at $\eta = 0$) and

$$\alpha^{\mu}_{b,k}(\Phi) \triangleq \sup_{\eta} |e^{-b|\omega|}\eta^{2k-\mu-1/2}\Phi(\eta)| < \infty \qquad k = 0, 1, 2, \ldots$$

The supremum is taken over the entire η plane. The topology of $\mathscr{Y}_{\mu,b}$ is the one generated by the countable multinorm $\{\alpha^{\mu}_{b,k}\}_{k=0}^{\infty}$. Here also, the following facts are true: $\mathscr{Y}_{\mu,b}$ is complete. If $b < c$, then $\mathscr{Y}_{\mu,b} \subset \mathscr{Y}_{\mu,c}$, and the topology induced on $\mathscr{Y}_{\mu,b}$ by $\mathscr{Y}_{\mu,c}$ is identical to the topology of $\mathscr{Y}_{\mu,b}$; in fact, $\alpha^{\mu}_{b,k}(\Phi) = \alpha^{\mu}_{c,k}(\Phi)$ for every $\Phi \in \mathscr{Y}_{\mu,b}$ and every k.

Again choosing $\{b_n\}_{n=1}^{\infty}$ as a monotonic sequence of real positive numbers tending to infinity (it doesn't matter which such sequence is chosen), we define $\mathscr{Y}_{\mu}$ as the strict countable-union space $\bigcup_{n=1}^{\infty} \mathscr{Y}_{\mu,b_n}$. $\mathscr{Y}_{\mu}'$ is the dual of $\mathscr{Y}_{\mu}$.

Now, by using a theorem due to Griffith [1], we can prove a result, which is crucial to this theory: *For $\mu \geq -\frac{1}{2}$, the conventional Hankel transformation $\mathfrak{H}_{\mu}$ is an isomorphism from $\mathscr{Y}_{\mu,b}$ onto $\mathscr{B}_{\mu,b}$ and therefore an isomorphism from $\mathscr{Y}_{\mu}$ onto $\mathscr{B}_{\mu}$.* Because of this, we can define the generalized Hankel transformation $\mathfrak{H}_{\mu}'$ on $\mathscr{B}_{\mu}'$ as the adjoint of $\mathfrak{H}_{\mu}$ on $\mathscr{Y}_{\mu}$. More specifically, for any $f \in \mathscr{B}_{\mu}'$ and $\mu \geq -\frac{1}{2}$, we define $\mathfrak{H}_{\mu}'f$ by

(1) $$\langle \mathfrak{H}_{\mu}'f, \Phi\rangle \triangleq \langle f, \mathfrak{H}_{\mu}\Phi\rangle \qquad \Phi \in \mathscr{Y}_{\mu}$$

By Theorem 1.10-2, $\mathfrak{H}_{\mu}'$ is an isomorphism from $\mathscr{B}_{\mu}'$ onto $\mathscr{Y}_{\mu}'$; moreover, its inverse is itself (i.e., $(\mathfrak{H}_{\mu}')^{-1} = \mathfrak{H}_{\mu}'$) since $\mathfrak{H}_{\mu}^{-1} = \mathfrak{H}_{\mu}$. It is also a fact that the generalized Hankel transformation, defined in Sec. 5.5, is a special case of our present Hankel transformation.

We finally point out one application of our present theory. Consider the differential equation:

(2) $$P(M_{\mu}N_{\mu})u = g$$

where $\mu \geq -\frac{1}{2}$, P is a polynomial such that $P(O) \neq 0$, $g \in \mathscr{B}_{\mu}'$, and u is unknown. The Hankel transformation $\mathfrak{H}_{\mu}'$, as defined by (1), can be used to prove that (2) always has solutions in $\mathscr{B}_{\mu}'$ and that any two of them differ by no more than a conventional solution of the homogeneous equation $P(M_{\mu}N_{\mu}) = 0$. See Zemanian [5], [7].

CHAPTER VI

The K Transformation

6.1. Introduction

The *conventional K transformation* is defined as follows: Let μ be a fixed complex number, let t, σ, and ω be real variables in $\mathscr{R}^1$, and set $s = \sigma + i\omega$. As is customary, $K_\mu(z)$ denotes the modified Bessel function of third kind and order μ (Jahnke, Emde, and Losch [1], p. 207). If $f(t)$ is a suitably restricted conventional function defined on $0 < t < \infty$, then its K transform of order μ is a function of the complex variable $s = \sigma + i\omega$, defined by

(1)
$$F(s) \triangleq \int_0^\infty f(t)\sqrt{st}\, K_\mu(st)\, dt$$

It turns out that $F(s)$ is an analytic function on a half-plane of the form $\{s\colon \operatorname{Re} s > b \geq 0\}$ where the abscissa b depends upon the function $f(t)$. Furthermore, it is a fact that $K_\mu(st) = K_{-\mu}(st)$; therefore, we lose no generality in (1) by restricting μ according to $0 \leq \operatorname{Re} \mu < \infty$.

C. S. Meijer [2] was apparently the first to investigate the K transformation. Indeed, it is sometimes called the Meijer transformation, but this latter name is also used for other integral transformations investigated by him. Other early investigations of (1) were made by Boas [1], [2] and Erdelyi [1].

The K transformation can be extended to certain generalized functions by combining some of the techniques used in generalizing the Laplace and Hankel transformations. More specifically, let us assume that μ is either the number zero or a complex number having a positive real part. For each such μ and each real positive number a, we construct a testing-function space $\mathscr{K}_{\mu,a}$ of smooth functions $\phi(t)$ on $0 < t < \infty$ which is closed with respect to the Bessel-type differential operator S_μ (defined by (3) below) and whose members tend to zero at least as fast as $e^{-at}t^{1/2-\mu}$ as $t \to \infty$. It turns out that the kernel function $\sqrt{st}\, K_\mu(st)$ is in $\mathscr{K}_{\mu,a}$ for $\operatorname{Re} s > a$. The dual space $\mathscr{K}'_{\mu,a}$ consists of those generalized functions to which we may apply our generalized K transformation of order μ. The transform $F(s)$ of $f \in \mathscr{K}'_{\mu, a}$ is defined by

$$F(s) \triangleq \langle f(t), \sqrt{st}\, K_\mu(st)\rangle \qquad \text{Re } s > a$$

Among the various properties of the generalized K transformation that we shall discuss are an analyticity theorem, some inversion formulas, and an operational calculus that is useful in solving certain Bessel-type differential equations. In the last section of this chapter we apply this transformation to analyze various time-varying electrical networks having generalized-function excitations.

The theory presented here is a simplification of the one given in Zemanian [9]. Moreover, we required in that work that $|\text{Re } \mu| \leq \frac{1}{2}$, as is also done by Meijer [2] and Boas [1], [2]. In the present discussion we allow $\text{Re } \mu$ to be greater than $\frac{1}{2}$. On the other hand, we do not permit μ to be purely imaginary because some of the arguments of this chapter do not hold for this case. For the theory pertaining to purely imaginary μ, see Zemanian [9].

Incidentally, there is another transformation that is similar to the K transformation, namely, the *I transformation:*

(2) $$F(s) = \langle f(t), \sqrt{st}\, I_\mu(st)\rangle$$

where I_μ denotes the modified Bessel function of first kind and order $\mu \geq -\frac{1}{2}$. (See Koh and Zemanian [1].) Actually, (2) can be shown to be a special case of the Hankel transformation discussed in the preceding chapter.

We shall make considerable use of the following differentiation operator of second order:

(3) $$S_\mu \triangleq t^{-\mu-1/2} D t^{2\mu+1} D t^{-\mu-1/2} \qquad D = \frac{\partial}{\partial t}$$

Thus, $S_\mu = M_\mu N_\mu$, where M_μ and N_μ are defined in Sec. 5.3. Sometimes, we shall replace S_μ by $S_{\mu,t}$ in order to indicate what the independent variable in the operator is. From the rule for the differentiation of products, we obtain

(4) $$S^k_{\mu,t}\phi(t) = a_{2k,0}\, t^{-2k}\phi + a_{2k,1}\, t^{-2k+1} D\phi + \cdots + a_{2k,2k} D^{2k}\phi$$

$$a_{2k,2k} = 1$$

where the $a_{2k,q}$ are constants depending on the value of μ. This type of computation also shows that

(5) $$S_{\mu,t}\,\phi(t) = D^2\phi + \frac{1 - 4\mu^2}{4t^2}\,\phi$$

Thus, $S_\mu = S_{-\mu}$.

Following our usual practice, we restrict all multivalued functions on the z plane that are analytic everywhere except for branch points at $z = 0$ and $z = \infty$ to their principal branches by requiring that $-\pi < \arg z \leq \pi$, unless the opposite is explicitly indicated. As is customary, $\Gamma(z)$ denotes the gamma function (Jahnke, Emde, and Losch [1], p. 4). Throughout this chapter, I again denotes the open interval $(0, \infty)$ in $\mathscr{R}^1$.

6.2. Some Classical Results

Let z be a complex variable and μ a fixed complex number. As before, $I_\mu(z)$ denotes the modified Bessel functions of first kind and order μ, and $K_\mu(z)$ denotes the modified Bessel function of third kind and order μ (Jahnke, Emde, and Losch [1], p. 207, and Erdelyi (Ed.) [1], Vol. II, pp. 5 and 9). These functions are analytic on the z plane except possibly for branch points at $z = 0$ and $z = \infty$. Under our convention of choosing the principal branch by requiring that $-\pi < \arg z \leq \pi$, both $I_\mu(z)$ and $K_\mu(z)$ are real-valued when μ is real and z is real and positive.

These functions possess the following series expansions, which converge for every nonzero value of z. For arbitrary μ,

(1)
$$I_\mu(z) = \sum_{k=0}^{\infty} \frac{z^{2k+\mu}}{k!\, 2^{2k+\mu}\, \Gamma(k+1+\mu)}$$

Also, for any μ not equal to an integer,

(2)
$$K_\mu(z) = \frac{\pi}{2 \sin \mu\pi} [I_{-\mu}(z) - I_\mu(z)]$$
$$= \frac{\pi}{2 \sin \mu\pi} \left[\sum_{k=0}^{\infty} \frac{z^{2k-\mu}}{k!\, 2^{2k-\mu}\, \Gamma(k+1-\mu)} - \sum_{k=0}^{\infty} \frac{z^{2k+\mu}}{k!\, 2^{2k+\mu}\, \Gamma(k+1+\mu)} \right]$$

On the other hand,

(3)
$$K_n(z) = \frac{1}{2} \sum_{k=0}^{n-1} \frac{(-1)^k (n-k-1)!}{k!} \left(\frac{z}{2}\right)^{2k-n}$$
$$+ (-1)^n \sum_{k=0}^{\infty} \frac{(z/2)^{n+2k}}{k!\,(n+k)!} \left[-\log \frac{Cz}{2} + \frac{1}{2} \left(\sum_{p=1}^{k} \frac{1}{p} + \sum_{p=1}^{n+k} \frac{1}{p} \right) \right]$$
$$n = 0, 1, 2, \ldots$$

where $C = e^\gamma$ and γ is Euler's constant $(0.5772 \cdots)$.

The asymptotic behaviors for these functions as $z \to \infty$, which hold for any value of μ, are the following (Watson [1], pp. 202–203, and Meijer [1], pp. 658–660). For any fixed $\varepsilon > 0$ and for $|z| \to \infty$,

$$\sqrt{z}\,K_\mu(z) = \sqrt{\frac{\pi}{2}}\,e^{-z}[1 + O(|z|^{-1})] \qquad -\frac{3\pi}{2} + \varepsilon < \arg z < \frac{3\pi}{2} - \varepsilon \tag{4}$$

$$\sqrt{z}\,I_\mu(z) = \frac{1}{\sqrt{2\pi}}\,(e^z + ie^{-z+i\mu\pi})[1 + O(|z|^{-1})] \qquad -\frac{\pi}{2} + \varepsilon < \arg z < \frac{3\pi}{2} - \varepsilon \tag{5}$$

$$\sqrt{z}\,I_\mu(z) = \frac{1}{\sqrt{2\pi}}\,(e^z - ie^{-z-i\mu\pi})[1 + O(|z|^{-1})] \qquad -\frac{3\pi}{2} + \varepsilon < \arg z < \frac{\pi}{2} - \varepsilon \tag{6}$$

uniformly for $\arg z$ in the indicated intervals.

Some differentiation formulas are (Erdelyi (Ed.) [1], Vol. II, p. 79):

$$Dz^\mu K_\mu(z) = -z^\mu K_{\mu-1}(z), \quad Dz^\mu I_\mu(z) = z^\mu I_{\mu-1}(z) \tag{7}$$

$$Dz^{-\mu} K_\mu(z) = -z^{-\mu} K_{\mu+1}(z), \quad Dz^{-\mu} I_\mu(z) = z^{-\mu} I_{\mu+1}(z) \tag{8}$$

and these yield the formulas:

$$S_{\mu,\,t}\sqrt{st}\,K_\mu(st) = s^2\sqrt{st}\,K_\mu(st) \tag{9}$$

$$S_{\mu,\,t}\sqrt{st}\,I_\mu(st) = s^2\sqrt{st}\,I_\mu(st) \tag{10}$$

Another result we shall need is the following indefinite integral:

$$\int tI_\mu(zt)\,K_\mu(st)\,dt = \frac{t}{z^2 - s^2}\,[zI_{\mu+1}(zt)\,K_\mu(st) + sI_\mu(zt)\,K_{\mu+1}(st)] \tag{11}$$

[See Erdelyi (Ed.) [1], Vol. II, Eqs. 7.14.1(9), 7.2.2(12), and 7.2.2(15).]

One of the inversion formulas for our generalized K transformation (see Sec. 6.7, Eq. (6)) is based upon the following theorem. The proof we present here is essentially that given by Meijer [2], pp. 709–710. Now, however, we only require that $\operatorname{Re}\mu \geq -\frac{1}{2}$, whereas Meijer imposed the restriction $|\operatorname{Re}\mu| \leq \frac{1}{2}$.

THEOREM 6.2-1. *Let μ be a fixed complex number such that* $\operatorname{Re}\mu \geq -\frac{1}{2}$. *Assume that, on the half-plane* $\{s\colon \operatorname{Re} s > b \geq 0\}$, *$F(s)$ is an analytic function and is bounded according to* $|F(s)| < M|s|^{-q}$, *where M and q are* *real constants not depending on s, and* $q > \operatorname{Re}\mu + \frac{3}{2}$. *Then, for any fixed real number* $c > b$ *and for* $\operatorname{Re} s > c$

$$F(s) = \int_0^\infty f(t)\sqrt{st}\,K_\mu(st)\,dt \tag{12}$$

where

$$f(t) = \frac{1}{i\pi}\int_{c-i\infty}^{c+i\infty} F(z)\sqrt{zt}\,I_\mu(zt)\,dz \tag{13}$$

Here, $f(t)$ *does not depend on the choice of* c.

PROOF. The last statement follows immediately from Cauchy's theorem, the bound on $F(z)$, and the fact that, for each fixed $t > 0$, $\sqrt{zt}\,I_\mu(zt)$ is bounded on every strip of the form $x_1 \le \operatorname{Re} z \le x_2$, where $0 < x_1 < x_2 < \infty$; see (5) and (6).

Next, let s be fixed and let $1 < T < \infty$; consider

$$\int_0^T f(t)\sqrt{st}\,K_\mu(st)\,dt = \frac{1}{i\pi}\int_0^T \sqrt{st}\,K_\mu(st)\int_{c-i\infty}^{c+i\infty} F(z)\sqrt{zt}\,I_\mu(zt)\,dz\,dt \tag{14}$$

Let Λ denote the domain $\{(t, \operatorname{Im} z);\ 0 < t < T,\ -\infty < \operatorname{Im} z < \infty\}$, and fix $\operatorname{Re} z = c > b$. In view of the series expansion and asymptotic behavior of $I_\mu(zt)$, it follows that

$$|\sqrt{zt}\,I_\mu(zt)| < C_1|(zt)^{\mu+1/2}|$$

on Λ, where C_1 is a constant. Consequently, by the series expansion of $K_\mu(st)$ and the bound on $F(z)$,

$$|\sqrt{st}\,K_\mu(st)\sqrt{zt}\,I_\mu(zt)\,F(z)| \le C_2\,|z^{-q+\mu+1/2}|\,(t + |t^{2\mu+1}|)$$

on Λ, where C_2 is another constant. The right-hand side is absolutely integrable on Λ. Therefore, by Fubini's theorem, we may change the order of integration in (14) and make use of (11) to get

$$\frac{1}{i\pi}\int_{c-i\infty}^{c+i\infty} F(z)\,\frac{\sqrt{sz}}{z^2 - s^2}\Big[ztI_{\mu+1}(zt)\,K_\mu(st) + stI_\mu(zt)\,K_{\mu+1}(st)\Big]_{t=0}^{t=T}\,dz \tag{15}$$

The series expansions (1), (2), and (3) and the identity (Jahnke, Emde, and Losch [1], p. 9):

$$\Gamma(1+\mu)\Gamma(-\mu) = \frac{\pi}{\sin(1+\mu)\pi}$$

show that (15) is equal to

$$\begin{aligned}\frac{1}{i\pi}\int_{c-i\infty}^{c+i\infty} \frac{F(z)}{z^2 - s^2}\,[z\sqrt{zT}\,I_{\mu+1}(zT)\sqrt{sT}\,K_\mu(sT)\\ + s\sqrt{zT}\,I_\mu(zT)\sqrt{sT}\,K_{\mu+1}(sT) - s^{1/2-\mu}z^{1/2+\mu}]\,dz\end{aligned} \tag{16}$$

Now, for $\operatorname{Re} s > c > 0$, $T > 1$, and $z = c + iy$, the asymptotic estimates (4), (5), and (6) show that there exists a constant N not depending on T or y such that

$$(17) \qquad \begin{aligned} &|\sqrt{sT}\,K_\mu(sT)| < Ne^{-\operatorname{Re} sT}, \; |\sqrt{sT}\,K_{\mu+1}(sT)| < Ne^{-\operatorname{Re} sT} \\ &|\sqrt{zT}\,I_\mu(zT)| < Ne^{cT}, \; |\sqrt{zT}\,I_{\mu+1}(zT)| < Ne^{cT} \end{aligned}$$

Therefore,

$$\begin{aligned} |z\sqrt{zT}\,I_{\mu+1}(zT)\sqrt{sT}\,K_\mu(sT) + s\sqrt{zT}\,I_\mu(zT)\sqrt{sT}\,K_{\mu+1}(sT)| \\ \leq N^2(|z| + |s|) \end{aligned}$$

By virtue of our assumption on $F(z)$, it follows that, for any fixed s with $\operatorname{Re} s > c > 0$, the integral (16) converges uniformly on $1 < T < \infty$, so that we may take the limit as $T \to \infty$ under the integral sign. Since $\operatorname{Re} s > c = \operatorname{Re} z$, the bounds (17) now show that

$$(18) \qquad \int_0^\infty f(t)\sqrt{st}\,K_\mu(st)\,dt = \frac{s^{1/2-\mu}}{i\pi}\int_{c-i\infty}^{c+i\infty} \frac{F(z)z^{1/2+\mu}}{s^2 - z^2}\,dz$$

To determine the right-hand side of (18), we evaluate the integral

$$(19) \qquad \oint \frac{F(z)z^{1/2+\mu}}{s^2 - z^2}\,dz$$

where the closed path of integration is in the positive direction around the rectangle whose corners are at the points $c - iY$, $Y - iY$, $Y + iY$, and $c + iY$, Y being a real number greater than $\operatorname{Re} s$. By Cauchy's residue theorem, (19) is equal to

$$-i\pi s^{\mu-1/2}F(s)$$

Moreover, the integral along the three sides extending from $c - iY$ to $Y - iY$ to $Y + iY$ to $c + iY$ tends to zero as $Y \to \infty$, again because of our hypothesis on $F(z)$. By combining these results, we find that the right-hand side of (18) is equal to $F(s)$. The proof is complete.

PROBLEM 6.2-1. Let $f(x)$ be a continuous function on $0 < x < \infty$ having a continuous first derivative and a compact support with respect to $0 < x < \infty$. Define its I transform of order μ by

$$F(s) = \int_0^\infty f(x)\sqrt{xs}\,I_\mu(xs)\,dx$$

By relating the function $I_\mu(xs)$ to $J_\mu(xs)$, derive an inversion formula for this transform.

6.3. The Testing-Function Space $\mathscr{K}_{\mu,a}$ and Its Dual

Throughout the rest of this chapter, a will always denote a real number, and μ will be either the number zero or a complex number satisfying $\operatorname{Re}\mu > 0$. Also, $h(t)$ will be the continuous function on $0 < t < \infty$ defined by

$$h(t) \triangleq \begin{cases} \log t & 0 < t < e^{-1} \\ -1 & e^{-1} \le t < \infty \end{cases}$$

We define the functionals $\rho^{\mu}_{a,k}$, $k = 0, 1, 2, \ldots$, on certain smooth functions $\phi(t)$ by

$$\rho^{\mu}_{a,k}(\phi) \triangleq \sup_{0<t<\infty} |e^{at} t^{\mu-1/2} S_{\mu}{}^{k} \phi(t)| \qquad \operatorname{Re}\mu > 0$$

$$\rho^{0}_{a,k}(\phi) \triangleq \sup_{0<t<\infty} \left| \frac{e^{at} S_0{}^{k} \phi(t)}{\sqrt{t}\, h(t)} \right|$$

If we set $j_{\mu}(t)$ equal to $t^{\mu-1/2}$ when $\operatorname{Re}\mu > 0$ and equal to $[\sqrt{t}\, h(t)]^{-1}$ when $\mu = 0$, then the two preceding equations can be written simultaneously as

$$\rho^{\mu}_{a,k}(\phi) \triangleq \sup_{0<t<\infty} |e^{at} j_{\mu}(t) S_{\mu}{}^{k} \phi(t)|$$

$\mathscr{K}_{\mu,a}$ is defined as the linear space of all complex-valued smooth functions $\phi(t)$ on $0 < t < \infty$ for which $\rho^{\mu}_{a,k}(\phi)$ exists (i.e., is finite) for every $k = 0, 1, 2, \ldots$. Each $\rho^{\mu}_{a,k}(\phi)$ is a seminorm on $\mathscr{K}_{\mu,a}$, and $\rho^{\mu}_{a,0}$ is a norm. Therefore, $\{\rho^{\mu}_{a,k}\}_{k=0}^{\infty}$ is a countable multinorm on $\mathscr{K}_{\mu,a}$. Henceforth, we always take $\mathscr{K}_{\mu,a}$ to be the countably multinormed space having the topology generated by $\{\rho^{\mu}_{a,k}\}_{k=0}^{\infty}$.

Note that the case $\mu = 0$ is a special one because the seminorms $\rho^{0}_{a,k}$ have a form different from the $\rho^{\mu}_{a,k}$ ($\operatorname{Re}\mu > 0$). This case will at times require an independent analysis.

For every fixed s such that $s \neq 0$ and $\operatorname{Re} s > a$, we have that $\sqrt{st}\, K_{\mu}(st) \in \mathscr{K}_{\mu,a}$. Indeed, by the analyticity of $\sqrt{z}\, K_{\mu}(z)$ for $z \neq 0$, it follows that $\sqrt{st}\, K_{\mu}(st)$ is smooth on $0 < t < \infty$. Also, the differentiation formula Sec. 6.2, Eq. (9), the series expansions Sec. 6.2, Eq. (2) and (3), and the asymptotic estimate Sec. 6.2, Eq. (4) imply that the quantities $\rho^{\mu}_{a,k}[\sqrt{st}\, K_{\mu}(st)]$ are finite for all $k = 0, 1, 2, \ldots$. (In fact, the seminorms were devised to provide just this property.) Thus, our assertion is verified.

LEMMA 6.3-1. *$\mathscr{K}_{\mu,a}$ is complete and therefore is a Fréchet space.*

PROOF. Let $\{\phi_\nu\}_{\nu=1}^{\infty}$ be a Cauchy sequence in $\mathscr{H}_{\mu,a}$, and let Ω denote an arbitrary compact subset of $I = \{t: 0 < t < \infty\}$. Also, let τ be a fixed point in I, and let D^{-1} denote the integration operator:

$$D^{-1} = \int_\tau^t \cdots \, dx$$

Thus, for any smooth function $\zeta(t)$ on I.

$$D^{-1} D\zeta(t) = \zeta(t) - \zeta(\tau)$$

By the definition of S_μ,

$$S_\mu \phi_\nu(t) = t^{-\mu-1/2} D t^{2\mu+1} D t^{-\mu-1/2} \phi_\nu(t) \tag{1}$$

and, in view of the seminorm $\rho^\mu_{a,1}$, $S_\mu \phi_\nu(t)$ converges uniformly on every Ω as $\nu \to \infty$. Moreover, we have

$$t^{-2\mu-1} D^{-1} t^{\mu+1/2} S_\mu \phi_\nu(t) = D_t t^{-\mu-1/2} \phi_\nu(t) - \left(\frac{\tau}{t}\right)^{2\mu+1} D_\tau \tau^{-\mu-1/2} \phi_\nu(\tau) \tag{2}$$

and

$$t^{\mu+1/2} D^{-1} t^{-2\mu-1} D^{-1} t^{\mu+1/2} S_\mu \phi_\nu(t) = \phi_\nu(t) - \left(\frac{t}{\tau}\right)^{\mu+1/2} \phi_\nu(\tau) + \ell_\tau(t) D_\tau \tau^{-\mu-1/2} \phi_\nu(\tau) \tag{3}$$

where

$$\ell_\tau(t) = \begin{cases} (2\mu)^{-1} t^{1/2} \tau [t^{-\mu} \tau^{2\mu} - t^\mu] & \operatorname{Re} \mu > 0 \\ t^{1/2} \tau [\log \tau - \log t] & \mu = 0 \end{cases}$$

The left-hand sides of (2) and (3) also converge uniformly on every Ω as $\nu \to \infty$ since multiplication by a power of t or the application of D^{-1} preserves this property. Moreover, $\phi_\nu(t)$ and $(t/\tau)^{\mu+1/2} \phi_\nu(\tau)$ converge in the same way, in view of the seminorm $\rho^\mu_{a,0}$. Since $\ell_\tau(t) \not\equiv 0$, it follows from (3) that $D_\tau \tau^{-\mu-1/2} \phi_\nu(\tau)$ must converge as $\nu \to \infty$. We now see from (2) that $D_t t^{-\mu-1/2} \phi_\nu(t)$ also converges uniformly on every Ω, which implies in turn that $D_t \phi_\nu(t)$ does the same. Next, by virtue of Sec. 6.1, Eq. (5), it follows that $D^2 \phi_\nu(t)$ also converges uniformly on every Ω.

We repeat this argument with ϕ_ν replaced by $S_\mu{}^k \phi_\nu$ and $S_\mu \phi_\nu$ replaced by $S_\mu^{k+1} \phi_\nu$. This shows that, for every nonnegative integer k, $D^k \phi_\nu(t)$ converges uniformly on every Ω. Consequently, there exists a smooth function $\phi(t)$ on I such that, for each k and t, $D^k \phi_\nu(t) \to D^k \phi(t)$ as $\nu \to \infty$ (Apostol [1], p. 402).

Next, the assumption that $\{\phi_\nu\}$ is a Cauchy sequence in $\mathscr{H}_{\mu,a}$ can be restated as follows: For each nonnegative integer k and any given $\varepsilon > 0$, there exists a real number N_k such that, for every ν, $\eta > N_k$, we have $\rho^{\mu}_{a,k}(\phi_\nu - \phi_\eta) < \varepsilon$. Therefore, when $\eta \to \infty$, we obtain

$$(4) \qquad \rho^{\mu}_{a,k}(\phi_\nu - \phi) \leq \varepsilon$$

for all $\nu > N_k$. In other words, $\rho^{\mu}_{a,k}(\phi_\nu - \phi) \to 0$ as $\nu \to 0$ for each k.

Finally, there exists a constant C_k not depending on ν such that $\rho^{\mu}_{a,k}(\phi_\nu) < C_k$. Therefore, by (4),

$$\rho^{\mu}_{a,k}(\phi) \leq \rho^{\mu}_{a,k}(\phi_\nu) + \rho^{\mu}_{a,k}(\phi - \phi_\nu) \leq C_k + \varepsilon$$

Thus, ϕ is in $\mathscr{K}_{\mu,a}$ and is the limit in $\mathscr{K}_{\mu,a}$ of $\{\phi_\nu\}$. The proof is complete.

Our discussion up to this point has shown that $\mathscr{K}_{\mu,a}$ is a testing-function space on I since the three conditions in Sec. 2.4 are satisfied.

$\mathscr{K}'_{\mu,a}$ is the dual of $\mathscr{K}_{\mu,a}$, and, by Theorem 1.8-3, it too is complete. The members of $\mathscr{K}'_{\mu,a}$, are generalized functions on I.

In a previous work (Zemanian [9]), the testing-function space $\mathscr{K}_{\mu,a}$ was defined in a somewhat more complicated way. In particular, certain seminorms involving differential operators of odd order were added to the $\rho^{\mu}_{a,k}$ to get the multinorm for the previous testing-function space, and, in addition, the order μ was restricted according to $0 \leq \operatorname{Re} \mu \leq \frac{1}{2}$. Consequently, when either $\mu = 0$ or $0 < \operatorname{Re} \mu \leq \frac{1}{2}$, the restrictions of our present generalized functions in $\mathscr{K}'_{\mu,a}$ to the previous testing-function space are generalized functions in the previous sense.

We now list some properties of $\mathscr{K}_{\mu,a}$ and $\mathscr{K}'_{\mu,a}$.

I. Clearly, $\mathscr{D}(I)$ is a subspace of $\mathscr{K}_{\mu,a}$, and convergence in $\mathscr{D}(I)$ implies convergence in $\mathscr{K}_{\mu,a}$. Consequently, the restriction of any $f \in \mathscr{K}'_{\mu,a}$ to $\mathscr{D}(I)$ is in $\mathscr{D}'(I)$.

On the other hand, $\mathscr{D}(I)$ is not dense in $\mathscr{K}_{\mu,a}$. Moreover, there exist two different members of $\mathscr{K}'_{\mu,a}$ whose restrictions to $\mathscr{D}(I)$ coincide. (See Problem 6.3-1.)

II. If $a < b$, then $\mathscr{K}_{\mu,b} \subset \mathscr{K}_{\mu,a}$, and the topology of $\mathscr{K}_{\mu,b}$ is stronger than the topology induced on it by $\mathscr{K}_{\mu,a}$. This follows from Lemma 1.6-3, and the inequalities $\rho^{\mu}_{a,k}(\phi) \leq \rho^{\mu}_{b,k}(\phi)$, $\phi \in \mathscr{K}_{\mu,b}$, $k = 0, 1, 2, \ldots$. Hence, the restriction of any $f \in \mathscr{K}'_{\mu,a}$ to $\mathscr{K}_{\mu,b}$ is in $\mathscr{K}'_{\mu,b}$.

Here again, however, $\mathscr{K}_{\mu,b}$ is not dense in $\mathscr{K}_{\mu,a}$. Also, there exist two different members of $\mathscr{K}'_{\mu,a}$ whose restrictions to $\mathscr{K}_{\mu,b}$ coincide. (See Problem 6.3-2.)

III. $\mathscr{K}_{\mu,a}$ is a dense subspace of $\mathscr{E}(I)$, whatever be the choices of μ and a. Indeed, $\mathscr{D}(I) \subset \mathscr{K}_{\mu,a} \subset \mathscr{E}(I)$, and, since $\mathscr{D}(I)$ is dense in $\mathscr{E}(I)$, so too is $\mathscr{K}_{\mu,a}$. Moreover, in the proof of Lemma 6.3-1 we established that the convergence of any sequence in $\mathscr{K}_{\mu,a}$ implies its convergence in $\mathscr{E}(I)$. Consequently, by Corollary 1.8-2a, $\mathscr{E}'(I)$ is a subspace of $\mathscr{K}'_{\mu,a}$ for any permissible values of μ and a.

IV. In our present context Theorem 1.8-1 states the following: For each $f \in \mathscr{K}'_{\mu,a}$, there exists a nonnegative integer r and a positive constant C such that

$$|\langle f, \phi \rangle| \leq C \max_{0 \leq k \leq r} \rho^{\mu}_{a,k}(\phi)$$

for every $\phi \in \mathscr{K}_{\mu,a}$. Here, C and r depend on f but not on ϕ.

V. The differential operator S_{μ} is a continuous linear mapping of $\mathscr{K}_{\mu,a}$ into $\mathscr{K}_{\mu,a}$, as is implied by the equation:

$$\rho^{\mu}_{a,k}(S_{\mu}\phi) = \rho^{\mu}_{a,k+1}(\phi) \qquad \phi \in \mathscr{K}_{\mu,a}$$

Thus, the adjoint S_{μ}' of S_{μ} is a generalized differential operator on $\mathscr{K}'_{\mu,a}$ into $\mathscr{K}'_{\mu,a}$ defined by

(5) $$\langle S_{\mu}'f, \phi \rangle \triangleq \langle f, S_{\mu}\phi \rangle$$

Since $S_{-\mu} = S_{\mu}$ (see Sec. 6.1, Eq. (5)), we also have that $S'_{-\mu} = S_{\mu}'$. Because of the symmetry of S_{μ} and our notational convention described in Sec. 2.5, we shall henceforth denote the generalized differential operator S_{μ}' by S_{μ}, thereby dropping the prime. As before, the symbol S_{μ} denotes either a conventional or a generalized differential operator depending on the way in which it is used. If it acts on a testing function in $\mathscr{K}_{\mu,a}$, it is a conventional operator; if it acts on a generalized function in $\mathscr{K}'_{\mu,a}$, it is a generalized operator (i.e., the adjoint of the conventional one).

VI. Let a be any real number, and let $f(t)$ be a locally integrable function on $0 < t < \infty$ such that $f(t)e^{-at}/j_{\mu}(t)$ is absolutely integrable on $0 < t < \infty$. ($j_{\mu}(t)$ is defined at the beginning of this section.) Then, $f(t)$ generates a regular member of $\mathscr{K}'_{\mu,a}$, which as usual we also denote by f, through the definition:

(6) $$\langle f, \phi \rangle \triangleq \int_0^\infty f(t)\phi(t)\,dt \qquad \phi \in \mathscr{K}_{\mu,a}$$

That f is truly a member of $\mathscr{K}'_{\mu,a}$ follows from the inequality:

(7) $$|\langle f, \phi \rangle| \leq \rho^{\mu}_{a,0}(\phi)\int_0^\infty \left| \frac{e^{-at}f(t)}{j_{\mu}(t)} \right| dt \qquad \phi \in \mathscr{K}_{\mu,a}$$

If $a > 0$ and either $\mu = 0$ or $0 < \operatorname{Re}\mu < 1$, then $\mathscr{K}_{\mu,a}$ can be identified with a subspace of $\mathscr{K}'_{\mu,a}$. Indeed, for any $\psi \in \mathscr{K}_{\mu,a}$,

(8) $$\frac{e^{-at}\psi(t)}{j_{\mu}(t)} = \frac{e^{-2at}}{[j_{\mu}(t)]^2} e^{at}j_{\mu}(t)\psi(t)$$

Under our assumptions on a and μ, $e^{-2at}/[j_{\mu}(t)]^2$ is absolutely integrable on $0 < t < \infty$. Moreover, $e^{at}j_{\mu}(t)\psi(t)$ is continuous and bounded on $0 < t < \infty$.

Therefore, (8) is also absolutely integrable on $0 < t < \infty$, so that each $\psi \in \mathscr{K}_{\mu, a}$ truly generates a unique regular member of $\mathscr{K}'_{\mu, a}$. On the other hand, two different members of $\mathscr{K}_{\mu, a}$ cannot generate the same member of $\mathscr{K}'_{\mu, a}$. For, if $\psi(t)$ and $\theta(t)$ are both in $\mathscr{K}_{\mu, a}$ and differ somewhere, we can choose a nonnegative testing function $\phi(t) \in \mathscr{D}(I)$, whose support is contained in an interval on which $\psi - \theta$ is never equal to zero. In this case, $\langle \psi, \phi \rangle \neq \langle \theta, \phi \rangle$ (assuming that $\phi(t) \not\equiv 0$). This verifies that ψ and θ, taken as members of $\mathscr{K}'_{\mu, a}$, must differ. So truly, if $a > 0$ and either $\mu = 0$ or $0 < \operatorname{Re} \mu < 1$, $\mathscr{K}_{\mu, a}$ can be identified with a subspace of $\mathscr{K}'_{\mu, a}$, and we write $\mathscr{K}_{\mu, a} \subset \mathscr{K}'_{\mu, a}$.

As a simple example, $\sqrt{t}\, K_\mu(t)$ is a regular member of $\mathscr{K}'_{\mu, a}$ under the stated restrictions on a and μ. (In this particular case, we need merely require that $a > -1$.)

PROBLEM 6.3-1. (*a*) Show that $\mathscr{D}(I)$ is not dense in $\mathscr{K}_{\mu, a}$. *Hint*: Let $\operatorname{Re} \mu > 0$, and let $\lambda(t)$ be a smooth function on $0 < t < \infty$ such that

$$\lambda(t) = \begin{cases} t^{1/2-\mu} & 0 < t < 1 \\ 0 & 2 < t < \infty \end{cases}$$

Show that λ is a member of $\mathscr{K}_{\mu, a}$ and that there is a balloon around λ which contains no members of $\mathscr{D}(I)$. Proceed similarly for $\mu = 0$.

(*b*) Show that there exist two different members of $\mathscr{K}'_{\mu, a}$ whose restrictions to $\mathscr{D}(I)$ coincide. *Hint*: See Example 3.2-1.

PROBLEM 6.3-2. Let $a < b$. Show that $\mathscr{K}_{\mu, b}$ is not dense in $\mathscr{K}_{\mu, a}$ and that there exist two different members of $\mathscr{K}'_{\mu, a}$ whose restrictions to $\mathscr{K}_{\mu, b}$ coincide. *Hint*: For the first part construct a member of $\mathscr{K}_{\mu, a}$ that is identically equal to $e^{-at}t^{-\mu+1/2}$ on $2 < t < \infty$.

PROBLEM 6.3-3. Let $0 < \operatorname{Re} \mu < 1$. Define the functionals g_k on $\mathscr{K}_{\mu, a}$ by

$$\langle g_k, \phi \rangle \triangleq \lim_{t \to 0+} \{ t^{-\mu+1} K_\mu(t) D[t^{\mu-1/2} S_\mu{}^k \phi(t)] + \sqrt{t} K_{\mu-1}(t) S_\mu{}^k \phi(t) \}$$

$$k = 0, 1, 2, \ldots; \qquad \phi \in \mathscr{K}_{\mu, a}$$

Show that these functionals exist and are members of $\mathscr{K}'_{\mu, a}$. Here, a may assume any real value. *Hint*: Let $\lambda(t)$ be a smooth function on $0 < t < \infty$ such that $\lambda(t) \equiv 1$ for $0 < t < 1$ and $\lambda(t) \equiv 0$ for $2 < t < \infty$. Apply two integrations by parts to the expression:

$$\int_\varepsilon^{\infty-} \sqrt{t}\, K_\mu(t) \lambda(t) S_\mu^{k+1} \phi(t)\, dt \qquad \varepsilon > 0$$

and use the fact that $S_\mu = S_{-\mu}$.

PROBLEM 6.3-4. Let a be any real number. Define the functionals h_k on $\mathscr{K}_{0,a}$, by

$$\langle h_k, \phi\rangle \triangleq \lim_{t\to 0+} \{tK_0(t)D[t^{-1/2}S_0{}^k\phi(t)] + t^{1/2}K_1(t)S_0{}^k\phi(t)\}$$

Show that the h_k exist and are members of $\mathscr{K}'_{0,a}$.

PROBLEM 6.3-5. Let $0 < \operatorname{Re}\mu < 1$. Demonstrate that the generalized differential operator S_μ is truly different from the corresponding conventional one in the following way. First note that in the conventional sense $S_\mu\sqrt{t}\,K_\mu(t) = \sqrt{t}\,K_\mu(t)$. (See Sec. 6.2, Eq. (9).) On the other hand, let $\sqrt{t}\,K_\mu(t)$ now denote a regular member of $\mathscr{K}'_{\mu,a}(a > 0)$ as discussed in note VI, and show that in the generalized sense $S_\mu\sqrt{t}\,K_\mu(t) = f + \sqrt{t}\,K_\mu(t)$. Here, f is a nonzero member of $\mathscr{K}'_{\mu,a}$ arising from the lower limit terms when two integrations by parts are applied to the right-hand side of the definition (5). Show that f truly is a member of $\mathscr{K}'_{\mu,a}$.

PROBLEM 6.3-6. Let $a > 0$ and either $\mu = 0$ or $0 < \operatorname{Re}\mu < 1$. Show that the topology of $\mathscr{K}_{\mu,a}$ is stronger than the topology induced on it by $\mathscr{K}'_{\mu,a}$.

6.4. The K Transformation

We shall call f a $\mathfrak{K}_\mu$-*transformable generalized function* if it is a member of $\mathscr{K}'_{\mu,a}$ for some real number a. According to Sec. 6.3, note II, f is then a member of $\mathscr{K}'_{\mu,b}$ for every $b > a$. This implies that there exists a real number σ_f (possibly $\sigma_f = -\infty$) such that $f \in \mathscr{K}'_{\mu,a}$ for every $a > \sigma_f$ and $f \notin \mathscr{K}'_{\mu,a}$ for every $a < \sigma_f$.

Since $\sqrt{st}\,K_\mu(st) \in \mathscr{K}_{\mu,a}$ for every fixed s such that $s \neq 0$ and $\operatorname{Re} s > a$, we may define the (*generalized*) μ*th order* K *transform* of f by

$$F(s) \triangleq (\mathfrak{K}_\mu f)(s) \triangleq \langle f(t), \sqrt{st}\,K_\mu(st)\rangle \qquad s \in \Omega_f \tag{1}$$

where Ω_f denotes the region:

$$\Omega_f \triangleq \{s\colon \operatorname{Re} s > \sigma_f,\ s \neq 0,\ -\pi < \arg s < \pi\} \tag{2}$$

When $\sigma_f < 0$, Ω_f is a cut half-plane obtained by deleting all real nonpositive values of s. The definition (1) has a sense as the application of $f(t) \in \mathscr{K}'_{\mu,a}$ to $\sqrt{st}\,K_\mu(st) \in \mathscr{K}_{\mu,a}$ where a is any real number such that $\sigma_f < a < \operatorname{Re} s$. The number σ_f will be called the *abscissa of definition*, and Ω_f the *region of definition* for the K transform of f. Moreover, we will refer to the operation $\mathfrak{K}_\mu\colon f \mapsto F$ as the (*generalized*) K *transformation of order* μ.

As usual, whenever, we write $F(s) = \mathfrak{K}_\mu f$ for $s \in \Omega_f$, it is understood that f is a $\mathfrak{K}_\mu$-transformable generalized function, where μ is either the number zero or a complex number with positive real part, and that $F(s)$ is defined by (1) and Ω_f by (2).

EXAMPLE 6.4-1. As a simple example of a generalized K transform, let us determine the transform of the generalized function $S_\mu{}^k\delta(t-\tau)$ where k is a nonnegative integer and τ is a fixed positive number. $S_\mu{}^k\delta(t-\tau)$ is a member of $\mathscr{E}'(I)$ and therefore of $\mathscr{K}'_{\mu,a}$ for every μ and a according to Sec. 6.3, note *III*. Thus,

$$\begin{aligned}[\mathfrak{K}_\mu S^k_{\mu,t}\delta(t-\tau)](s) &= \langle S^k_{\mu,t}\delta(t-\tau), \sqrt{st}\,K_\mu(st)\rangle \\ &= \langle \delta(t-\tau), S^k_{\mu,t}\sqrt{st}\,K_\mu(st)\rangle \\ &= s^{2k}\sqrt{s\tau}\,K_\mu(s\tau) \qquad s\neq 0, \qquad -\pi<\arg s<\pi\end{aligned}$$

Note that, on any vertical line (i.e., with $\operatorname{Re} s$ fixed), the magnitude of this transform grows according to $|s|^{2k}$ by virtue of the asymptotic relation Sec. 6.2, Eq. (4).

Under certain conditions, the generalized K transformation contains the conventional K transformation as a special case. In particular, if $f(t)$ is a locally integrable function such that, for some a, $f(t)e^{-at}/j_\mu(t)$ is absolutely integrable on $0<t<\infty$, then, according to Sec. 6.3, note *VI*, f generates a regular member of $\mathscr{K}'_{\mu,a}$ whose K transform (1) takes on the form:

$$F(s)=\int_0^\infty f(t)\sqrt{st}\,K_\mu(st)\,dt \qquad \operatorname{Re} s>a, \quad s\neq 0, \quad -\pi<\arg s<\pi$$

THEOREM 6.4-1. *If $F(s)=\mathfrak{K}_\mu f$ for $s\in\Omega_f$, then, for any positive integer k,*

$$\mathfrak{K}_\mu S_\mu{}^k f=\mathfrak{K}_\mu S^k_{-\mu}f=s^{2k}F(s) \qquad s\in\Omega_f$$

PROOF. By Sec. 6.3, note *V* and Sec. 6.2, Eq. (9),

$$\begin{aligned}\mathfrak{K}_\mu S_\mu{}^k f &= \langle f(t), S_\mu{}^k\sqrt{st}\,K_\mu(st)\rangle \\ &= \langle f(t), s^{2k}\sqrt{st}\,K_\mu(st)\rangle \\ &= s^{2k}F(s) \qquad s\in\Omega_f\end{aligned}$$

Since $S_\mu=S_{-\mu}$, the proof is complete.

PROBLEM 6.4-1. Compute the K transform of the generalized functions specified in Problems 6.3-3 and 6.3-4.

PROBLEM 6.4-2. Let

$$f(t)=\begin{cases}0 & 0<t<T\\ t^{\mu+1/2} & T\le t<\infty\end{cases}$$

Show that

$$(\mathfrak{K}_\mu f)(s) = T^{\mu+1}s^{-1/2}K_{\mu+1}(sT) \qquad \operatorname{Re} s > 0$$

by using Sec. 6.2, Eq. (7). Then, use Theorem 6.4-1 with $k = 1$ to establish another formula.

PROBLEM 6.4-3. Show that

$$\sqrt{st}\, K_{1/2}(st) = \sqrt{\frac{\pi}{2}}\, e^{-st}$$

This allows us to define a generalized Laplace transformation corresponding to the conventional one-sided transform:

$$F(s) = \int_0^\infty f(t)e^{-st}\, dt$$

by means of the definition:

$$F(s) = \sqrt{\frac{2}{\pi}}\, \mathfrak{K}_{1/2} f$$

whenever f is a $\mathfrak{K}_{1/2}$-transformable generalized function. How does the testing function space $\mathscr{K}_{1/2,\,a}$ compare with the space $\mathscr{L}_{+,\,a}$ introduced in Problem 3.10-13?

PROBLEM 6.4-4. Let α be a complex number not equal to zero. Show that

$$[\mathfrak{K}_\mu \sqrt{\alpha t}\, I_\mu(\alpha t)](s) = \frac{\alpha^\mu s^{-\mu}}{s^2 - \alpha^2} \qquad \operatorname{Re} s > \operatorname{Re} \alpha$$

PROBLEM 6.4-5. Let a be a fixed positive number. Show that

$$\mathfrak{K}_\mu\, Pf\, \frac{1_+(t-a)}{t-a} = \int_a^{a+1} \frac{\sqrt{st}\, K_\mu(st) - \sqrt{sa}\, K_\mu(sa)}{t-a}\, dt + \int_{a+1}^\infty \frac{\sqrt{st}\, K_\mu(st)}{t-a}\, dt$$

where Pf denotes as usual a pseudofunction (see Zemanian [1], Secs. 1.4 and 2.5).

PROBLEM 6.4-6. Let $f \in \mathscr{E}'(I)$. Show that, for all s not equal to real nonpositive values,

$$(a) \qquad [\mathfrak{K}_\mu t^{-1} f(t)](s) = -\frac{s}{2\mu}[(\mathfrak{K}_{\mu-1} f)(s) - (\mathfrak{K}_{\mu+1} f)(s)]$$

$$(b) \qquad [\mathfrak{K}_\mu t^{-1/2} D_t t^{1/2} f(t)](s) = -\frac{s}{2}[(\mathfrak{K}_{\mu-1} f)(s) + (\mathfrak{K}_{\mu+1} f)(s)]$$

(c) $$[\mathfrak{K}_\mu t^{\mu-1/2} D_t t^{-\mu+1/2} f(t)](s) = -s(\mathfrak{K}_{\mu-1} f)(s)$$

(d) $$[\mathfrak{K}_\mu t^{-\mu-1/2} D_t t^{\mu+1/2} f(t)](s) = -s(\mathfrak{K}_{\mu+1} f)(s)$$

6.5. The Analyticity of a K Transform

A K transform is analytic on its region of definition. To show this we shall need some inequalities.

LEMMA 6.5-1. *Let $\mu_R \triangleq \operatorname{Re} \mu > 0$, and let a and b be real numbers with $a < b$. Then, for $\operatorname{Re} \zeta \geq b$, $\zeta \neq 0$, $-\pi < \arg \zeta \leq \pi$, and $0 < t < \infty$, we have that*

(1) $$|e^{at}(\zeta t)^\mu K_\mu(\zeta t)| < A_\mu(1 + |\zeta|^{\mu_R})$$

where A_μ is a constant with respect to ζ and t.

PROOF. Recall that $K_\mu(z)$ is analytic everywhere except at $z = 0$ and $z = \infty$. In this proof assume that $z \neq 0$ and $-\pi < \arg z \leq \pi$. By the series expansions Sec. 6.2, Eqs. (2) and (3), there exists a constant B_μ such that

$$|z^\mu K_\mu(z)| < B_\mu \qquad |z| \leq 1$$

On the other hand, by the asymptotic expansion Sec. 6.2, Eq. (4), there exists a constant C_μ such that

$$|z^\mu K_\mu(z)| = |z^{\mu-1/2}||\sqrt{z}\, K_\mu(z)| < C_\mu |z|^{\mu_R} e^{-\operatorname{Re} z} \qquad |z| > 1$$

Consequently, for all permissible z,

$$|z^\mu K_\mu(z)| < E_\mu(1 + |z|^{\mu_R}) e^{-\operatorname{Re} z}$$

where E_μ is still another constant. So, for ζ and t restricted as stated,

$$|e^{at}(\zeta t)^\mu K_\mu(\zeta t)| < E_\mu(1 + |\zeta t|^{\mu_R}) e^{(a - \operatorname{Re} \zeta)t}$$
$$< E_\mu(1 + |\zeta|^{\mu_R})(1 + t^{\mu_R}) e^{(a - \operatorname{Re} \zeta)t}$$

Since $b > a$, $(1 + t^{\mu_R}) e^{(a - \operatorname{Re} \zeta)t}$ is bounded on $0 < t < \infty$ uniformly for all $\operatorname{Re} \zeta \geq b$. This proves our lemma.

LEMMA 6.5-2. *Let a and ε be fixed real numbers with $\varepsilon > 0$. Then, for all ζ such that $|\zeta| \geq \varepsilon$, $\operatorname{Re} \zeta \geq a$, and $-\pi < \arg \zeta \leq \pi$, and for $0 < t < \infty$, we have that*

(2) $$\left| \frac{e^{at} K_0(\zeta t)}{h(t)} \right| < A_0$$

where A_0 is a constant with respect to ζ and t.

PROOF. The series expansion Sec. 6.2, Eq. (3), the asymptotic behavior as $z \to \infty$ Sec. 6.2, Eq. (4), and the analyticity of $K_0(z)$ show that

$$\left|\frac{K_0(z)}{h(|z|)}\right| < Ce^{-\operatorname{Re} z} \qquad z \neq 0, \qquad -\pi < \arg z \leq \pi \tag{3}$$

where C is a constant with respect to z. Moreover, it follows from the definition of $h(t)$ that

$$\left|\frac{h(|\zeta|t)}{h(t)}\right| < B \qquad |\zeta| \geq \varepsilon, \qquad 0 < t < \infty \tag{4}$$

where B is also a constant with respect to ζ and t. Consequently, for ζ and t restricted as stated,

$$\left|\frac{e^{at}K_0(\zeta t)}{h(t)}\right| = \left|\frac{e^{at}K_0(\zeta t)}{h(|\zeta|t)} \cdot \frac{h(|\zeta|t)}{h(t)}\right|$$
$$< CBe^{(a-\operatorname{Re}\zeta)t} \leq CB = A_0 \qquad \text{Q.E.D}$$

THEOREM 6.5-1. *Let $F(s) = \mathfrak{K}_\mu f$ for $s \in \Omega_f$. Then, $F(s)$ is an analytic function on Ω_f, and*

$$D_sF(s) = \langle f(t), D_s\sqrt{st}\,K_\mu(st)\rangle \qquad s \in \Omega_f \tag{5}$$

PROOF. Let s be an arbitrary but fixed point in Ω_f. Choose the real positive numbers a, b, r, and r_1 such that

$$\sigma_f < a < b = \operatorname{Re} s - r_1 < \operatorname{Re} s - r < \operatorname{Re} s.$$

Also, let C denote the circle whose center is at s and whose radius is r_1. Restrict r_1 (and thereby h and r) still further by requiring that C lie entirely within Ω_f (i.e., C does not intersect the real nonpositive axis). Finally, let Δs be a nonzero complex increment such that $|\Delta s| < r$, and consider the expression:

$$\frac{F(s+\Delta s) - F(s)}{\Delta s} - \langle f(t), D_s\sqrt{st}\,K_\mu(st)\rangle = \langle f(t), \psi_{\Delta s}(t)\rangle \tag{6}$$

where

$$\psi_{\Delta s}(t) = \frac{\sqrt{st + \Delta st}\,K_\mu(st + \Delta st) - \sqrt{st}\,K_\mu(st)}{\Delta s} - D_s\sqrt{st}\,K_\mu(st)$$

The series expansion and asymptotic behavior of $K_\mu(st)$ show that $D_s\sqrt{st}\,K_\mu(st)$ is a member of $\mathscr{K}_{\mu,a}$ so that (5) and (6) have a sense.

Since

$$S_{\mu,t}^k\sqrt{st}\,K_\mu(st) = s^{2k}\sqrt{st}\,K_\mu(st)$$

and

$$S_{\mu,t}^k D_s \sqrt{st}\, K_\mu(st) = D_s s^{2k} \sqrt{st}\, K_\mu(st)$$

$S_{\mu,t}^k \psi_{\Delta s}(t)$ can be written as a closed integral on C by using Cauchy's integral formulas. This yields

$$S_{\mu,t}^k \psi_{\Delta s}(t) = \frac{1}{2\pi i} \int_C \zeta^{2k} \sqrt{\zeta t}\, K_\mu(\zeta t) \left[\frac{1}{\Delta s} \left(\frac{1}{\zeta - s - \Delta s} - \frac{1}{\zeta - s} \right) - \frac{1}{(\zeta - s)^2} \right] d\zeta$$

$$= \frac{\Delta s}{2\pi i} \int_C \frac{\zeta^{2k} \sqrt{\zeta t}\, K_\mu(\zeta t)}{(\zeta - s)^2 (\zeta - s - \Delta s)}\, d\zeta$$

Next, assume that $\operatorname{Re} \mu > 0$, and let Q_μ be a constant bound on $e^{at}(\zeta t)^\mu K_\mu(\zeta t)$ for $0 < t < \infty$ and for all $\zeta \in C$ (see Lemma 6.5-1). We may write

$$|e^{at} t^{\mu - 1/2} S_\mu{}^k \psi_{\Delta s}(t)| \leq \frac{|\Delta s| Q_\mu}{2\pi} \int_C \left| \frac{\zeta^{2k+1/2-\mu}}{(\zeta - s)^2 (\zeta - s - \Delta s)} \right| d\zeta$$

$$\leq \frac{|\Delta s| Q_\mu}{r_1(r_1 - r)} \sup_{\zeta \in C} |\zeta^{2k+1/2-\mu}|$$

$$\to 0 \qquad |\Delta s| \to 0$$

This proves that $\psi_{\Delta s}(t)$ converges in $\mathscr{K}_{\mu,a}$ to zero as $|\Delta s| \to 0$. Since $f \in \mathscr{K}'_{\mu,a}$, (6) implies (5), and the theorem is established for $\operatorname{Re} \mu > 0$.

In the case where $\mu = 0$, the argument presented in the preceding paragraph is almost the same. Now, the factor $t^{\mu-1/2}$ is replaced by $[\sqrt{t}\, h(t)]^{-1}$, and Lemma 6.5-1 is replaced by Lemma 6.5-2 with ε chosen small enough.

Q.E.D.

PROBLEM 6.5-1. Let $f \in \mathscr{E}'(I)$. Show that, for all s not equal to real nonpositive values,

$$[\mathfrak{K}_\mu t f(t)](s) = -s^{\mu-1/2} D_s s^{-\mu+1/2} (\mathfrak{K}_{\mu-1} f)(s)$$
$$= -s^{-\mu-1/2} D_s s^{\mu+1/2} (\mathfrak{K}_{\mu+1} f)(s)$$

6.6. Inversion

The bulk of this section is devoted to proving an inversion formula, which determines the restriction to $\mathscr{D}(I)$ of any $\mathfrak{K}_\mu$-transformable generalized function from its K transform. From this we will obtain an incomplete version of a uniqueness theorem, which states that two $\mathfrak{K}_\mu$-transformable generalized functions having the same transform must have the same restriction to $\mathscr{D}(I)$.

LEMMA 6.6-1. *Let $\mathfrak{K}_\mu f = F(s)$ for $s \in \Omega_f$, let $\phi \in \mathscr{D}(I)$, and set*

$$\Phi(s) = \int_0^\infty \phi(t)\sqrt{st}\, I_\mu(st)\, dt \qquad \operatorname{Re} s > 0$$

Then, for any fixed real number r with $0 < r < \infty$,

$$(1) \quad \int_{-r}^{r} \Phi(s)\langle f(\tau), \sqrt{s\tau}\, K_\mu(s\tau)\rangle d\omega = \left\langle f(\tau), \int_{-r}^{r} \Phi(s)\sqrt{s\tau}\, K_\mu(s\tau)\, d\omega \right\rangle$$

where $s = \sigma + i\omega$ and $\sigma = \operatorname{Re} s$ is fixed with $\sigma > \max(0, \sigma_f)$.

(As always, σ_f denotes the abscissa of definition for $\mathfrak{K}_\mu f$.)

PROOF. Our conclusion is obvious when $\phi(t) \equiv 0$. So, assume $\phi(t) \not\equiv 0$. We shall first show that, for every real number a such that $\max(0, \sigma_f) < a < \sigma$,

$$(2) \qquad V(\tau) \triangleq \int_{-r}^{r} \Phi(s)\sqrt{s\tau}\, K_\mu(s\tau)\, d\omega$$

is a member of $\mathscr{K}_{\mu,a}$; this will insure that the right-hand side of (1) has a sense. We assume that $\operatorname{Re} \mu > 0$, the argument for the case $\mu = 0$ being almost the same. By the smoothness of the integrand of (2), we may carry the operator $S_{\mu,\tau}^k$ under the integral sign in (2) to write

$$(3) \qquad |e^{a\tau}\tau^{\mu-1/2}S_\mu{}^k V(\tau)| = \left| \int_{-r}^{r} \Phi(s)s^{2k+(1/2)-\mu}e^{a\tau}(s\tau)^\mu K_\mu(s\tau)\, d\omega \right|$$

$$\leq B\int_{-r}^{r} |\Phi(s)s^{2k+(1/2)-\mu}|\, d\omega < \infty$$

where B is a constant bound on $e^{a\tau}(s\tau)^\mu K_\mu(s\tau)$ for $0 < \tau < \infty$ and s on the straight line connecting the points $\sigma - ir$ and $\sigma + ir$ (see Lemma 6.5-1). This proves that $V(\tau) \in \mathscr{K}_{\mu,a}$. (When $\mu = 0$, Lemma 6.5-2 should be used here.)

Next, we construct the following Riemann sum for the integral (2).

$$J(\tau, m) \triangleq \frac{r}{m}\sum_{p=-m}^{m-1} \Phi\left(\sigma + \frac{ipr}{m}\right)\sqrt{\sigma\tau + \frac{ipr\tau}{m}}\, K_\mu\left(\sigma\tau + \frac{ipr\tau}{m}\right)$$

Upon applying $f(\tau)$ to this term by term, we get another Riemann sum which converges to the left-hand side of (1) as $m \to \infty$ by virtue of the continuity of the integrand on $-r \leq \omega \leq r$.

Since $f \in \mathscr{K}'_{\mu,a}$, our lemma will therefore be proven when we show that $J(\tau, m)$ converges in $\mathscr{K}_{\mu,a}$ to (2) as $m \to \infty$. Again, we do this only for the case where Re $\mu > 0$ and leave it to the reader to supply the few modifications needed for the case $\mu = 0$. Set

$$H(\tau, m) \triangleq e^{a\tau}\tau^{\mu-1/2}S^k_{\mu,\tau}[V(\tau) - J(\tau, m)] \tag{4}$$

$$= e^{a\tau}\int_{-r}^{r} \Phi(s)s^{2k+(1/2)-\mu}(s\tau)^{\mu}K_{\mu}(s\tau)\,ds$$

$$- e^{a\tau}\frac{r}{m}\sum_{p=-m}^{m-1} \Phi(s)s^{2k+(1/2)-\mu}(s\tau)^{\mu}K_{\mu}(s\tau)|_{s=\sigma+\frac{ipr}{m}}$$

We have to show that $H(\tau, m)$ converges uniformly to zero on $0 < \tau < \infty$ as $m \to \infty$.

Observe that

$$e^{a\tau}s^{2k+(1/2)-\mu}(s\tau)^{\mu}K_{\mu}(s\tau) \tag{5}$$

tends uniformly to zero on $-r \leq \omega \leq r$ as $\tau \to \infty$ by virtue of the asymptotic behavior of $K_\mu(s\tau)$ and the condition $\sigma =$ Re $s > a$. Consequently, given an $\varepsilon > 0$, there exists a T such that, for $\tau > T$ and $-r \leq \omega \leq r$, (5) is bounded by

$$\frac{\varepsilon}{3}\left[\int_{-r}^{r} |\Phi(s)|\,d\omega\right]^{-1}$$

which is a finite quantity since $\phi(t) \not\equiv 0$. It follows that

$$\sup_{\tau > T} |e^{a\tau}\tau^{\mu-1/2}S_\mu{}^k V(\tau)| < \frac{\varepsilon}{3}$$

Also, for all m,

$$\sup_{\tau > T} |e^{a\tau}\tau^{\mu-1/2}S_\mu{}^k J(\tau, m)| < \frac{\varepsilon}{3}\left[\int_{-r}^{r} |\Phi(s)|\,d\omega\right]^{-1}\frac{r}{m}\sum_{p=-m}^{m-1}\left|\Phi\left(\sigma + \frac{ipr}{m}\right)\right|$$

Thus, there exists an m_0 such that, for $m > m_0$, the right-hand side is bounded by $2\varepsilon/3$. We have thus shown that, for $m > m_0$ and $\tau > T$, $|H(\tau, m)| < \varepsilon$.

Next, we consider the range $0 < \tau \leq T$. It is a fact that, with σ fixed as before, $(s\tau)^\mu K_\mu(s\tau)$ is a uniformly continuous function of (τ, ω) for $0 < \tau \leq T$ and $-r \leq \omega \leq r$. Indeed, $K_\mu(z)$ is analytic for Re $z > 0$. Moreover, by its series expansion, $(s\tau)^\mu K_\mu(s\tau)$ converges, as $\tau \to 0+$, to $2^{\mu-1}\Gamma(\mu)$ when Re $\mu > 0$, this being so uniformly on $-r \leq \omega \leq r$. Thus, our assertion follows from the standard result that a function that is continuous on a closed bounded domain is uniformly continuous there. This result coupled with (4) shows that there exists an m_1 such that, for all $m > m_1$, $|H(\tau, m)| < \varepsilon$ on $0 < \tau \leq T$ as well. Thus, when $m > \max(m_0, m_1)$, $|H(\tau, m)| < \varepsilon$ on $0 < \tau < \infty$. This completes the proof of Lemma 6.6-1.

LEMMA 6.6-2. *Assume that* $\phi \in \mathscr{D}(I)$ *and that either* $\mu = 0$ *or* $\operatorname{Re} \mu > 0$. *For any real number* $a \geq 0$, *fix* $\sigma \triangleq \operatorname{Re} s$ *such that* $\sigma > a$. *Set* $s = \sigma + i\omega$ *and*

$$M_r(\tau) \triangleq \frac{1}{\pi} \int_{-r}^{r} \sqrt{s\tau}\, K_\mu(s\tau) \int_0^\infty \phi(t) \sqrt{st}\, I_\mu(st)\, dt\, d\omega \tag{6}$$

Then, $M_r(\tau)$ *converges in* $\mathscr{K}_{\mu,a}$ *to* $\phi(\tau)$ *as* $r \to \infty$.

PROOF. By the smoothness of the integrand of (6) and the fact that $\phi \in \mathscr{D}(I)$, we may repeatedly differentiate under the integral sign and make use of Sec. 6.2, Eqs. (9) and (10) to write

$$\begin{aligned} S^k_{\mu,\tau} M_r(\tau) &= \frac{1}{\pi} \int_{-r}^{r} S^k_{\mu,\tau} \sqrt{s\tau}\, K_\mu(s\tau) \int_0^\infty \phi(t) \sqrt{st}\, I_\mu(st)\, dt\, d\omega \\ &= \frac{1}{\pi} \int_{-r}^{r} \sqrt{s\tau}\, K_\mu(s\tau) \int_0^\infty \phi(t) s^{2k} \sqrt{st}\, I_\mu(st)\, dt\, d\omega \\ &= \frac{1}{\pi} \int_{-r}^{r} \sqrt{s\tau}\, K_\mu(s\tau) \int_0^\infty \phi(t) S^k_{\mu,t} \sqrt{st}\, I_\mu(st)\, dt\, d\omega \end{aligned} \tag{7}$$

Upon integrating by parts the last inner integral $2k$ times and noting that the limit terms are always equal to zero, we get that (7) is equal to

$$\frac{1}{\pi} \int_{-r}^{r} \sqrt{s\tau}\, K_\mu(s\tau) \int_0^\infty \sqrt{st}\, I_\mu(st) S^k_{\mu,t} \phi(t)\, dt\, d\omega$$

We may now reverse the order of integration and make use of Sec. 6.2, Eq. (11) to finally obtain

$$S^k_{\mu,\tau} M_r(\tau) = \int_0^\infty L_r(t, \tau) S^k_{\mu,t} \phi(t)\, dt \tag{8}$$

where

$$\begin{aligned} L_r(t, \tau) = \frac{1}{i\pi(t^2 - \tau^2)} &[t\sqrt{st} I_{\mu+1}(st) \sqrt{s\tau}\, K_\mu(s\tau) \\ &+ \tau \sqrt{st}\, I_\mu(st) \sqrt{s\tau}\, K_{\mu+1}(s\tau)]\Big|_{s=\sigma-ir}^{s=\sigma+ir} \end{aligned} \tag{9}$$

For a simpler notation we shall denote $S^k_{\mu,t}\phi(t)$ by $\phi_k(t)$.

Now assume that the support of $\phi(t)$ is contained in the closed interval $[A, B]$ where $0 < A < B < \infty$. Let the real number δ be such that $0 < \delta < A$ and let us break the integral (8) into

$$\begin{aligned} S^k_{\mu,\tau} M_r(\tau) &= V_1(\tau) + V_2(\tau) + V_3(\tau) \\ &= \int_0^{\tau-\delta} + \int_{\tau-\delta}^{\tau+\delta} + \int_{\tau+\delta}^{\infty} \end{aligned}$$

where V_1, V_2, and V_3 denote respectively integrals on the intervals $0 < t < \tau - \delta$, $\tau - \delta < t < \tau + \delta$, and $\tau + \delta < t < \infty$. We shall first show that

$$N_r(\tau) \triangleq e^{a\tau} j_\mu(\tau)[V_2(\tau) - \phi_k(\tau)]$$

converges uniformly to zero on $0 < \tau < \infty$ as $r \to \infty$. If either $\tau + \delta \leq A$ or $\tau - \delta \geq B$, then $V_2(\tau) \equiv 0$ and $\phi_k(\tau) \equiv 0$. Thus, we need merely consider the interval $A - \delta < \tau < B + \delta$. Moreover, we can restrict the interval of integration in (8) to $A < t < B$. Since $s = \sigma \pm ir$ where $\sigma > 0$ is fixed and since $r \to \infty$, we have that $|st| \geq |\sigma \pm ir| A \to \infty$ and $|s\tau| \geq |\sigma \pm ir|(A - \delta) \to \infty$. So, we may employ the asymptotic expressions Sec. 6.2, Eqs. (4), (5), and (6) to estimate $N_r(\tau)$. (Actually, when $s = \sigma + ir$, Sec. 6.2, Eq. (5) should be used, and when $s = \sigma - ir$, Sec. 6.2, Eq. (6) should be used.) This yields after some simplification

$$
\begin{aligned}
(10)\quad N_r(\tau) = {} & \frac{e^{a\tau} j_\mu(\tau)}{\pi} \int_{\tau-\delta}^{\tau+\delta} \frac{\sin(rt - r\tau)}{t - \tau}\, e^{\sigma t - \sigma\tau} \phi_k(t)\, dt \\
& + \frac{e^{a\tau} j_\mu(\tau)}{\pi} \int_{\tau-\delta}^{\tau+\delta} \frac{\sin(rt - r\tau)}{t - \tau}\, e^{\sigma t - \sigma\tau} \phi_k(t) \left[O\left(\frac{1}{|st|}\right) + O\left(\frac{1}{|s\tau|}\right) \right. \\
& \qquad \left. + O\left(\frac{1}{|st|}\right) O\left(\frac{1}{|s\tau|}\right) \right] dt \\
& - \frac{e^{a\tau} j_\mu(\tau)}{\pi} \left[1 + O\left(\frac{1}{|s\tau|}\right) \right] \int_{\tau-\delta}^{\tau+\delta} \frac{\cos(rt + r\tau - \mu\pi)}{t + \tau} \\
& \times e^{-\sigma t - \sigma\tau} \phi_k(t) \left[1 + O\left(\frac{1}{|st|}\right) \right] dt \\
& - e^{a\tau} j_\mu(\tau) \phi_k(\tau)
\end{aligned}
$$

Observe that the quantities represented by the order notations $O(|st|^{-1})$ and $O(|s\tau|^{-1})$ are bounded by $C/|\sigma \pm ir|$ for $t > A$ and $\tau > A - \delta$, where C is a sufficiently large constant.

Let $\Lambda \triangleq \{(t, \tau) : A < t < B,\ A - \delta < \tau < B + \delta\}$. For all $r > 1$, the integrand of the third term on the right-hand side of (10) is bounded on Λ by a constant independent of r. Therefore, given an $\varepsilon > 0$, we can choose δ so small, say, $\delta = \delta_1$, that this term is uniformly bounded on Λ by $\varepsilon/3$ for all $r > 1$.

Next,

$$\left| \frac{\sin(rt - r\tau)}{t - \tau}\, O(|st|^{-1}) \right| \leq \left| \frac{\sin(rt - r\tau)}{rt - r\tau} \right| \frac{Cr}{|\sigma \pm ir|} \leq C$$

and similarly

$$\left| \frac{\sin(rt - r\tau)}{t - \tau} O(|s\tau|^{-1}) \right| \leq C$$

It follows that the second term on the right-hand side of (10) is also uniformly bounded on Λ by $\varepsilon/3$ for all $r > 1$ when δ is chosen sufficiently small, say $\delta = \delta_2$.

Now, consider the sum of the first and last terms in (10). Through some more manipulation, this sum can be written as

$$(11) \quad \frac{1}{\pi} \int_{-\delta}^{\delta} H(x, \tau) \sin rx \, dx + e^{a\tau} j_\mu(\tau) \phi_k(\tau) \left[\frac{1}{\pi} \int_{-\delta r}^{\delta r} \frac{\sin y}{y} \, dy - 1 \right]$$

where

$$H(x, \tau) = e^{a\tau} j_\mu(\tau) \frac{\phi_k(x + \tau) e^{\sigma x} - \phi_k(\tau)}{x}$$

$H(x, \tau)$ is a continuous function of (x, τ) for $x + \tau > 0$ and $\tau > 0$ (assuming it has been defined by continuity at $x = 0$). Since supp $\phi(\tau) \subset [A, B]$, $H(x, \tau)$ is therefore bounded on the domain:

$$\{(x, \tau): -A/2 < x < A/2, \ A/2 < \tau < B + A/2\}$$

It follows that the magnitude of the first term in (11) can be made less than $\varepsilon/3$ for all $r > 1$ by choosing δ small enough, say, $\delta = \delta_3$. Henceforth in this proof, we fix $\delta = \min(\delta_1, \delta_2, \delta_3)$.

Clearly, the second term in (11) converges uniformly to zero on $0 < \tau < \infty$ as $r \to \infty$. Thus, we have shown that

$$(12) \qquad \overline{\lim_{r \to \infty}} N_r(\tau) \leq \varepsilon$$

Since $\varepsilon > 0$ is arbitrary, we conclude that $N_r(\tau)$ converges uniformly to zero on $0 < \tau < \infty$ as $r \to \infty$.

We next discuss

$$\begin{aligned} P_r(\tau) &\triangleq e^{a\tau} j_\mu(\tau) V_1(\tau) \\ &= e^{a\tau} j_\mu(\tau) \int_0^{\tau - \delta} L_r(t, \tau) \phi_k(t) \, dt \end{aligned}$$

For $\tau - \delta \leq A$, $P_r(\tau) \equiv 0$. So, we need merely consider the range $\tau - \delta > A$ Here again we may use the asymptotic expressions since

$$|st| \geq |\sigma \pm ir| A \to \infty$$

and

$$|s\tau| \geq |\sigma \pm ir|(A + \delta) \to \infty$$

as $r \to \infty$. As before, we obtain

(13)

$$P_r(\tau) = \frac{e^{a\tau - \sigma\tau} j_\mu(\tau)}{\pi} \left\{ \int_A^{\min(B,\tau-\delta)} \frac{e^{\sigma t}\phi_k(t)}{t-\tau} \sin(rt - r\tau)\,dt \right.$$

$$- \int_A^{\min(B,\tau-\delta)} \frac{e^{-\sigma t}\phi_k(t)}{t+\tau} \cos(rt + r\tau - \mu\pi)\,dt$$

$$+ \int_A^{\min(B,\tau-\delta)} \left[\frac{e^{\sigma t}\phi_k(t)}{t-\tau} \sin(rt - r\tau) + \frac{e^{-\sigma t}\phi_k(t)}{t+\tau} \cos(rt + r\tau - \mu\pi) \right]$$

$$\left. \times \left[O\left(\frac{1}{|st|}\right) + O\left(\frac{1}{|s\tau|}\right) + O\left(\frac{1}{|st|}\right) O\left(\frac{1}{|s\tau|}\right) \right] dt \right\}$$

First note that, since $a < \sigma$, $e^{(a-\sigma)\tau} j_\mu(\tau)$ is a bounded function for $\tau - \delta > A$. Similarly, the quantity:

$$\frac{e^{\sigma t}\phi_k(t)}{t-\tau} \sin(rt - r\tau) + \frac{e^{-\sigma t}\phi_k(t)}{t+\tau} \cos(rt + r\tau - \mu\pi)$$

is bounded on the domain

$$\Theta \triangleq \{(t, \tau) \colon A < \tau - \delta,\ A < t < \min(B, \tau - \delta)\}$$

Therefore, the last term within the braces in (13) (i.e., the term containing the order notations) converges uniformly to zero for $A < \tau - \delta$ as $r \to \infty$.

As the next step, we integrate by parts the first term within the braces in (13) to get

$$(14) \qquad - \left. \frac{e^{\sigma t}\phi_k(t) \cos(rt - r\tau)}{rt - r\tau} \right|_A^{\min(B,\tau-\delta)}$$

$$+ \frac{1}{r} \int_A^{\min(B,\tau-\delta)} \left[D_t \frac{e^{\sigma t}\phi_k(t)}{t-\tau} \right] \cos(rt - r\tau)\,dt$$

The lower limit term is zero. So is the upper limit term if $B \le \tau - \delta$. On the other hand, if $\tau - \delta < B$, the upper limit term is bounded by

$$(r\delta)^{-1} \sup_{A<t<B} |e^{\sigma t}\phi_k(t)|$$

Consequently, this upper limit term converges uniformly to zero for $A < \tau - \delta$ as $r \to \infty$. Moreover,

$$D_t \frac{e^{\sigma t}\phi_k(t)}{t-\tau}$$

is also bounded on the domain Θ, which implies that the second term in (14) also converges uniformly to zero for $A < \tau - \delta$ as $r \to \infty$.

A similar argument using an integration by parts may be applied to the second term within the braces in (13). All this proves that, as $r \to \infty$, $P_r(\tau)$ converges uniformly to zero on $0 < \tau < \infty$,

Finally, we discuss

$$Q_r(\tau) \triangleq e^{a\tau} j_\mu(\tau) V_3(\tau)$$
$$= e^{a\tau} j_\mu(\tau) \int_{\tau+\delta}^{\infty} L_r(t, \tau) \phi_k(t)\, dt$$

For $\tau + \delta \geq B$, $Q_r(\tau) \equiv 0$. So, we need merely consider the range $0 < \tau < B - \delta$. Set

$$Q_r(\tau) = Q_{1,r}(\tau) + Q_{2,r}(\tau) - Q_{3,r}(\tau) - Q_{4,r}(\tau)$$

where

(15) $Q_{1,r}(\tau) =$

$$e^{a\tau} j_\mu(\tau) \sqrt{s\tau}\, K_\mu(s\tau) \frac{1}{i\pi} \int_{\max(A,\, \tau+\delta)}^{B} \frac{t\phi_k(t)}{t^2 - \tau^2} \sqrt{st}\, I_{\mu+1}(st)\, dt \Big|_{s=\sigma+ir}$$

(16) $Q_{2,r}(\tau) =$

$$e^{a\tau} j_\mu(\tau) \tau \sqrt{s\tau}\, K_{\mu+1}(s\tau) \frac{1}{i\pi} \int_{\max(A,\, \tau+\delta)}^{B} \frac{\phi_k(t)}{t^2 - \tau^2} \sqrt{st}\, I_\mu(st)\, dt \Big|_{s=\sigma+ir}$$

and $Q_{3,r}$ and $Q_{4,r}$ are respectively the same as $Q_{1,r}$ and $Q_{2,r}$ except that $s = \sigma - ir$.

According to Lemma 6.5-1, for Re $s = \sigma > a \geq 0$, Im $s = r > 1$, $\mu_R \triangleq$ Re $\mu > 0$, and $0 < \tau < \infty$,

$$(17) \qquad |e^{a\tau} j_\mu(\tau) \sqrt{s\tau}\, K_\mu(s\tau)| = |s^{1/2-\mu} e^{a\tau} (s\tau)^\mu K_\mu(s\tau)|$$
$$< B_\mu |s|^{1/2-\mu_R} (1 + |s|^{\mu_R}) < C_\mu \sqrt{r}$$

where B_μ and C_μ are sufficiently large constants. Similarly, under the same restrictions on s, σ, a, and τ, but with $\mu = 0$, Lemma 6.5-2 shows that

$$(18) \qquad |e^{a\tau} j_0(\tau) \sqrt{s\tau}\, K_0(s\tau)| = \left| \sqrt{s}\, \frac{e^{a\tau} K_0(s\tau)}{h(\tau)} \right| < C_0 \sqrt{r}$$

where C_0 is also a sufficiently large constant. Furthermore, these results imply that, for Re $s = \sigma > a \geq 0$, Im $s = r > 1$, $\mu = 0$ or Re $\mu > 0$, and $0 < \tau < \infty$

$$(19) \qquad |e^{a\tau} j_\mu(\tau) \tau \sqrt{s\tau}\, K_{\mu+1}(s\tau)| \leq C_{\mu+1} \sqrt{r}$$

because $j_\mu(\tau)\tau = j_{\mu+1}(\tau)$ if Re $\mu > 0$, and $|j_0(\tau)\tau| \leq |j_1(\tau)|$.

We shall now show that the integral in (15) is of the order $O(r^{-1})$ as $r \to \infty$ uniformly on $0 < \tau < B - \delta$. In view of (17) and (18), this will prove that $Q_{1,r}(\tau)$ tends uniformly to zero on $0 < \tau < \infty$ as $r \to \infty$. Since $|st| \geq |\sigma + ir|A \to \infty$, we may use the asymptotic expression Sec. 6.2, Eq. (15). Thus, for the integral in (15) we may write

$$(20) \quad \frac{1}{\sqrt{2\pi}} \int_{\max(A,\, \tau+\delta)}^{B} \frac{t\phi_k(t)}{t^2 - \tau^2} (e^{st} - ie^{-st+i\mu\pi})\, dt$$
$$+ \frac{1}{\sqrt{2\pi}} \int_{\max(A,\, \tau+\delta)}^{B} \frac{t\phi_k(t)}{t^2 - \tau^2} (e^{st} - ie^{-st+i\mu\pi})\, O\left(\frac{1}{|st|}\right) dt$$

Now, on the domain:

$$\{(t, \tau) : 0 < \tau < B - \delta,\ \max(A, \tau + \delta) < t < B\}$$

the integrand of the first term in (20) is bounded when Re s is fixed. This implies that the second term in (20) is of the order $O(r^{-1})$ uniformly on $0 < \tau < B - \delta$ as $r \to \infty$. Upon applying an integration by parts to the first term in (20) and using similar reasoning, we can come to the same conclusion for the first term. So truly, $Q_{1,r}(\tau)$ converges uniformly to zero on $0 < \tau < \infty$ as $r \to \infty$.

The same arguments applied to $Q_{2,r}(\tau)$, $Q_{3,r}(\tau)$, and $Q_{4,r}(\tau)$ show that these quantities and therefore $Q_r(\tau)$ also converge uniformly to zero on $0 < \tau < \infty$ as $r \to \infty$. By virtue of our conclusions concerning $N_r(\tau)$, $P_r(\tau)$, and $Q_r(\tau)$, Lemma 6.6-2 has been proven.

The inversion formula we have been preparing for is stated by

THEOREM 6.6-1. *Let* $F(s) = \mathfrak{K}_\mu f$ *for* $s \in \Omega_f$. *Then, in the sense of convergence in* $\mathscr{D}'(I)$,

$$(21) \qquad f(t) = \lim_{r \to \infty} \frac{1}{i\pi} \int_{\sigma - ir}^{\sigma + ir} F(s) \sqrt{st}\, I_\mu(st)\, ds$$

where σ *is any fixed real positive number in* Ω_f.

PROOF. Let $\phi \in \mathscr{D}(I)$. We shall show that

$$(22) \qquad \left\langle \frac{1}{i\pi} \int_{\sigma - ir}^{\sigma + ir} F(s) \sqrt{st}\, I_\mu(st)\, ds,\ \phi(t) \right\rangle$$

tends to $\langle f, \phi \rangle$ as $r \to \infty$. From the analyticity of $F(s)$ on Ω_f and the fact that the support of $\phi(t)$ is a compact subset of I, it follows that (22) is really a repeated integral on (t, ω) having a continuous integrand on a closed bounded domain of integration. Thus, we may change the order of integration to obtain for (22)

$$\frac{1}{\pi} \int_{-r}^{r} \langle f(\tau), \sqrt{s\tau}\, K_\mu(s\tau) \rangle \int_0^\infty \phi(t) \sqrt{st}\, I_\mu(st)\, dt d\omega \qquad s = \sigma + i\omega$$

By Lemma 6.6-1, this is equal to

$$\left\langle f(\tau), \frac{1}{\pi}\int_{-r}^{r} \sqrt{s\tau}\, K_\mu(s\tau) \int_0^\infty \phi(t)\sqrt{st}\, I_\mu(st)\, dt d\omega \right\rangle \tag{23}$$

Now, choose a such that $\max(0, \sigma_f) < a < \sigma$ where σ_f is the abscissa of definition for $\mathfrak{K}_\mu f$. Then, $f \in \mathscr{K}'_{\mu, a}$, and according to Lemma 6.6-2, the testing function inside (23) converges in $\mathscr{K}_{\mu, a}$ to $\phi(\tau)$ as $r \to \infty$. Therefore, (23) tends to $\langle f, \phi \rangle$, and the theorem is proven.

An immediate consequence of Theorem 6.6-1 is the following weak version of a uniqueness theorem.

COROLLARY 6.6-1a. *Let $F(s) = \mathfrak{K}_\mu f$ for $s \in \Omega_f$, let $G(s) = \mathfrak{K}_\mu g$ for $s \in \Omega_g$, and assume that $F(s) = G(s)$ for $s \in \Omega_f \cap \Omega_g$. Then, in the sense of equality in $\mathscr{D}'(I), f = g$.*

There is another inversion formula for our K transformation. It is a generalization of a result due to Boas [1] and can be stated as follows: *Let $F(s) = \mathfrak{K}_\mu f$ for $s \in \Omega_f$, and assume that f is concentrated on an interval of the form $T \le t < \infty$, $T > 0$. Then, in the sense of convergence in $\mathscr{D}'(I)$,*

$$f(t) = \lim_{k \to \infty} \sqrt{\frac{2}{\pi}}\, \frac{1}{(2k)!} \left(\frac{2k}{t}\right)^{2k+1} [S^k_{\mu, s} F(s)|_{s=2k/t}] \tag{24}$$

See Zemanian [9] for a proof; the condition Re $\mu \le \frac{1}{2}$ assumed throughout that paper is nowhere used in the proof.

PROBLEM 6.6-1. Prove Lemma 6.6-1 for the case where $\mu = 0$.

6.7. Characterization of K Transforms

We shall prove in this section that every K transform can be characterized as follows:

A necessary and sufficient condition for a function $F(s)$ to be the μth order K transform of some generalized function according to our definition (*Sec.* 6.4, *Eq.* (1)) *is that there be a half-plane* $\{s\colon \text{Re } s \ge b > 0\}$ *on which $F(s)$ is analytic and bounded according to*

$$|F(s)| \le P_b(|s|) \tag{1}$$

where $P_b(|s|)$ is a polynomial in $|s|$.

As we shall see, b can be any real positive point in the region of definition for the K transform. However, $P_b(|s|)$ will depend in general on the choice of b. We first prove the necessity part in a more precise form, wherein b may be a nonpositive number.

THEOREM 6.7-1. *Let $F(s) = \mathfrak{K}_\mu f$ for $s \in \Omega_f$, and let b be any real number such that $b > \sigma_f$. Also, let Λ_b denote the subset of Ω_f defined by*

$$\Lambda_b \triangleq \{s\colon s \in \Omega_f,\ \operatorname{Re} s \geq b,\ |s| \geq b - \sigma_f\}$$

Then, $F(s)$ is analytic on Λ_b and bounded according to (1). *$P_b(|s|)$ depends in general on the choice of b.*

PROOF. Theorem 6.5-1 asserts that $F(s)$ is analytic for $s \in \Omega_f$ and therefore for $s \in \Lambda_b$. Next, choose two real numbers a and b such that $\sigma_f < a < b$. (As always, σ_f denotes the abscissa of definition for $\mathfrak{K}_\mu f$.) As was shown in Sec. 6.3, for each fixed $s \neq 0$ in the half-plane $\{s\colon \operatorname{Re} s \geq b\}$, $\sqrt{st}\,K_\mu(st)$ as a function of t is a member of $\mathscr{K}_{\mu,a}$. Moreover, $f \in \mathscr{K}'_{\mu,a}$, and so, by Sec. 6.3, note IV, there exists a positive constant C and a nonnegative integer r such that

$$|F(s)| \leq C \max_{0 \leq k \leq r} \rho^\mu_{a,k}[\sqrt{st}\,K_\mu(st)] \qquad s \neq 0, \qquad \operatorname{Re} s \geq b$$

Setting $\mu_R \triangleq \operatorname{Re}\mu$ and $\mu_X \triangleq \operatorname{Im}\mu$, we may use Sec. 6.2, Eq. (9) and the condition $s \in \Omega_f$ to write

$$\rho^\mu_{a,k}[\sqrt{st}\,K_\mu(st)] \leq \begin{cases} |s|^{2k+(1/2)-\mu_R} e^{|\mu_X|\pi} \sup\limits_{0<t<\infty} |e^{at}(st)^\mu K_\mu(st)| & \operatorname{Re}\mu > 0 \\ |s|^{2k+1/2} \sup\limits_{0<t<\infty} \left|\dfrac{e^{at}K_0(st)}{h(t)}\right| & \mu = 0 \end{cases}$$

(Remember that the condition $s \in \Omega_f$ implies that $-\pi < \arg s < \pi$.) The inequality (1) now follows from Lemmas 6.5-1 and 6.5-2. That $P_b(|s|)$ depends in general on b can be seen by noting that, for the transform $F(s)$ specified in Problem 6.4-2. $\sigma_f = 0$ whereas $|F(s)| \to \infty$ as $s \to 0$.

The sufficiency part of our characterization can be restated as follows.

THEOREM 6.7-2. *Assume that, on the half-plane $\{s\colon \operatorname{Re} s \geq b > 0\}$, $F(s)$ is an analytic function that is bounded according to $|F(s)| \leq P(|s|)$ where $P(|s|)$ is a polynomial in $|s|$. Then, for each μ (as always, either $\mu = 0$ or $\operatorname{Re}\mu > 0$), $F(s)$ is a μth order K transform of some generalized function f according to the definition Sec. 6.4, Eq. (1). Moreover, the half-plane $\{s\colon \operatorname{Re} s > b\}$ is contained in the region of definition for $\mathfrak{K}_\mu f$.*

PROOF. Let q be a real number such that $q > \operatorname{Re}\mu + \frac{3}{2}$. Also, let m be an even integer such that $m - q$ is no less than the degree of $P(|s|)$. Then, $s^{-m}F(s)$ satisfies the hypothesis of Theorem 6.2-1. Consequently, for $\operatorname{Re} s > c > b$,

$$s^{-m}F(s) = \int_0^\infty g(t)\sqrt{st}\,K_\mu(st)\,dt \tag{2}$$

where

$$(3) \qquad g(t) = \frac{1}{i\pi}\int_{c-i\infty}^{c+i\infty} z^{-m}F(z)\sqrt{zt}\,I_\mu(zt)\,dz \triangleq \mathfrak{K}_{\mu,c}^{-1} z^{-m}F(z)$$

Here, $g(t)$ does not depend on the choice of c so long as $c > b$.

Now, consider the expression (wherein either Re $\mu > 0$ or $\mu = 0$):

$$(4) \qquad g(t)e^{-ct}t^{-\mu-1/2} = \frac{1}{i\pi}\int_{c-i\infty}^{c+i\infty} z^{-m+\mu+1/2}F(z)e^{-ct}(zt)^{-\mu}I_\mu(zt)\,dz$$

From Sec. 6.2, Eq. (1) we see that

$$(5) \qquad |e^{-ct}(zt)^{-\mu}I_\mu(zt)|$$

is bounded when $|zt| \le 1$, $0 < t < \infty$. The same is true on the domain described by $|zt| \ge 1$, Re $z = c$, $0 < t < \infty$; indeed, from Sec. 6.2, Eqs. (5) and (6) we see that, for some sufficiently large constant A_μ, (5) is bounded by

$$e^{-ct}|(zt)^{-\mu-1/2}|A_\mu e^{\text{Re}\,zt} = A_\mu|zt|^{-\mu_R-1/2}e^{|\mu_X|\pi/2}$$
$$\le A_\mu e^{|\mu_X|\pi/2} \qquad \mu_R \triangleq \text{Re}\,\mu,\ \mu_X \triangleq \text{Im}\,\mu$$

Moreover,

$$|z^{-m+\mu+1/2}F(z)| \le |z|^{-q+\mu_R+1/2}e^{|\mu_X|\pi/2}P(|z|)/|z|^{m-q}$$
$$\le C|z|^{-q+\mu_R+1/2}$$

where C is still another constant. Since $-q + \mu_R + \frac{1}{2} < -1$, all this shows that the integral in (4) converges uniformly for $0 < t < \infty$ and that $g(t)e^{-ct}t^{-\mu-1/2}$ is continuous and bounded on $0 < t < \infty$. It follows that, for any $d > c$ and for both of the cases Re $\mu > 0$ and $\mu = 0$, $g(t)e^{-dt}/j_\mu(t)$ is absolutely integrable on $0 < t < \infty$. By Sec. 6.3, note *VI*, $g(t)$ generates a regular member of $\mathscr{K}'_{\mu,d}$.

Thus, we can conclude that (2) is a generalized K transform whose region of definition contains the half-plane $\{s: \text{Re}\,s > d\}$. Set $f = S_\mu^{m/2}g$. By Theorem 6.4-1, $\mathfrak{K}_\mu f = s^m\mathfrak{K}_\mu g = F(s)$ for at least Re $s > d$. Since c and therefore d can be chosen arbitrarily close to b, the half-plane $\{s: \text{Re}\,s > b\}$ is contained in the region of definition for $\mathfrak{K}_\mu f$. This completes the proof.

Our original assertion characterizing a K transform can now be obtained by combining Theorems 6.7-1 and 6.7-2.

The proof of Theorem 6.7-2 has also established still another inversion formula, namely, (6) below.

COROLLARY 6.7-2a. *Assume that $F(s)$ satisfies the condition stated in Theorem 6.7-2. Then, for any fixed μ (restricted in the usual way), choose m*

as an even integer such that $m - \text{Re}\,\mu - \frac{3}{2}$ *is greater than the degree of* $P(|s|)$. *Set*

$$f = S_\mu^{m/2}[\mathfrak{K}_{\mu,c}^{-1} s^{-m} F(s)] \tag{6}$$

where $c > b$, $\mathfrak{K}_{\mu,c}^{-1}$ *is defined by* (3), *and* S_μ *denotes the generalized differential operator. Then,* f *is a member of* $\mathscr{K}'_{\mu,d}$ *for every* $d > b$ *and is independent of the choice of* $c > b$; *moreover,* $\mathfrak{K}_\mu f = F(s)$ *for at least* $\text{Re}\, s > b$.

Incidentally, Theorem 6.7-1 and Corollaries 6.6-1a and 6.7-2a imply that we may place a hypothesis on $f(t)$, rather than on $F(s)$, to obtain the following assertion: *If* $F(s) = \mathfrak{K}_\mu f$ *for* $\text{Re}\, s > \sigma_f$, *then* (6) *holds true at least in the sense of equality in* $\mathscr{D}'(I)$.

PROBLEM 6.7-1. Prove the following: $F(s)$ is the μth order K transform of a generalized function $f(t)$ that is concentrated on the interval $T \le t < \infty$ ($T > 0$) if and only if $F(s)$ is analytic on some half-plane $\{s: \text{Re}\, s \ge b > 0\}$ and there exists a polynomial $P(|s|)$ such that

$$|F(s)| \le e^{-\text{Re}\, sT} P(|s|) \qquad \text{Re}\, s \ge b \tag{7}$$

Hint: For necessity, apply Sec. 6.3, note IV to

$$F(s) = \langle f(t), \lambda[|s|(t-T)] \sqrt{st}\, K_\mu(st)\rangle$$

where $\lambda(t)$ is a smooth function on $-\infty < t < \infty$, $\lambda(t) \equiv 1$ on $-\frac{1}{2} < t < \infty$, and $\lambda(t) \equiv 0$ on $-\infty < t < -1$. For sufficiency, choose q, m, and $g(t)$ as in the proof of Theorem 6.7-2 so that $f(t) = S_\mu^{m/2} g(t)$; then prove that $g(t) \equiv 0$ on $0 < t < T$ by closing the path of integration in (3) to the right.

PROBLEM 6.7-2. Show that Theorem 6.6-1 remains valid when the lower limit in the integral of Sec. 6.6, Eq. (21) tends to $\sigma - i\infty$ independently from the way the upper limit tends to $\sigma + i\infty$.

6.8. An Operational Calculus

Let P be any polynomial and consider the differential equation:

$$P(S_\mu)u = g \tag{1}$$

where g is a given $\mathfrak{K}_\mu$-transformable generalized function and the operator S_μ is understood in the generalized sense. The K transformation generates an operational calculus through which a solution u for (1) can be found. By applying $\mathfrak{K}_\mu$ to (1) and invoking Theorem 6.4-1, we obtain

$$\mathfrak{K}_\mu u = \frac{G(s)}{P(s^2)} \tag{2}$$

where $G(s) = \mathfrak{K}_\mu g$ for $\text{Re } s > \sigma_g$. Let σ_p be the largest of the real parts of the roots of $P(s^2)$. Then, Theorem 6.7-1 implies that (2) satisfies the hypothesis of Theorem 6.7-2 on some half-plane $\{s\colon \text{Re } s \geq b > \max(0, \sigma_p, \sigma_g)\}$ Thus, by Theorem 6.7-2, (2) is truly a K transform of order μ. We may apply either of the inversion formulas Sec. 6.6, Eq. (21) or Sec. 6.7, Eq. (6) to determine u (at least in principle—the computations of these formulas may in particular cases be very difficult to carry out). The solution u is hereby determined as a distribution on I, which satisfies (1) in the sense of equality in $\mathscr{D}'(I)$.

The extension of this technique to simultaneous differential equations of the form (1) is straightforward.

(Furthermore, if g is concentrated on the interval $T \leq t < \infty$ where $T > 0$, then $G(s)/P(s^2)$ satisfies the conditions stated in Problem 6.7-1. It follows that we may apply either of the aforementioned inversion formulas or also Sec. 6.6, Eq. (24) to $G(s)/P(s^2)$ to determine u as a member of $\mathscr{D}'(I)$; this u will also be concentrated on $T \leq t < \infty$. Moreover, we may add to u any solution $v \in \mathscr{D}'(I)$ of the homogeneous equation $P(S_\mu)v = 0$ to obtain still another solution to (1), but such (nonzero) $v \in \mathscr{D}'(I)$ will not be concentrated on $T \leq t < \infty$. (See Schwartz [1], Vol. I, p. 130 and Hurewicz [1], p. 46.)

Let us compare our $\mathfrak{K}_\mu$ operational calculus with the $\mathfrak{H}_\mu$ operational calculus discussed in Sec. 5.7. The $\mathfrak{K}_\mu$ operational calculus solves the same type of differential equation as does the $\mathfrak{H}_\mu$ operational calculus. However, $\mathfrak{H}_\mu$ is defined only for real μ. (For the case where $\mu < -\frac{1}{2}$, see Sec. 5.10.) On the other hand, $\mathfrak{K}_\mu$ is defined for all real and complex μ. (We have used the fact that $K_\mu(t) = K_{-\mu}(t)$ to restrict μ to $\text{Re } \mu \geq 0$ without loss of generality. For the case where $\text{Re } \mu = 0$ and $\mu \neq 0$, see Zemanian [10].) Also, there is no restriction on the roots of the polynomial P in (1) in the present case, whereas for the $\mathfrak{H}_\mu$ operational calculus $P(x)$ was required to have no roots on the real semiaxis $-\infty < x < 0$.

Moreover, the allowable generalized-function solutions in the $\mathfrak{K}_\mu$ operational calculus may be of exponential growth as $t \to \infty$, whereas they must be at most of slow growth in the $\mathfrak{H}_\mu$ operational calculus. Hence, so far as the behavior of the solutions as $t \to \infty$ is concerned, the $\mathfrak{K}_\mu$ transformation is an extension of the $\mathfrak{H}_\mu$ transformation in much the same way as is the Laplace transformation an extension of the Fourier transformation.

On the other hand, the situation concerning the behavior of the allowable solutions as $t \to 0+$ is somewhat different as is seen from the following facts. The testing functions in $\mathscr{H}_\mu$ behave as does the function $\sqrt{t}J_\mu(t)$ when $t \to 0+$. In contrast to this, the testing functions in $\mathscr{K}_{\mu, a}$ are of the order of $\sqrt{t}\,K_\mu(t)$ when $t \to 0+$. Note that $K_\mu(t)$ becomes unbounded as

$t \to 0+$, whereas $J_\mu(t)$ tends to zero if $\mu > 0$ or $\mu = -1, -2, -3, \ldots$ and tends to 1 if $\mu = 0$. It follows that under these conditions on μ the allowable generalized-function solutions under the $\mathfrak{K}_\mu$ operational calculus are considerably more restricted as $t \to 0+$ than are the generalized-function solutions under the $\mathfrak{H}_\mu$ operational calculus.

PROBLEM 6.8-1. In the sense of ordinary differentiation, $u(t) = \sqrt{t}J_\mu(t)$ is a solution of the homogeneous differential equation $(S_\mu + 1)u = 0$. Prove that this is still so in the sense of differentiation and equality in $\mathscr{D}'(I)$ but is no longer so when S_μ is taken as a generalized differential operator in the space $\mathscr{K}'_{\mu,a}$ $(a > 0)$. In the course of your proof, also show that $\sqrt{t}\, J_\mu(t)$ is a regular member of $\mathscr{K}'_{\mu,a}$ for every $a > 0$.

PROBLEM 6.8-2. Describe the technique one may use to solve simultaneous differential equations of the form of (1) by using the generalized K transformation. State what conditions must be imposed on the differential equations in order to insure the success of the method.

PROBLEM 6.8-3. Let β be a complex number not equal to zero. Find the $\mathfrak{K}_\mu$-transformable generalized function $u(t)$ that satisfies the following differential equation on I.

$$(S_\mu - \beta)u(t) = t^{\mu-1}$$

Hint: From Grobner and Hofreiter [1], Vol. 2, p. 197, we have that

$$\int_0^\infty z^{\rho-1}K_\mu(z)\, dz = 2^{\rho-2}\Gamma\left(\frac{\rho+\nu}{2}\right)\Gamma\left(\frac{\rho-\nu}{2}\right) \qquad \operatorname{Re}\rho > \operatorname{Re}\mu$$

6.9. Applications to Certain Time-Varying Electrical Networks

In this section we use our K transformation to analyze three time-varying electrical networks. These examples, as well as several others, were suggested by Aseltine [1] and Gerardi [1].

EXAMPLE 6.9-1. Consider the electrical network shown in Fig. 6.9-1 for the time interval $0 < t < \infty$. The network consists of a voltage source $v(t)$, two inductors $L_1(t)$ and $L_2(t)$, and a capacitor $C(t)$. The inductors and capacitor vary with time as indicated in the diagram. The symbols a, b, and c denote real positive constants and μ any real constant. (We could allow μ to be complex, but this would not be physically meaningful.)

Let $q_1(t)$ and $q_2(t)$ be the mesh charges as shown. Upon applying a mesh analysis, we obtain the simultaneous differential equations:

$$
(1) \qquad
\begin{aligned}
v(t) &= D[L_1(t)Dq_1(t)] + \frac{q_1(t) - q_2(t)}{C(t)} \\
0 &= \frac{q_2(t) - q_1(t)}{C(t)} + D[L_2(t)Dq_2(t)]
\end{aligned}
$$

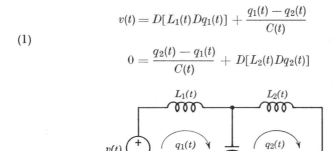

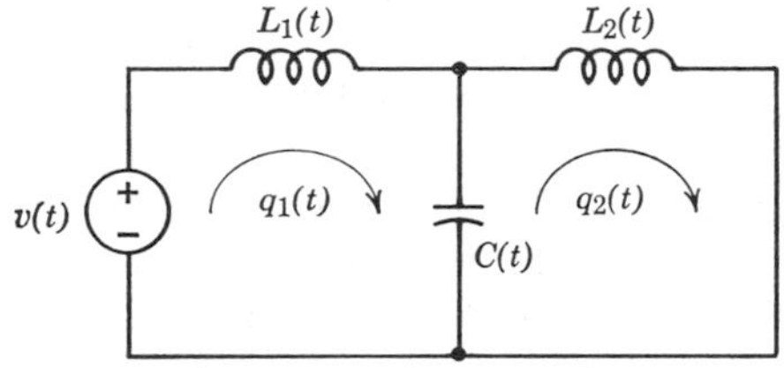

FIGURE 6.9-1

By setting $v = t^{\mu+1/2}g$, $q_1 = t^{-\mu-1/2}u_1$, and $q_2 = t^{-\mu-1/2}u_2$, inserting the time variations for $L_1(t)$, $L_2(t)$, and $C(t)$, and multiplying each equation by $t^{-\mu-1/2}$, we obtain

$$
(2) \qquad
\begin{aligned}
g &= aS_\mu u_1 + c(u_1 - u_2) \\
0 &= c(u_2 - u_1) + bS_\mu u_2
\end{aligned}
$$

These equations have the right form for an analysis via the K transformation, assuming that g is a $\mathfrak{K}_{|\mu|}$-transformable generalized function.

If $\mu \geq 0$, apply $\mathfrak{K}_\mu$. If $\mu < 0$, apply $\mathfrak{K}_{-\mu}$. This yields

$$
(3) \qquad
\begin{aligned}
G(s) &= (as^2 + c)U_1(s) - cU_2(s) \\
0 &= -cU_1(s) + (bs^2 + c)U_2(s)
\end{aligned}
$$

where G, U_1, and U_2 denote the K transforms of g, u_1, and u_2 respectively. The determinant of the coefficient matrix for the simultaneous equations (3) has all its zeros on the imaginary axis of the s plane. Upon solving (3) for U_1 and U_2, we get

$$
U_1(s) = \frac{bs^2 + c}{abs^4 + c(a + b)s^2} G(s)
$$

$$
U_2(s) = \frac{c}{abs^4 + c(a + b)s^2} G(s) \qquad \operatorname{Re} s > \max(0, \sigma_g)
$$

where σ_g is the abscissa of definition for the transform $G(s)$. We may now apply one of our inversion formulas to obtain (at least in principle) two distributions u_1 and u_2 which satisfy (2) in the sense of equality in $\mathscr{D}'(I)$. Thus, q_1 and q_2 satisfy (1) in the sense of equality in $\mathscr{D}'(I)$.

Actually, we could obtain another solution by adding to the pair q_1, q_2 some free oscillation of the network in Fig. 6.9-1 (i.e., any solution of the homogeneous system obtained by setting $v = 0$ in (1)). However, such free oscillations can be prohibited by requiring that the voltage v and the mesh charges be the zero distribution on some initial interval of the form $0 < t < T$. Indeed, according to the result stated in Problem 6.7-1, if v is the zero distribution on $0 < t < T$, then $G(s)$ and therefore $U_1(s)$ and $U_2(s)$ as well will satisfy the conditions stated in Problem 6.7-1. Consequently, the solution q_1, q_2 obtained in the preceding paragraph must also be the zero distribution on $0 < t < T$. Therefore, if the mesh charges are also to satisfy this condition, there can be no free oscillation in the network. This is because any free oscillation must vanish on $0 < t < \infty$ if it vanishes on $0 < t < T$ (see Schwartz [1], Vol. I, p. 130 and Hurewicz, p. 46). In summary, the mesh charges q_1 and q_2, obtained by using the K transformation, are the only members of $\mathscr{D}'(I)$ that satisfy the system (1) and the requirement that they be the zero distribution on $0 < t < T$, assuming that v also satisfies this requirement.

Finally, we note that one can analyze in the very same way any network consisting only of time-varying inductors and capacitors whose time variations are proportional to $t^{2\mu+1}$ and $t^{-2\mu-1}$ respectively and also of suitably restricted voltage sources. Upon making the change of variables mentioned above, one obtains differential equations containing only the differential operator S_μ.

EXAMPLE 6.9-2. Consider the series circuit of Fig. 6.9-2 consisting of a voltage source $v(t)$, a fixed inductor L, a fixed capacitor C, and a time-varying resistor $R(t) = r/t$. Here, L, C, and r are fixed real numbers, and we

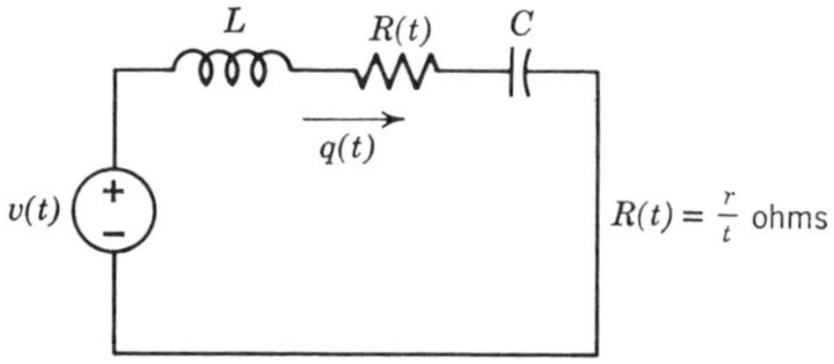

FIGURE 6.9-2

again restrict our attention to the time interval $0 < t < \infty$. In terms of the circulating charge $q(t)$, Kirchhoff's voltage law yields

$$LD^2q + \frac{r}{t}Dq + \frac{q}{C} = v \tag{4}$$

Upon setting $q(t) = t^{\mu-1/2}u(t)$, $g(t) = L^{-1}t^{-\mu+1/2}v(t)$, and $r/L = -2\mu + 1$, this can be rewritten as

$$S_\mu u + \frac{u}{LC} = g$$

Assuming that g is a $\mathfrak{K}_{|\mu|}$-transformable generalized function, we can apply the K transformation of order $|\mu|$ to get

$$U(s) = \frac{G(s)}{s^2 + \dfrac{1}{LC}} \qquad \operatorname{Re} s > \max(\sigma_g, \pm \operatorname{Re} \sqrt{-LC})$$

As usual, the capital letters denote the transforms of the corresponding lower case letters, and σ_g is the abscissa of definition of $\mathfrak{K}_{|\mu|}g$. Applying one of the inversion formulas of Secs. 6.6 and 6.7, we determine a distribution $u(t) \in \mathscr{D}'(I)$. This in turn determines a distribution $q(t) = t^{\mu-1/2}u(t)$, which satisfies (4) in the sense of equality in $\mathscr{D}'(I)$. Here again, we could add any free oscillation of the network of Fig. 6.9-2 to obtain still another solution to (4). But, the imposition of the condition that both $v(t)$ and $q(t)$ be the zero distribution on some interval $0 < t < T$ insures that no such free oscillation can exist so that $q(t)$ is the only solution in $\mathscr{D}'(I)$ of (4).

EXAMPLE 6.9-3. As a last example, let $0 < t < \infty$, and consider Fig. 6.9-3, which shows a system that can be approximately realized by appropriately connecting certain components of an analog computer. There are two ideal voltage amplifiers whose time-variable gains are μ^2/t and $1/t$, two ideal

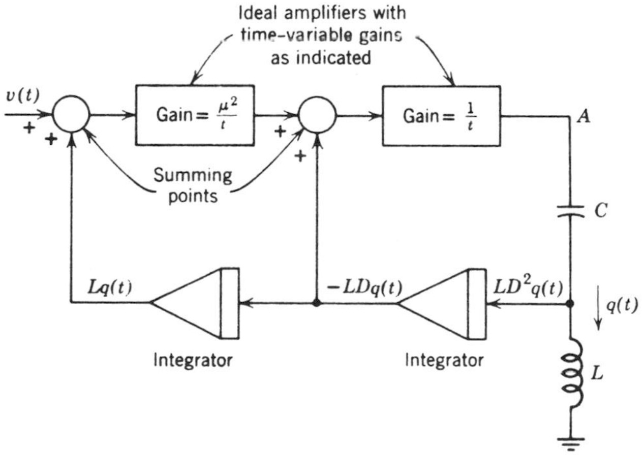

FIGURE 6.9-3

integrators, two summing points, a capacitor C, and an inductor L; μ^2, L, and C have fixed positive values. The input voltage $v(t)$ is assumed known. We wish to determine the charge $q(t)$ flowing through the inductor L and capacitor C during the time interval $0 < t < \infty$. The voltage drop from the point A to ground is equal to

$$LD^2q + \frac{q}{C} \tag{5}$$

On the other hand, the voltage applied between point A and ground resulting from the rest of the system is

$$-\frac{L}{t}\,Dq + \frac{L\mu^2}{t^2}\,q + \frac{\mu^2}{t^2}\,v \tag{6}$$

Upon equating (5) and (6) and dividing by L, we obtain the differential equation:

$$D^2q + \frac{1}{t}\,Dq + \left(\frac{1}{LC} - \frac{\mu^2}{t^2}\right)q = \frac{\mu^2}{Lt^2}\,v \tag{7}$$

A little computation shows that the operator:

$$D^2 + t^{-1}D - \mu^2t^{-2}$$

is equal to the operator $t^{-1/2}S_\mu t^{1/2}$, where μ can be chosen positive. Therefore, upon setting $u(t) = \sqrt{t}\,q(t)$, we may rewrite (7) as

$$S_\mu u(t) + \frac{u(t)}{LC} = \frac{\mu^2}{L}\,t^{-3/2}v(t) \tag{8}$$

Once again we have a differential equation that can be analyzed by means of the K transformation of order μ, assuming that $t^{-3/2}v(t)$ is a $\mathfrak{K}_\mu$-transformable generalized function. Setting $G(s) = \mathfrak{K}_\mu[t^{-3/2}v(t)]$ for $\operatorname{Re} s > \sigma_g$, we obtain

$$U(s) = \frac{\mu^2G(s)}{Ls^2 + C^{-1}} \qquad \operatorname{Re} s > \max(0, \sigma_g) \tag{9}$$

where $U(s) = \mathfrak{K}_\mu u$. An application of an inversion formula to (9) determines a distribution $u(t)$; this in turn determines $q(t) = t^{-1/2}u(t)$, which satisfies (7) in the sense of equality in $\mathscr{D}'(I)$. As before, this will be the only solution in $\mathscr{D}'(I)$ if we require that both v and q be the zero distribution on an interval of the form $0 < t < T$.

PROBLEM 6.9-1. Verify our assertion that every network consisting solely of voltage sources and time-varying inductors and capacitors whose time variations are proportional to $t^{2\mu+1}$ and $t^{-2\mu-1}$ respectively generate (after the indicated change of variables) differential equations involving only the operator S_μ.

CHAPTER VII

The Weierstrass Transformation

7.1. Introduction

The *conventional Weierstrass transformation* is defined by

$$(1) \qquad F(s) \triangleq \frac{1}{\sqrt{4\pi}} \int_{-\infty}^{\infty} f(\tau) \exp[-(s-\tau)^2/4]\, d\tau$$

where $f(\tau)$ is a suitably restricted conventional function on $-\infty < \tau < \infty$ and s is a complex variable. Hirschman and Widder [1], Chapter VIII, discuss it at some length. Other names for it are the Gauss transformation (Rooney [1], [2]), the Gauss-Weierstrass transformation (Hille and Phillips [1], p. 578), and the Hille transformation (Gonzalez Dominguez [1]).

The heat equation for one-dimensional flow:

$$(2) \qquad \kappa \frac{\partial^2 u}{\partial x^2} = \frac{\partial u}{\partial t} \qquad u = u(x, t), \qquad -\infty < x < \infty, \qquad 0 < t < \infty$$

where κ is a positive constant, has as its Green's function:

$$(3) \qquad k(x, \kappa t) = \frac{e^{-x^2/4\kappa t}}{\sqrt{4\pi\kappa t}}$$

Upon replacing x by $s - \tau$ and setting $\kappa t = 1$, we obtain the kernel of (1), and thus (1) can be rewritten as

$$(4) \qquad F(s) = \int_{-\infty}^{\infty} f(\tau) k(s-\tau, 1)\, d\tau$$

Because of this, the Weierstrass transformation arises naturally in problems involving the heat equation (2). This is discussed in Sec. 7.5 wherein the Cauchy problem for (2) with a generalized-function initial condition is solved.

Note that (4) is a convolution. In fact, the Weierstrass transformation is a singular case of the convolution transformation (Hirschman and Widder [1], p. 170), a subject we will discuss in the next chapter.

Our procedure for extending (4) to generalized functions (Zemanian [10]) is quite similar to the technique used in Chapter III for the two-sided Laplace transformation. The basic idea is to construct a testing-function space $\mathscr{W}(\sigma_1, \sigma_2)$, which is a countable-union space and contains the kernel $k(s-\tau, 1)$ for $\sigma_1 < \operatorname{Re} s < \sigma_2$. Then, the Weierstrass transform $F(s)$ of any member f of the dual space $\mathscr{W}'(\sigma_1, \sigma_2)$ is obtained by applying $f(\tau)$ to the kernel $k(s-\tau, 1)$:

$$(5) \qquad F(s) \triangleq \frac{1}{\sqrt{4\pi}} \langle f(\tau), \exp[-(s-\tau)^2/4] \rangle \qquad \sigma_1 < \operatorname{Re} s < \sigma_2$$

It turns out that this transformation can be related to the two-sided Laplace transformation, and, as a result, several properties of the latter transformation can be carried over to the present case. Indeed, in Sec. 7.3, we obtain an analyticity theorem, an inversion formula, and a characterization of Weierstrass transforms by converting the corresponding results of the Laplace transformation. A second inversion formula, which is in fact an extension to generalized functions of Hirschman and Widder's complex inversion formula (Hirschman and Widder [1], p. 191), is established in Sec. 7.4. There is still a third inversion theorem for our generalized Weierstrass transformation (Queen [1]). It extends a formula due to Rooney [1], [2] but does not hold for as wide a class of generalized functions as do the inversion formulas of Secs. 7.3 and 7.4.

One can also define an n-dimensional Weierstrass transformation wherein $\tau \in \mathscr{R}^n$ and $s \in \mathscr{C}^n$. Its properties are quite similar to those of the one-dimensional transformation discussed here. See Queen [1].

7.2. The Testing-Function Spaces $\mathscr{W}_{a,b}$ and $\mathscr{W}(w, z)$ and Their Duals

Let a and b be fixed numbers in $\mathscr{R}^1$, let τ be a variable in $\mathscr{R}^1$, and let $\rho_{a,b}(\tau)$ denote the function:

$$\rho_{a,b}(\tau) = \begin{cases} e^{-a\tau/2} & -\infty < \tau < 0 \\ e^{-b\tau/2} & 0 \leq \tau < \infty \end{cases}$$

We define $\mathscr{W}_{a,b}$ as the linear space of all complex-valued smooth functions $\phi(\tau)$ on $-\infty < \tau < \infty$ such that for each $p = 0, 1, 2, \ldots$

$$\chi_p(\phi) \triangleq \chi_{a,b,p}(\phi) \triangleq \sup_{-\infty < \tau < \infty} |e^{\tau^2/4} \rho_{a,b}(\tau) D^p \phi(\tau)| < \infty$$

Moreover, we assign to $\mathscr{W}_{a,b}$ the topology generated by the sequence of seminorms $\{\chi_p\}_{p=0}^{\infty}$. Since χ_0 is a norm, $\{\chi_p\}_{p=0}^{\infty}$ is a multinorm on $\mathscr{W}_{a,b}$, and $\mathscr{W}_{a,b}$ is a countably multinormed space. By the usual argument (see, for example, the proof of lemma 3.2-1), it can be shown that $\mathscr{W}_{a,b}$ is complete and therefore a Fréchet space. Also, note that differentiation is a continuous linear mapping of $\mathscr{W}_{a,b}$ into itself.

Throughout this chapter, $k(v, t)$ will always denote the function

$$k(v, t) = \frac{e^{-v^2/4t}}{\sqrt{4\pi t}} \tag{1}$$

where $v \in \mathscr{C}^1$ and $t \in \mathscr{R}^1$ with $0 < t \leq 1$. A crucial result is that, for any fixed $s \in \mathscr{C}^1$, the kernel $k(s - \tau, 1)$ as a function of τ is a member of $\mathscr{W}_{a,b}$ if and only if $a < \operatorname{Re} s < b$. Indeed, some computation shows that

$$D_\tau^p k(s - \tau, t) = \frac{e^{-(s-\tau)^2/4t}}{\sqrt{4\pi t}} P_p\left(s - \tau, \frac{1}{t}\right) \tag{2}$$

where $P_p(s - \tau, 1/t)$ is a polynomial of degree p in $s - \tau$ and $1/t$ separately. Therefore,

$$\chi_p[k(s - \tau, t)] = \sup_{-\infty<\tau<\infty} \left| \frac{e^{-s^2/4t}}{\sqrt{4\pi t}} e^{\tau^2(1-1/\tau)/4} \rho_{a,b}(\tau) e^{s\tau/2t} P_p\left(s - \tau, \frac{1}{t}\right) \right| \tag{3}$$

By setting $t = 1$ in (3) and noting the definition of $\rho_{a,b}(\tau)$, we obtain our assertion. Eq. (3) also shows that, for any fixed t such that $0 < t < 1$ and for every fixed $s \in \mathscr{C}^1$, the kernel $k(s - \tau, t)$ as a function of τ is a member of $\mathscr{W}_{a,b}$, whatever be the values of a and b.

The dual $\mathscr{W}'_{a,b}$ of $\mathscr{W}_{a,b}$ is also a complete space under its usual (weak) topology, according to Theorem 1.8-3. Also, in agreement with Sec. 2.5, we define (generalized) differentiation on $\mathscr{W}'_{a,b}$ by

$$\langle Df, \phi \rangle \triangleq \langle f, -D\phi \rangle \qquad f \in \mathscr{W}'_{a,b}, \qquad \phi \in \mathscr{W}_{a,b}$$

Such differentiation is a continuous linear mapping of $\mathscr{W}'_{a,b}$ into itself.

If $a \leq c$ and $d \leq b$, then $\mathscr{W}_{c,d}$ is a subspace of $\mathscr{W}_{a,b}$, and the topology of $\mathscr{W}_{c,d}$ is stronger than that induced on it by $\mathscr{W}_{a,b}$. Indeed, $0 < \rho_{a,b}(\tau) \leq \rho_{c,d}(\tau)$ on $-\infty < \tau < \infty$ so that $\chi_{a,b,p}(\phi) \leq \chi_{c,d,p}(\phi)$; this fact and Lemma 1.6-3 now verify our assertion. As a result, the restriction of any $f \in \mathscr{W}'_{a,b}$ to $\mathscr{W}_{c,d}$ is a member of $\mathscr{W}'_{c,d}$.

Following the general scheme used for the Laplace transformation, we now construct a countable-union space out of the $\mathscr{W}_{a,b}$ spaces. As before, let w denote either a (finite) real number or $-\infty$, and let z denote either a (finite) real number or $+\infty$. Choose two monotonic sequences of real numbers $\{a_\nu\}_{\nu=1}^{\infty}$ and $\{b_\nu\}_{\nu=1}^{\infty}$ such that $a_\nu \to w+$ and $b_\nu \to z-$. We define $\mathscr{W}(w, z)$ as the countable-union space $\bigcup_{\nu=1}^{\infty} \mathscr{W}_{a_\nu, b_\nu}$; thus, a sequence converges in $\mathscr{W}(w, z)$ if and only if it converges in one of the $\mathscr{W}_{a_\nu, b_\nu}$ spaces. It is readily shown that this definition is independent of the choices of $\{a_\nu\}$ and $\{b_\nu\}$. By Sec. 1.7, $\mathscr{W}(w, z)$ is complete. The dual $\mathscr{W}'(w, z)$ of $\mathscr{W}(w, z)$ is also complete according to Theorem 1.9-2.

The kernel $k(s-\tau, 1)$ as a function of τ is a member of $\mathscr{W}(w, z)$ if and only if $w < \operatorname{Re} s < z$; also, if t is fixed with $0 < t < 1$, then $k(s-\tau, t)$ as a function of τ is a member of $\mathscr{W}(w, z)$ for every fixed $s \in \mathscr{C}^1$ and every w and z. Furthermore, conventional differentiation is a continuous linear mapping of $\mathscr{W}(w, z)$ into itself; generalized differentiation is a continuous linear mapping of $\mathscr{W}'(w, z)$ into itself.

We now relate the spaces $\mathscr{W}(w, z)$ and $\mathscr{W}'(w, z)$ to $\mathscr{L}(-z/2, -w/2)$ and $\mathscr{L}'(-z/2, -w/2)$, respectively. In the next section this will lead to a connection between the Laplace and Weierstrass transformations.

THEOREM 7.2-1. *The mapping* $\theta(\tau) \leftrightarrow e^{-\tau^2/4}\theta(\tau)$ *is an isomorphism from* $\mathscr{L}(-z/2, -w/2)$ *onto* $\mathscr{W}(w, z)$.

PROOF. Let $\theta \in \mathscr{L}(-z/2, -w/2)$. This means that $\theta \in \mathscr{L}_{-d/2, -c/2}$ for some $c > w$ and some $d < z$. Choose a and b such that $w < a < c$ and $d < b < z$. We first prove that $\theta(\tau) \leftrightarrow e^{-\tau^2/4}\theta(\tau)$ is a continuous linear mapping of $\mathscr{L}_{-d/2, -c/2}$ into $\mathscr{W}_{a, b}$.

Note that $D^p e^{-\tau^2/4} = e^{-\tau^2/4} P_p(\tau) \qquad (p = 0, 1, 2, \ldots)$ where P_p is a polynomial. Moreover, the weighting functions $\kappa_{a, b}(\tau)$ and $\rho_{a, b}(\tau)$ in the seminorms for the $\mathscr{L}_{a, b}$ and $\mathscr{W}_{a, b}$ spaces are related by

$$\rho_{a, b}(\tau) = \kappa_{-b/2, -a/2}(\tau) \tag{4}$$

Therefore, we may write

$$e^{\tau^2/4}\rho_{a, b}(\tau) D_\tau^{\,p}[e^{-\tau^2/4}\theta(\tau)] =$$

$$\sum_{q=0}^{p} \binom{p}{q} \frac{\kappa_{-b/2, -a/2}(\tau)}{\kappa_{-d/2, -c/2}(\tau)} P_{p-q}(\tau) \kappa_{-d/2, -c/2}(\tau) D^q \theta(\tau)$$

But,

$$\frac{\kappa_{-b/2, -a/2}(\tau)}{\kappa_{-d/2, -c/2}(\tau)} P_{p-q}(\tau)$$

is bounded on $-\infty < \tau < \infty$ by a constant, say, B_{p-q} because $a < c$ and $d < b$. Consequently.

$$\chi_{a, b, p}[e^{-\tau^2/4}\theta(\tau)] \le \sum_{q=0}^{p} \binom{p}{q} B_{p-q} \gamma_{-d/2, -c/2, q}[\theta(\tau)] \qquad p = 0, 1, 2, \ldots \tag{5}$$

So truly, our mapping, which is clearly linear, is also continuous from $\mathscr{L}_{-d/2, -c/2}$ into $\mathscr{W}_{a, b}$. It immediately follows that $\theta(\tau) \leftrightarrow e^{-\tau^2/4}\theta(\tau)$ is also a continuous linear mapping of $\mathscr{L}(-z/2, -w/2)$ into $\mathscr{W}(w, z)$.

The unique inverse mapping is $\phi(\tau) \leftrightarrow e^{\tau^2/4}\phi(\tau)$. Upon choosing $w < a < c$ and $d < b < z$ again, we can apply almost the same argument to conclude that

(6) $\gamma_{-b/2, -a/2, \, p}[e^{\tau^2/4}\phi(\tau)] \leq \sum_{q=0}^{p} C_{p, \, q} \chi_{c, \, d, \, q}(\phi)$ $\phi \in \mathscr{W}_{c, \, d}$,

$$p = 0, 1, 2, \ldots$$

where the $C_{p, \, q}$ are constants not depending on ϕ. Hence, $\phi(\tau) \mapsto e^{\tau^2/4}\phi(\tau)$ is a continuous linear mapping from $\mathscr{W}_{c, \, d}$ into $\mathscr{L}_{-b/2, -a/2}$ and therefore from $\mathscr{W}(w, z)$ into $\mathscr{L}(-z/2, -w/2)$.

These results coupled with the fact that $\theta(\tau) \mapsto e^{-\tau^2/4}\theta(\tau)$ is a one-to-one mapping imply that the mapping is also onto. This completes the proof of the theorem.

Now, in the dual spaces the mapping $f(\tau) \mapsto e^{-\tau^2/4}f(\tau)$ is defined as usual as the adjoint of $\theta(\tau) \mapsto e^{-\tau^2/4}\theta(\tau)$. That is,

(7)

$$\langle e^{-\tau^2/4}f(\tau), \, \theta(\tau)\rangle \triangleq \langle f(\tau), \, e^{-\tau^2/4}\theta(\tau)\rangle \qquad \theta \in \mathscr{L}(-z/2, \, -w/2),$$

$$f \in \mathscr{W}'(w, z)$$

Theorems 1.10-2 and 7.2-1 immediately imply

THEOREM 7.2-2.   *The mapping* $f(\tau) \mapsto e^{-\tau^2/4}f(\tau)$ *is an isomorphism from* $\mathscr{W}'(w, z)$ *onto* $\mathscr{L}'(-z/2, -w/2)$.

Finally, we list the usual remarks concerning our spaces.

I. $\mathscr{D}$ is clearly a subspace of $\mathscr{W}(w, z)$ for every choice of w and z, and convergence in $\mathscr{D}$ implies convergence in $\mathscr{W}(w, z)$, Moreover, $\mathscr{D}$ is dense in $\mathscr{W}(w, z)$. (Prove this.) Theorem 1.9-1 now shows that $\mathscr{W}'(w, z)$ is a subspace of $\mathscr{D}'$. Thus, the members of $\mathscr{W}'(w, z)$ are distributions. Moreover, the values that $f \in \mathscr{W}'(w, z)$ assigns to the elements of $\mathscr{D}$ uniquely determine the values that f assigns to the elements of $\mathscr{W}(w, z)$.

Now, let $w \leq u$ and $v \leq z$. This implies that $\mathscr{W}(u, v) \subset \mathscr{W}(w, z)$, and that convergence in $\mathscr{W}(u, v)$ implies convergence in $\mathscr{W}(w, z)$. Since $\mathscr{D} \subset \mathscr{W}(u, v) \subset \mathscr{W}(w, z)$, it follows that $\mathscr{W}(w, z)$ is also dense in $\mathscr{W}(u, v)$. By Theorem 1.9-1 again, $\mathscr{W}'(w, z)$ is a subspace of $\mathscr{W}'(u, v)$.

II. $\mathscr{W}(w, z)$ is a dense subspace of $\mathscr{E}$, and convergence in $\mathscr{W}(w, z)$ implies convergence in $\mathscr{E}$, whatever be the choices of w and z. (Prove this, too.) Therefore, Theorem 1.9-1 implies that $\mathscr{E}'$ is a subspace of $\mathscr{W}'(w, z)$.

III. In the present context, Theorem 1.8-1 asserts the following: For any given $f \in \mathscr{W}'_{a, \, b}$, there exists a positive constant C and a nonnegative integer r such that, for all $\phi \in \mathscr{W}_{a, \, b}$,

$$|\langle f, \phi\rangle| \leq C \max_{0 \leq p \leq r} \chi_{a, \, b, \, p}(\phi)$$

IV. Assume that f is a functional on $\mathscr{W}_{a, \, a} \cup \mathscr{W}_{b, \, b}(a < b)$ such that it has continuous linear restrictions to $\mathscr{W}_{a, \, a}$ and $\mathscr{W}_{b, \, b}$. Let $\lambda(\tau)$ be a smooth function on $-\infty < \tau < \infty$ such that $\lambda(\tau) = 0$ for $\tau < -1$ and $\lambda(\tau) = 1$ for $\tau > 1$. Then, f can be extended into a member of $\mathscr{W}'_{a,b}$ through the definition

$$\langle f, \phi \rangle \triangleq \langle f, \lambda\phi \rangle + \langle f, (1 - \lambda)\phi \rangle \qquad \phi \in \mathscr{W}_{a,b}$$

The right-hand side has a sense because $\lambda\phi \in \mathscr{W}_{b,b}$ and $(1 - \lambda)\phi \in \mathscr{W}_{a,a}$ whenever $\phi \in \mathscr{W}_{a,b}$. Moreover, this extension is unique; that is, two members of $\mathscr{W}'_{a,b}$ having the same restrictions to $\mathscr{W}_{a,a}$ and $\mathscr{W}_{b,b}$ must be identical. The proofs of these assertions, being almost identical to those of Sec. 3.2, note *IV*, are also omitted.

V. Let $f(\tau)$ be a locally integrable function such that $f(\tau)/e^{\tau^2/4}\rho_{a,b}(\tau)$ is absolutely integrable on $-\infty < \tau < \infty$ for every a and b satisfying $a > w$ and $b < z$. Then, $f(\tau)$ generates a regular member f of $\mathscr{W}'(w, z)$ through the definition:

$$\langle f, \phi \rangle \triangleq \int_{-\infty}^{\infty} f(\tau)\phi(\tau)\, d\tau \qquad \phi \in \mathscr{W}(w, z)$$

PROBLEM 7.2-1. Prove that $\mathscr{W}_{a,b}$ is complete.

PROBLEM 7.2-2. Prove that $\mathscr{D}$ is a dense subspace of $\mathscr{W}(w, z)$ for every choice of w and z.

PROBLEM 7.2-3. Prove note *II*.

PROBLEM 7.2-4. Prove note *IV*.

PROBLEM 7.2-5. Prove note *V*.

7.3. The Weierstrass Transformation

As is the case with so much of this chapter, our discussion of a Weierstrass-transformable generalized function is quite analogous to the corresponding discussion in Chapter III. We define f to be a *Weierstrass-transformable generalized function* if it possesses the following four properties:

(i) f is a functional on some domain $d(f)$ of conventional functions.

(ii) f is additive; that is, if ϕ, θ, $\phi + \theta \in d(f)$, then $\langle f, \phi + \theta \rangle = \langle f, \phi \rangle + \langle f, \theta \rangle$.

(iii) $\mathscr{W}_{a,b} \subset d(f)$ for at least one pair of real numbers a and b with $a < b$.

(iv) For every $\mathscr{W}_{c,d} \subset d(f)$, the restriction of f to $\mathscr{W}_{c,d}$ is a member of $\mathscr{W}'_{c,d}$.

Corresponding to f there exists a unique set Λ_f in $\mathscr{R}^1$ defined as follows: A point σ is in Λ_f if and only if there exist two real numbers a_σ and b_σ with $a_\sigma < \sigma < b_\sigma$ such that $\mathscr{W}_{a_\sigma, b_\sigma} \subset d(f)$. Let σ_1 be the infimum of Λ_f and σ_2 the supremum of Λ_f. Here, $\sigma_1 = -\infty$ and $\sigma_2 = \infty$ are possible. By using Sec. 7.2, note *IV* and the technique described at the beginning of Sec. 3.3, we can extend f into a functional f_1 on $d(f) \cup \mathscr{W}(\sigma_1, \sigma_2)$ having the following two properties:

(*A*) *The restriction of* f_1 *to* $\mathscr{W}(\sigma_1, \sigma_2)$ *is a member of* $\mathscr{W}'(\sigma_1, \sigma_2)$.

(*B*) *The restriction of* f_1 *to* $d(f)$ *coincides with* f.

This extension is unique in that there is no other functional on $d(f) \cup \mathscr{W}(\sigma_1, \sigma_2)$ with these two properties.

We shall always assume that every Weierstrass-transformable generalized function f has been extended this way into the functional f_1, but we shall always denote f_1 simply by f. Under this convention, we have that *for every Weierstrass-transformable generalized function* f *there exists a unique nonvoid open interval* (σ_1, σ_2) *such that* f *has a continuous linear restriction to* $\mathscr{W}(\sigma_1, \sigma_2)$ *and is not defined on all of* $\mathscr{W}(w, z)$ *if either* $w < \sigma_1$ *or* $z > \sigma_2$.

We are now ready to define our *generalized Weierstrass transformation* $\mathfrak{W}$. Given a Weierstrass-transformable generalized function f, the *strip of definition* Ω_f *for* $\mathfrak{W}f$ is a set in $\mathscr{C}^1$ defined by

$$\Omega_f \triangleq \{s: \sigma_1 < \operatorname{Re} s < \sigma_2\}$$

Here, σ_1 and σ_2 are the infimum and supremum of Λ_f, as above, and are called the *abscissas of definition for* $\mathfrak{W}f$. We noted in Sec. 7.2 that, for each fixed $s \in \Omega_f$, $k(s - \tau, 1)$ as a function of τ is a member of $\mathscr{W}(\sigma_1, \sigma_2)$. Therefore, since $f \in \mathscr{W}'(\sigma_1, \sigma_2)$, we can define the *Weierstrass transform* $\mathfrak{W}f$ *of* f as a conventional function $F(s)$ on Ω_f through the expression:

$$F(s) \triangleq (\mathfrak{W}f)(s) \triangleq \langle f(\tau), k(s - \tau, 1)\rangle \qquad s \in \Omega_f \tag{1}$$

Henceforth, whenever we write $\mathfrak{W}f$ it is always understood that f is a Weierstrass-transformable generalized function that has been extended as stated above.

The conventional Weierstrass transformation (Sec. 7.1, Eq. (4)) is a special case of our generalized transformation (1) whenever $f(\tau)$ is a locally integrable function such that $f(\tau)/e^{\tau^2/4}\rho_{a,b}(\tau)$ is absolutely integrable on $-\infty < \tau < \infty$ for every $a > \sigma_1$ and every $b < \sigma_2$. This assertion is an immediate consequence of Sec. 7.2, note *V*.

The Weierstrass and Laplace transformations are related as follows:

THEOREM 7.3-1. *$f(\tau)$ is a Weierstrass-transformable generalized function with the strip of definition $\{s: \sigma_1 < \operatorname{Re} s < \sigma_2\}$ for $(\mathfrak{W}f)(s)$ if and only if*

$e^{-\tau^2/4}f(\tau)$ *is a Laplace-transformable generalized function with the strip of definition* $\{z\colon -\sigma_2/2 < \operatorname{Re} z < -\sigma_1/2\}$ *for* $[\mathfrak{L}e^{-\tau^2/4}f(\tau)](z)$. *In this case,*

$$(2) \qquad (\mathfrak{W}f)(s) = \frac{e^{-s^2/4}}{\sqrt{4\pi}}\,[\mathfrak{L}e^{-\tau^2/4}f(\tau)]\left(-\frac{s}{2}\right) \qquad \sigma_1 < \operatorname{Re} s < \sigma_2$$

PROOF. The first statement follows directly from Theorem 7.2-2. On the other hand, (2) can be obtained by a straightforward computation using Sec. 7.2, Eq. (7).

By virtue of this theorem, a number of properties of the Laplace transformation can be converted into properties of the Weierstrass transformation. For example, we have the analyticity property of a Weierstrass transform:

THEOREM 7.3-2. *Let* $F(s) = \mathfrak{W}f$ *for* $s \in \Omega_f = \{s\colon \sigma_1 < \operatorname{Re} s < \sigma_2\}$. *Then,* $F(s)$ *is analytic on* Ω_f, *and*

$$(3) \quad D^m F(s) = \langle f(\tau), D_s{}^m k(s-\tau, 1)\rangle \qquad s \in \Omega_f, \qquad m = 1, 2, 3, \ldots$$

PROOF. $F(s)$ is analytic on Ω_f because the right-hand side of (2) is analytic there according to Theorems 3.3-1 and 7.3-1. Moreover, (3) can be obtained from (2) and Sec. 3.4, Eq. (5) through an easy computation.

Similarly, Theorem 3.5-1 is converted via Theorem 7.3-1 into an inversion theorem for the Weierstrass transformation. Indeed, by setting $z = -s/2$, we can rewrite (2) as

$$(4) \qquad [\mathfrak{L}e^{-\tau^2/4}f(\tau)](z) = \sqrt{4\pi}\, e^{z^2} F(-2z) \qquad -\frac{\sigma_2}{2} < \operatorname{Re} z < -\frac{\sigma_1}{2}$$

where $F = \mathfrak{W}f$. Upon applying the inversion formula of theorem 3.5-1 to this and making the change of variable $s = -2z$ in the result, we arrive at

THEOREM 7.3-3. *Let* $\mathfrak{W}f = F(s)$ *for* $\sigma_1 < \operatorname{Re} s < \sigma_2$, *and let* r *be a real variable, Then, in the sense of convergence in* $\mathscr{D}'$.

$$(5) \qquad f(\tau) = \lim_{r\to\infty} \frac{1}{i\sqrt{4\pi}} \int_{\sigma - ir}^{\sigma + ir} F(s) e^{(s-\tau)^2/4}\, ds$$

$$= \lim_{r\to\infty} \int_{-r}^{r} F(\sigma + i\omega) k(\omega + i\tau - i\sigma, 1)\, d\omega$$

where σ *is any fixed real number such that* $\sigma_1 < \sigma < \sigma_2$.

Actually, we can take $r \to \infty$ independently in the upper and lower limits of the above integrals; see Problem 3.5-1.

As a consequence of Theorem 7.3-3, we have

THEOREM 7.3-4 (*The Uniqueness Theorem*). *If* $\mathfrak{W}f = F(s)$ *for* $s \in \Omega_f$ *and* $\mathfrak{W}h = H(s)$ *for* $s \in \Omega_h$, *if* $\Omega_f \cap \Omega_h$ *is not empty, and if* $F(s) = H(s)$ *for* $s \in \Omega_f \cap \Omega_h$, *then* $f = h$ *in the sense of equality in* $\mathscr{W}'(w, z)$, *where the interval* $w < \sigma < z$ *is the intersection of* $\Omega_f \cap \Omega_h$ *with the real axis.*

PROOF. By the inversion formula (5), wherein we choose σ such that $w < \sigma < z$, both f and h must assign the same value to each $\phi \in \mathscr{D}$. But, $\mathscr{D}$ is dense in $\mathscr{W}(w, z)$, and therefore f and h must also assign the same value to each $\phi \in \mathscr{W}(w, z)$. Q.E.D.

Finally, we can characterize Weierstrass transforms by their growth properties as follows:

THEOREM 7.3-5. *Necessary and sufficient conditions for a function* $F(s)$ *to be the Weierstrass transform of a generalized function (according to the definition* (1)) *and for the corresponding strip of definition to be* $\Omega_f = \{s: \sigma_1 < \operatorname{Re} s < \sigma_2\}$ *are that* $F(s)$ *be analytic on* Ω_f *and, for each closed substrip* $\{s: a \leq \operatorname{Re} s \leq b\}$ *of* Ω_f $(\sigma_1 < a < b < \sigma_2)$, *there be a polynomial* B *such that*

$$|F(\sigma + i\omega)| \leq e^{\omega^2/4} B(|\omega|)$$

for $a \leq \operatorname{Re} s \leq b$. *The polymonial* B *will depend in general on the choices of* a *and* b.

This theorem is also obtained by combining Theorem 7.3-1 with the analogous result for the Laplace transformation, namely, Theorem 3.6-1.

Theorems 7.3-4 and 7.3-5 imply that, for every choice of σ_1 and σ_2, the generalized Weierstrass transformation is a one-to-one mapping of $\mathscr{W}'(\sigma_1, \sigma_2)$ onto the space of functions that are analytic on the strip $\{s: \sigma_1 < \operatorname{Re} s < \sigma_2\}$ and satisfy the growth requirements stated in Theorem 7.3-5.

PROBLEM 7.3-1. Show that a Weierstrass-transformable generalized function can be extended as described in the beginning of this section.

PROBLEM 7.3-2. Prove Theorem 7.3-2 directly without resorting to Theorem 7.3-1.

PROBLEM 7.3-3. Prove Theorem 7.3-5.

PROBLEM 7.3-4. Let $x \in \mathscr{R}^1$ be fixed. Prove that the mapping $\phi(\tau) \mapsto \phi(\tau + x)$ is an isomorphism from $\mathscr{W}_{a+x,\, b+x}$ onto $\mathscr{W}_{a,\, b}$. Then, assuming that $\mathfrak{W}[f(\tau)] = F(s)$ for $s \in \Omega_f$, establish the following operation-transform formula:

$$\mathfrak{W}[f(\tau - x)] = F(s - x) \qquad s - x \in \Omega_f$$

PROBLEM 7.3-5. Assuming that $\mathfrak{W}f = F(s)$ for $s \in \Omega_f$, establish the following operation-transform formula:

$$\mathfrak{W}[D_\tau{}^m f(\tau)] = D_s{}^m F(s) \qquad s \in \Omega_f, \qquad m = 1, 2, 3, \ldots$$

PROBLEM 7.3-6. Prove that the differential operator $\tau - 2D_\tau$ is a continuous linear mapping of $\mathscr{W}'(w, z)$ into $\mathscr{W}'(w, z)$. Then, establish the following operation-transform formula wherein $F(s) = \mathfrak{W}f$ for $s \in \Omega_f$.

$$[\mathfrak{W}(\tau - 2D_\tau)f(\tau)](s) = sF(s) \qquad s \in \Omega_f$$

Finally, describe an operational calculus for differential equations of the form:

$$P(\tau - 2D_\tau)u(\tau) = g(\tau) \qquad -\infty < \tau < \infty$$

where P is a polynomial, $g \in \mathscr{W}'(\sigma_1, \sigma_2)$ and u is the unknown.

PROBLEM 7.3-7. By using the operational calculus indicated in the preceding problem, find a Weierstrass-transformable generalized function $u(\tau)$ that satisfies the differential equation:

$$[(\tau - 2D_\tau)^2 + \alpha^2]u(\tau) = \delta(\tau)$$

where α is a complex number. Can you find another solution?

PROBLEM 7.3-8. State necessary and sufficient conditions on a function $F(s)$ in order for it to be the Weierstrass transform of a Weierstrass-transformable generalized function $f(t)$ that is concentrated on the interval $T \le t < \infty$ $(T > -\infty)$.

PROBLEM 7.3-9. Prove the following assertion: Let $\mathfrak{W}f = F(s)$ for $s \in \Omega_f$. Choose three fixed real numbers a, σ, and b in Ω_f such that $a < \sigma < b$ Also, choose a polynomial $Q(s)$ that has no zeros for $a \le \operatorname{Re} s \le b$ and which satisfies for some constant K the inequality:

$$\left|\frac{F(s)e^{s^2}}{Q(s)}\right| \le \frac{K}{|s|^2} \qquad a < \operatorname{Re} s < b$$

Then, in the sense of equality in $\mathscr{W}'(a, b)$,

$$f(\tau) = Q(\tau - 2D_\tau)\frac{1}{i\sqrt{4\pi}}\int_{\sigma - i\infty}^{\sigma + i\infty} \frac{F(s)}{Q(s)} e^{(s-\tau)^2/4}\, ds$$

where D_τ represents generalized differentiation in $\mathscr{W}'(a, b)$ and the integral converges in the conventional sense to a continuous function that generates a regular member of $\mathscr{W}'(a, b)$.

PROBLEM 7.3-10. An n-dimensional Weierstrass transformation can be constructed by relying on the results of Sec. 3.11. In this problem we use the notation:

$$e^{az^2} \triangleq \exp(az_1{}^2 + \cdots + az_n{}^2)$$

$$e^{az\tau} \triangleq \exp(az_1\tau_1 + \cdots + az_n\tau_n)$$

where $z \in \mathscr{C}^n$, $\tau \in \mathscr{R}^n$, and $a \in \mathscr{R}^1$. We say that f is a Weierstrass-transformable generalized function and that its transform $\mathfrak{W}f$ has the tube of definition $\Omega_f \subset \mathscr{C}^n$ if and only if $e^{-\tau^2/4}f(\tau)$ is a Laplace-transformable generalized function and the tube of definition for $\mathfrak{L}e^{-\tau^2/4}f(\tau)$ is $\{s: -s/2 \in \Omega_f\}$. In this case, define $(\mathfrak{W}f)(s)$ by

$$(\mathfrak{W}f)(s) = \frac{e^{-s^2/4}}{(4\pi)^{n/2}}[\mathfrak{L}e^{-\tau^2/4}f(\tau)]\left(-\frac{s}{2}\right) \qquad s \in \Omega_f$$

By referring to Sec. 3.11 and especially Problem 3.11-9, determine the n-dimensional analogs to Theorems 7.3-2, 7.3-4, and 7.3-5.

7.4. Another Inversion Formula

The inversion formula given by Sec. 7.3, Eq. (5) states that a given Weierstrass-transformable generalized function $f(\tau)$ is the limit in $\mathscr{D}'$ as $r \to \infty$ of the following conventional functions depending on the parameter r:

$$\int_{-r}^{r} F(\sigma + i\omega)k(\omega + i\tau - i\sigma, 1)\, d\omega$$

Here, $F(s) = \mathfrak{W}f$ for $s \in \Omega_f$ and σ is any fixed real number in Ω_f. The reason we have to take a generalized limit is that the integral does not always converge in the conventional sense when r is set equal to ∞.

There is another way to generate $f(\tau)$ as the limit in $\mathscr{D}'$ of a directed set of conventional functions. That is to replace the above integral by

$$\int_{-\infty}^{\infty} F(\sigma + i\omega)k(\omega + i\tau - i\sigma, t)\, d\omega$$

and then to take the limit as $t \to 1-$. That the latter integral converges for each positive $t < 1$ follows from the bound of Theorem 7.3-5 and the definition of the kernel k. Thus, our objective in this section is to prove the following alternative inversion formula for our generalized Weierstrass transformation wherein convergence in $\mathscr{D}'$ is understood.

$$(1) \qquad f(\tau) = \lim_{t\to 1-} \frac{1}{i\sqrt{4\pi t}} \int_{\sigma - i\infty}^{\sigma + i\infty} F(s) e^{(s-\tau)^2/4t}\, ds$$

$$= \lim_{t \to 1-} \int_{-\infty}^{\infty} F(\sigma + i\omega)k(\omega + i\tau - i\sigma, t)\, d\omega$$

This formula was established by Hirschman and Widder [1], p. 191, for certain classes of conventional functions and measures, wherein the limit was taken in a conventional sense.

We start with three lemmas.

LEMMA 7.4-1. *Let $f(\tau) \in \mathscr{W}'_{a,b}$, let $\psi(x)$ be a smooth function on the finite interval $A \leq x \leq B$, and let $\theta(\tau, x)$ be a smooth function on the domain:*

$$\{(\tau, x): -\infty < \tau < \infty, A \leq x \leq B\}$$

such that for every nonnegative integer p

$$\lim_{|\tau| \to \infty} e^{\tau^2/4} \rho_{a,b}(\tau) D_\tau^{\,p} \theta(\tau, x) = 0 \tag{2}$$

uniformly for $A \leq x \leq B$.

Then, the following three assertions are true.

(i)

$$\int_A^B \psi(x)\theta(\tau, x)\, dx \in \mathscr{W}_{a,b}$$

(ii) $\langle f(\tau), \theta(\tau, x)\rangle$ *is a continuous function on $A \leq x \leq B$.*

(iii)

$$\left\langle f(\tau), \int_A^B \psi(x)\theta(\tau, x)\, dx\right\rangle = \int_A^B \psi(x)\langle f(\tau), \theta(\tau, x)\rangle\, dx \tag{3}$$

PROOF. First note that

$$I(\tau) \triangleq \int_A^B \psi(x)\theta(\tau, x)\, dx$$

is a smooth function of τ and may be differentiated under the integral sign any number of times. Moreover, for each $p = 0, 1, 2, \ldots$

$$\chi_{a,b,p}\left[\int_A^B \psi(x)\theta(\tau, x)\, dx\right]$$

$$= \sup_{-\infty<\tau<\infty} \left| e^{\tau^2/4} \rho_{a,b}(\tau) \int_A^B \psi(x) D_\tau^{\,p} \theta(\tau, x)\, dx \right|$$

$$\leq \int_A^B |\psi(x)|\, dx \sup_{-\infty<\tau<\infty} [e^{\tau^2/4} \rho_{a,b}(\tau) \sup_{A<x<B} |D_\tau^{\,p} \theta(\tau, x)|]$$

The last expression is finite because of (2), and this establishes (i).

To prove (ii), fix x as a point in the closed interval $[A, B]$, and restrict Δx so that $x + \Delta x$ is also a point in $[A, B]$. We will have proven that $\langle f(\tau), \theta(\tau, x)\rangle$ is a continuous function on $A \leq x \leq B$ when we show that

$$\theta(\tau, x + \Delta x) - \theta(\tau, x) \to 0$$

in $\mathscr{W}_{a,b}$ as $\Delta x \to 0$. In view of (2), given an $\varepsilon > 0$ and a nonnegative integer p, we can choose T so large that, for all $|\tau| > T$ and all permissible Δx,

$$|e^{\tau^2/4} \rho_{a,b}(\tau) D_\tau^{\,p}] \theta(\tau, x + \Delta\tau) - \theta(\tau, x)]| < \frac{\varepsilon}{2} \tag{4}$$

Fix T this way. Since the left-hand side of (4) is a continuous function on all of the $(\tau, \Delta x)$ plane, it tends to zero as $\Delta x \to 0$ uniformly on

$$-T \leq \tau \leq T.$$

Thus, $\theta(\tau, x + \Delta x) - \theta(\tau, x)$ does tend to zero in $\mathscr{W}_{a,b}$ as $\Delta x \to 0$.

We now prove assertion (iii) by making use of certain "Riemann sums" for the integrals therein. Set $X = B - A$. If we can show that, as $m \to \infty$,

$$J(\tau, m) \triangleq \frac{X}{m} \sum_{\nu=1}^{m} \psi\left(A + \frac{\nu X}{m}\right) \theta\left(\tau, A + \frac{\nu X}{m}\right)$$

converges in $\mathscr{W}_{a,b}$ to $I(\tau)$, then we can write

$$\text{(5)} \quad \left\langle f(\tau), \frac{X}{m} \sum_{\nu=1}^{m} \psi\left(A + \frac{\nu X}{m}\right) \theta\left(\tau, A + \frac{\nu X}{m}\right)\right\rangle$$
$$\to \left\langle f(\tau), \int_A^B \psi(x)\theta(\tau, x)\, dx\right\rangle \qquad m \to \infty$$

On the other hand, we may apply $f(\tau)$ term by term to the sum in the left-hand side of (5), and, in view of (ii), the result tends to

$$\int_A^B \psi(x)\langle f(\tau), \theta(\tau, x)\rangle\, dx$$

as $m \to \infty$. Assertion (iii) will thereby be proven.

Throughout the following, p is fixed, and $\psi(x) \not\equiv 0$. (There is nothing to prove if $\psi(x) \equiv 0$.) Set

$$H(\tau, m) \triangleq e^{\tau^2/4} \rho_{a,b}(\tau) D_\tau^{\,p}[I(\tau) - J(\tau, m)]$$

We have to show that $H(\tau, m)$ tends to zero as $m \to \infty$ uniformly on $-\infty < \tau < \infty$.

In view of (2), given an $\varepsilon > 0$, there exists a T such that for $|\tau| > T$ and $A \leq x \leq B$,

$$|e^{\tau^2/4} \rho_{a,b}(\tau) D_\tau^{\,p} \theta(\tau, x)| < \frac{\varepsilon}{3}\left[\int_A^B |\psi(x)|\, dx\right]^{-1}$$

Consequently,

$$\sup_{|\tau| > T} |e^{\tau^2/4} \rho_{a,b}(\tau) D_\tau^{\,p} I(\tau)| < \frac{\varepsilon}{3}$$

Also, for all m,

$$\text{(6)} \quad \sup_{|\tau| > T} |e^{\tau^2/4} \rho_{a,b}(\tau) D_\tau^{\,p} J(\tau, m)|$$
$$< \frac{\varepsilon}{3}\left[\int_A^B |\psi(x)|\, dx\right]^{-1} \frac{X}{m} \sum_{\nu=1}^{m} \left|\psi\left(A + \frac{\nu X}{m}\right)\right|$$

Moreover, there exists an m_0 such that for all $m > m_0$ the right-hand side of (6) is bounded by $2\varepsilon/3$. Thus, for $m > m_0$ and $|\tau| > T$, we have $|H(\tau, m)| < \varepsilon$.

Next, set

$$K \triangleq \sup_{|\tau| < T} |e^{\tau^2/4}\rho_{a,b}(\tau)|$$

Then, for $|\tau| \leq T$,

$$(7) \quad |H(\tau, m)| \leq$$
$$K \left| \int_A^B \psi(x) D_\tau^{\,p} \theta(\tau, x)\, dx - \frac{X}{m} \sum_{\nu=1}^{m} \psi\left(A + \frac{\nu X}{m}\right) D_\tau^{\,p} \theta\left(\tau, A + \frac{\nu X}{m}\right) \right|$$

Since $\psi(x)D_\tau^{\,p}\theta(\tau, x)$ is uniformly continuous for all (τ, x) such that $-T \leq \tau \leq T$ and $A \leq x \leq B$, there exists an m_1 such that for all $m > m_1$, the right hand side of (7) is bounded by ε on $-T \leq \tau \leq T$. This completes the proof.

LEMMA 7.4-2. *Let $\mathfrak{W}f = F(s)$ for $s \in \Omega_f = \{s: \sigma_1 < \operatorname{Re} s < \sigma_2\}$, and let t and σ be fixed real numbers with $0 < t < 1$ and $\sigma_1 < \sigma < \sigma_2$. Given any $\varepsilon > 0$ and any compact subset Ξ of $\mathscr{R}^1$, the positive numbers ω_1 and ω_2 can be so chosen that, for all $x \in \Xi$,*

$$(8) \quad \left| \left(\int_{-\infty}^{-\omega_1} + \int_{\omega_2}^{\infty} \right) k(\omega + ix - i\sigma, t) \langle f(\tau), k(\sigma + i\omega - \tau, 1) \rangle d\omega \right| < \varepsilon$$

and simultaneously

$$(9) \quad \left| \left\langle f(\tau), \left(\int_{-\infty}^{-\omega_1} + \int_{\omega_2}^{\infty} \right) k(\omega + ix - i\sigma, t) k(\sigma + i\omega - \tau, 1)\, d\omega \right\rangle \right| < \varepsilon$$

PROOF. By Theorem 7.3-5, for s restricted to a strip $\{s: a \leq \operatorname{Re} s = \sigma \leq b\}$ where $\sigma_1 < a < b < \sigma_2$, we have

$$(10) \quad |k(\omega + ix - i\sigma, t) F(\sigma + i\omega)| \leq \frac{e^{(x-\sigma)^2/t4}}{\sqrt{4\pi t}} e^{\omega^2(1-1/t)/4} B(|\omega|)$$

where B is a polynomial. The inequality (8) follows immediately from (10).

To establish (9), first note that we may differentiate

$$(11) \quad \left(\int_{-\infty}^{-\omega_1} + \int_{\omega_2}^{\infty} \right) k(\omega + ix - i\sigma, t) k(\sigma + i\omega - \tau, 1)\, d\omega$$

with respect to τ under the integral sign any number of times. This is because, if τ is confined to any bounded interval, Sec. 7.2, Eq. (2) implies that

$$|k(\omega + ix - i\sigma, t) D_\tau^{\,p} k(\sigma + i\omega - \tau, 1)| \leq e^{\omega^2(1-1/t)/4} Q(|\omega|)$$

where Q is a polynomial depending on x, t, σ, and the said τ interval. Thus, every differentiation with respect to τ of (11) under the integral sign leads to an integral which converges uniformly on every bounded τ interval.

By differentiating (11) with respect to τ under the integral sign p times and by again using Sec. 7.2, Eq. (2), we obtain

$$\left| e^{\tau^2/4}\rho_{a,\,b}(\tau)D_\tau{}^p \left(\int_{-\infty}^{-\omega_1} + \int_{\omega_2}^{\infty}\right) k(\omega + ix - i\sigma,\, t)k(\sigma + i\omega - \tau,\, 1)\, d\omega\right| \tag{12}$$

$$\leq \frac{e^{(x-\sigma)^2/4t}}{\sqrt{t}} \sum_\nu{}^* C_\nu e^{\sigma\tau/2}\rho_{a,\,b}(\tau)R_\nu(|\tau|)\left(\int_{-\infty}^{-\omega_1} + \int_{\omega_2}^{\infty}\right) e^{\omega^2(1-1/t)/4}S_\nu(|\omega|)d\omega$$

Here, $\sum^*$ denotes a finite sum, R_ν and S_ν denote monomials, and the C_ν are constants depending on σ. With $\sigma_1 < a < \sigma < b < \sigma_2$, we have that $e^{\sigma\tau/2}\rho_{a,\,b}(\tau)R_\nu(|\tau|)$ is bounded on $-\infty < \tau < \infty$. Also, since $0 < t < 1$, the integral in the right-hand side of (12) is finite. This proves that (11) as a function of τ is a member of $\mathscr{W}_{a,\,b}$. Moreover, the integral in the right-hand side of (12) can be made arbitrarily small by choosing ω_1 and ω_2 sufficiently large. In view of Sec. 7.2, note *III* and the restriction on x, this verifies (9) and completes the proof of Lemma 7.4-2.

LEMMA 7.4-3. *Let x, σ, t, and τ be fixed real numbers with $0 < t < 1$. Then,*

$$\int_{-\infty}^{\infty} k(\omega + ix - i\sigma,\, t)k(\sigma + i\omega - \tau,\, 1)\, d\omega = k(x - \tau,\, 1 - t) \tag{13}$$

PROOF. We start from the well-known Fourier transform:

$$\int_{-\infty}^{\infty} e^{-y^2/4}e^{-i\eta y/2}\, dy = \sqrt{4\pi}\, e^{-\eta^2/4} \tag{14}$$

(A derivation for it is indicated in Zemanian [1], pp. 180–181.) Set

$$y = \omega\sqrt{\frac{1-t}{t}}\,,\ \eta = \sqrt{\frac{t}{1-t}}\left(\frac{x-\sigma}{t} + \sigma - \tau\right) \tag{15}$$

Upon multiplying the resulting equation by

$$\frac{1}{4\pi\sqrt{1-t}}\exp\left\{\frac{1}{4}\left[\frac{(x-\sigma)^2}{t} - (\sigma-\tau)^2\right]\right\}$$

we obtain (13) after some simplification.

We are at last ready to prove the inversion formula (1) for our generalized Weierstrass transformation.

THEOREM 7.4-1. *Let* $F(s) = \mathfrak{W}f$ *for* $s \in \Omega_f = \{s: \sigma_1 < \text{Re } s < \sigma_2\}$. *Let* σ *be any fixed real number such that* $\sigma_1 < \sigma < \sigma_2$. *Then,* (1) *is true in the sense of convergence in* $\mathscr{D}'$; *in other words, for each* $\phi \in \mathscr{D}$,

$$\lim_{t \to 1-} \left\langle \int_{-\infty}^{\infty} k(\omega + i\tau - i\sigma, t) F(\sigma + i\omega)\, d\omega, \phi(\tau) \right\rangle = \langle f(\tau), \phi(\tau) \rangle \tag{16}$$

PROOF. Formally, the proof runs as follows. For $0 < t < 1$ and for a and b chosen such that $\sigma_1 < a < \sigma < b < \sigma_2$,

$$\left\langle \int_{-\infty}^{\infty} k(\omega + ix - i\sigma, t) F(\sigma + i\omega)\, d\omega, \phi(x) \right\rangle \tag{17}$$

$$= \left\langle \int_{-\infty}^{\infty} k(\omega + ix - i\sigma, t) \langle f(\tau), k(\sigma + i\omega - \tau, 1)\rangle\, d\omega, \phi(x) \right\rangle \tag{18}$$

$$= \left\langle \left\langle f(\tau), \int_{-\infty}^{\infty} k(\omega + ix - i\sigma, t) k(\sigma + i\omega - \tau, 1)\, d\omega \right\rangle, \phi(x) \right\rangle \tag{19}$$

$$= \langle \phi(x), \langle f(\tau), k(x - \tau, 1 - t)\rangle\rangle \tag{20}$$

$$= \langle f(\tau), \langle \phi(x), k(x - \tau, 1 - t)\rangle\rangle \tag{21}$$

Since $f \in \mathscr{W}'_{a,b}$, the proof is completed by showing that

$$\langle \phi(x), k(x - \tau, 1 - t)\rangle \to \phi(\tau)$$

in $\mathscr{W}_{a,b}$ as $t \to 1-$.

We now justify the various steps of this formal argument. For fixed t with $0 < t < 1$, we see from (10) that

$$\int_{-\infty}^{\infty} k(\omega + ix - i\sigma, t) F(\sigma + i\omega)\, d\omega \tag{22}$$

converges uniformly for x restricted to any compact set, which implies that (22) is a continuous function of x. (By Cauchy's theorem, (10) also shows that we may change the value of σ in the range $\sigma_1 < \sigma < \sigma_2$ without altering (22) as a function of x. Thus, (17) and (18) have a sense as the integral of the product of (22) and $\phi(x)$.

Next, by Sec. 7.2, Eq. (2) once again, we see that

$$|e^{\tau^2/4} \rho_{a,b}(\tau) D_\tau^{\,p} k(\sigma + i\omega - \tau, 1)| =$$

$$\frac{1}{\sqrt{4\pi}} e^{(\omega^2 - \sigma^2)/4} \rho_{a,b}(\tau) e^{\sigma\tau/2} |P_p(\sigma + i\omega - \tau, 1)|$$

and, since $a < \sigma < b$, this quantity tends to zero as $|\tau| \to \infty$ uniformly on $-\omega_1 \le \omega \le \omega_2$, where ω_1 and ω_2 are any finite positive numbers. Thus, from Lemma 7.4-1 we have that

$$\int_{-\omega_1}^{\omega_2} k(\omega + ix - i\sigma, t)\langle f(\tau), k(\sigma + i\omega - \tau, 1)\rangle \, d\omega$$

$$= \left\langle f(\tau), \int_{-\omega_1}^{\omega_2} k(\omega + ix - i\sigma, t)k(\sigma + i\omega - \tau, 1)\, d\omega \right\rangle$$

By combining this equation with Lemma 7.4-2 and recalling that $\phi(x)$ is a fixed member of $\mathscr{D}$, we see that the difference between (18) and (19) can be made arbitrarily small. Hence, (18) is truly equal to (19) when $\sigma_1 < \sigma < \sigma_2$ and $0 < t < 1$.

That (19) is equal to (20) is asserted by Lemma 7.4-3.

Next, the support of $\phi(x) \in \mathscr{D}$ is contained in a finite interval, say, $A \leq x \leq B$. Moreover,

$$e^{\tau^2/4}\rho_{a,\,b}(\tau)D_\tau{}^p k(x - \tau, 1 - t) =$$

$$\frac{e^{-x^2/4(1-t)}}{\sqrt{4\pi(1 - t)}}\, \rho_{a,\,b}(\tau)e^{x\tau/2(1-t)}e^{\tau^2 t/4(t-1)} P_p\left(x - \tau, \frac{1}{1 - t}\right)$$

Since $t/(t - 1) < 0$ for every choice of a and b, this quantity tends to zero as $|\tau| \to \infty$ uniformly on $A \leq x \leq B$. Therefore, we can invoke Lemma 7.4-1 again, which proves that (20) is equal to (21) when $\sigma_1 < a < b < \sigma_2$ and $0 < t < 1$.

As the last step, we prove that, as $t \to 1-$,

$$\langle \phi(x), k(x - \tau, 1 - t)\rangle \to \phi(\tau)$$

in $\mathscr{W}_{a,\,b}$ for every a and b. Set $x = \tau + 2y\sqrt{1 - t}$ where τ and t are fixed $(0 < t < 1)$. Then,

$$(23) \qquad \frac{1}{\sqrt{4\pi(1 - t)}} \int_{-\infty}^{\infty} \exp \frac{(x - \tau)^2}{4(t - 1)}\, dx = \pi^{-1/2} \int_{-\infty}^{\infty} e^{-y^2}\, dy = 1$$

Using this equation and the fact that $\phi \in \mathscr{D}$, we may differentiate under the integral sign and integrate by parts n times to get

$$(24) \quad e^{\tau^2/4}\rho_{a,\,b}(\tau)D_\tau{}^p[\langle \phi(x), k(x - \tau, 1 - t)\rangle - \phi(\tau)]$$

$$= \frac{e^{\tau^2/4}\rho_{a,\,b}(\tau)}{\sqrt{4\pi(1 - t)}} \int_{-\infty}^{\infty} [\phi^{(p)}(x) - \phi^{(p)}(\tau)] \exp \frac{(x - \tau)^2}{4(t - 1)}\, dx$$

$$= I_1 + I_2 + I_3 \qquad \phi^{(p)}(x) \triangleq D_x{}^p\phi(x)$$

Here, I_1, I_2, and I_3 denote respectively the expressions obtained after the integration on $-\infty < x < \infty$ is broken up into integrations on $-\infty < x < \tau - \delta$, $\tau - \delta \leq x \leq \tau + \delta$, and $\tau + \delta < x < \infty$ $(\delta > 0)$.

Consider $I_2(\tau)$. In view of (23), we have

$$|I_2(\tau)| \leq e^{\tau^2/4}\rho_{a,b}(\tau) \sup_{\tau-\delta<x<\tau+\delta} |\phi^{(p)}(x) - \phi^{(p)}(\tau)|$$

$$\leq \delta e^{\tau^2/4}\rho_{a,b}(\tau) \sup_{\tau-\delta<y<\tau+\delta} |\phi^{(p+1)}(y)|$$

Since $\phi \in \mathscr{D}$, the right-hand side is dominated by δB where B is a constant with respect to t, τ, and δ ($0 < t < 1$, $-\infty < \tau < \infty$, and $0 < \delta < 1$). There, given an $\varepsilon > 0$, we have $|I_2(\tau)| < \varepsilon$ for $\delta = \min(1, \varepsilon/B)$. Fix δ this way.

Next, consider

$$(25) \quad I_1(\tau) = \frac{e^{\tau^2/4}\rho_{a,b}(\tau)}{\sqrt{4\pi(1-t)}} \int_{-\infty}^{\tau-\delta} \phi^{(p)}(x) \exp \frac{(x-\tau)^2}{4(t-1)}\, dx$$

$$- \frac{e^{\tau^2/4}\rho_{a,b}(\tau)}{\sqrt{4\pi(1-t)}} \phi^{(p)}(\tau) \int_{-\infty}^{\tau-\delta} \exp \frac{(x-\tau)^2}{4(t-1)}\, dx$$

By the change of variable $x = \tau + 2y\sqrt{1-t}$, we obtain

$$\frac{1}{\sqrt{4\pi(1-t)}} \int_{-\infty}^{\tau-\delta} \exp \frac{(x-\tau)^2}{4(t-1)}\, dx = \pi^{-1/2} \int_{-\infty}^{-\delta/2\sqrt{1-t}} e^{-y^2}\, dy \to 0 \qquad t \to 1-$$

Since the support of ϕ is bounded, this shows that the second term on the right-hand side of (25) tends uniformly to zero on $-\infty < \tau < \infty$ as $t \to 1-$.

Let $J_1(\tau)$ denote the first term on the right-hand side of (25), and assume that the support of $\phi(x)$ is contained in the finite interval $A \leq x \leq B$. Then, for $-\infty < \tau - \delta < A$, $J_1(\tau) \equiv 0$. For $A \leq \tau - \delta \leq B$, let K be a constant bound on $e^{\tau^2/4}\rho_{a,b}(\tau)$. Therefore.

$$|J_1(\tau)| \leq \frac{Ke^{\delta^2/4(t-1)}}{\sqrt{4\pi(1-t)}} \int_A^B |\phi^{(p)}(x)|\, dx \to 0 \qquad t \to 1-$$

Hence, as $t \to 1-$, $J_1(\tau) \to 0$ uniformly on $A \leq \tau - \delta \leq B$.

Finally, consider the range $B < \tau - \delta < \infty$. There exists a constant M such that $\rho_{a,b}(\tau) < Me^{-b\tau/2}$ for $B < \tau - \delta < \infty$. Consequently since $0 < t < 1$,

$$(26) \quad |J_1(\tau)| \leq \frac{M}{\sqrt{4\pi(1-t)}} \exp\left[\frac{\tau^2}{4} - \frac{b\tau}{2} + \frac{(B-\tau)^2}{4(t-1)}\right] \int_A^B |\phi^{(p)}(x)|\, dx$$

The quantity:

$$(27) \qquad \exp\left[\frac{\tau^2}{4} - \frac{b\tau}{2} + \frac{(B-\tau)^2}{4(t-1)}\right]$$

tends to zero as $|\tau| \to \infty$ and has a single maximum which occurs at the point:

$$\tau = \frac{B}{t} + b\left(1 - \frac{1}{t}\right)$$

Therefore there exists a t_1 such that, for $t_1 < t < 1$, the single maximum of (27) lies within the interval $(B - \delta, B + \delta)$. Since $B < \tau - \delta < \infty$, we may set $\tau = B + \delta$ to get, for $t_1 < t < 1$,

$$|J_1(\tau)| \leq$$

$$M \exp\left[\frac{(B+\delta)^2}{4} - \frac{b(B+\delta)}{2}\right] \int_A^B |\phi^{(p)}(x)|\, dx \frac{e^{\delta^2/4(t-1)}}{\sqrt{4\pi(1-t)}} \to 0 \qquad t \to 1-$$

Thus, as $t \to 1-$, $|J_1(\tau)|$ converges uniformly to zero on $B < \tau - \delta < \infty$ as well.

Altogether, we have demonstrated that, as $t \to 1-$, $I_1(\tau)$ converges to zero uniformly on $-\infty < \tau < \infty$. A similar argument shows that $I_3(\tau)$ also converges to zero uniformly on $-\infty < \tau < \infty$ as $t \to 1-$. In view of (24) and our bound on $I_2(\tau)$, this proves that for each $p = 0, 1, 2, \ldots$

$$\lim_{t \to 1-} \chi_{a,\, b,\, p}[\langle \phi(x),\, k(x - \tau,\, 1 - t)\rangle - \phi(\tau)] \leq \varepsilon$$

Since ε is arbitrary, the proof of Theorem 7.4-1 is complete.

7.5 The Cauchy Problem for the Heat Equation for One-Dimensional Flow

Let $v(x, y)$, where $-\infty < x < \infty$ and $0 < y < Y$, denote the temperature at time y in an infinitely long rod lying along the x axis. Thus, it is assumed that the temperature varies only with time y and the spacial dimension x but not with any other spacial dimension; this implies that the surface of the rod is perfectly insulated. Under the further assumptions that the rod material is uniform throughout and that the rod possesses no heat sources, the differential equation governing the temperature $v(x, y)$ is the so-called homogeneous heat equation for one-dimensional flow:

$$(1) \qquad \kappa D_x^2 v(x, y) = D_y v(x, y) \qquad -\infty < x < \infty, \qquad 0 < y < Y$$

Here, κ is a positive constant depending on the properties of the rod material. The Cauchy problem for (1) is to solve for $v(x, y)$ under an initial condition on $v(x, y)$ at $y = 0$.

More specifically, let the temperature at $y = 0$ be a distribution $g(x)$ having the following property:

(i) There exists a positive constant A such that $g(Ax) \in \mathscr{W}'(w, z)$ for for some w and some z.

Here, $g(Ax)$ denotes that unique member of $\mathscr{W}'(w, z)$ whose restriction to $\mathscr{D}$ is given by

$$(2) \qquad \langle g(Ax), \phi(x) \rangle \triangleq \left\langle g(x), \frac{1}{A} \phi\left(\frac{x}{A}\right) \right\rangle \qquad \phi(x) \in \mathscr{D}$$

(See Zemanian [1], p. 27.) As the initial condition, we require that the temperature $v(x, y)$, treated as a distribution on $-\infty < x < \infty$ depending parametrically on y, converges in $\mathscr{D}'$ to $g(x)$ as $y \to 0+$. As we shall see, the property (i) allows us to use our generalized Weierstrass transformation to determine $v(x, y)$. Moreover, $v(x, y)$ will turn out to be (a regular distribution corresponding to) a smooth function of x for each fixed y in $0 < y < Y$, and Y will depend on the value of A. Also, Y may be set equal to ∞ if condition (i) holds for every $A > 0$.

Here is a result from classical analysis: If $g(x)$ is a suitably restricted function, then $v(x, y)$ is given by the integral transformation:

$$(3) \qquad v(x, y) = \int_{-\infty}^{\infty} g(\eta) \frac{e^{-(x-\eta)^2/4\kappa y}}{\sqrt{4\pi\kappa y}} \, d\eta$$

The kernel of (3) is the Green's function for our problem. Moreover, (3) can be converted into an integral transformation whose kernel is $k(\sigma - \tau, t)$ by normalizing the variables as follows. Let α be a fixed real number satisfying $0 < \alpha < A$, and set

$$(4) \qquad \eta = \alpha\tau, \; x = \alpha\sigma, \; y = \frac{\alpha^2}{\kappa} t, \; u(\sigma, t) = v(x, y)$$

Thus, τ and σ are normalized space variables, whereas t is the normalized time variable; the temperature is now represented by u. The differential equation (1) becomes

$$(5) \qquad D_\sigma^2 u(\sigma, t) = D_t u(\sigma, t)$$

and the initial condition becomes

$$(6) \qquad u(\sigma, t) \to g(\alpha\sigma) \text{ in } \mathscr{D}' \text{ as } t \to 0+$$

Finally, the generalized analog to (3) under the change-of-variable formula (2) is

$$(7) \qquad v(x, y) = u(\sigma, t) = \langle g(\alpha\tau), k(\sigma - \tau, t) \rangle_\tau \qquad 0 < \sqrt{\kappa y} \leq \alpha < A$$

The subscript τ at the end of this expression indicates that τ is the independent variable for the testing function $k(\sigma - \tau, t)$; thus, α, σ, and t are to be treated as parameters. The restriction on y indicated in (7) insures that $0 < t \leq 1$ so that $k(\sigma - \tau, t) \in \mathscr{W}(-\infty, \infty)$. The restriction on α therein insures that $g(\alpha\tau) \in \mathscr{W}'(-\infty, \infty)$, as will be proven in the next lemma. As a consequence, (7) has a sense.

LEMMA, 7.5-1. *Let $g(\sigma) \in \mathscr{D}'$ be such that $g(A\sigma) \in \mathscr{W}'(w, z)$ for some $A > 0$, some w, and some z. If $0 < \alpha < A$, then $g(\alpha\sigma) \in \mathscr{W}'(-\infty, \infty)$.*

PROOF. By the definition (2), we have for any $\phi \in \mathscr{D}$

$$\langle g(\alpha\sigma), \phi(\sigma)\rangle = \langle g(A\sigma), r\phi(r\sigma)\rangle \tag{8}$$

where $r = A/\alpha > 1$. We shall first show that the right-hand side of (8) has a sense when $\phi(\sigma) \in \mathscr{W}(-\infty, \infty)$ (i.e., when $\phi(\sigma) \in \mathscr{W}_{c,\, d}$ for some $c > -\infty$ and $d < \infty$). To do this, we will first prove that $\phi(\sigma) \mapsto r\phi(r\sigma)$ is a continuous linear mapping of $\mathscr{W}_{c,\, d}$ into $\mathscr{W}_{a,\, b}$ whatever be the values of a, b, c, and d. Bearing in mind that σ is the independent variable for the testing function $\phi(r\sigma)$ and setting $\xi = r\sigma$, we may write

$$\chi_{a,\, b,\, p}[r\phi(r\sigma)] = r \sup_{-\infty<\sigma<\infty} |e^{\sigma^2/4}\rho_{a,\, b}(\sigma) D_\sigma{}^p \phi(r\sigma)|$$

$$= r^{p+1} \sup_{-\infty<\xi<\infty} \left| \left[\frac{e^{\xi^2/4r^2}\rho_{a,\, b}\left(\dfrac{\xi}{r}\right)}{e^{\xi^2/4}\rho_{c,\, d}(\xi)} \right] [e^{\xi^2/4}\rho_{c,\, d}(\xi) D_\xi{}^p \phi(\xi)] \right|$$

Now, the first factor under the last supremum sign is a bounded function of ξ. Therefore, for any choices of a, b, c, and d.

$$\chi_{a,\, b,\, p}[r\phi(r\sigma)] \leq K\chi_{c,\, d,\, p}[\phi(\sigma)]$$

where K is a constant not depending on ϕ, hence our assertion concerning $\phi(\sigma) \mapsto r\phi(r\sigma)$. This in turn implies that $\phi(\sigma) \mapsto r\phi(r\sigma)$ is a continuous linear mapping of $\mathscr{W}(-\infty, \infty)$ into $\mathscr{W}(w, z)$ for every w and z. So truly, the right-hand side of (8) has a sense as the application of $g(A\sigma) \in \mathscr{W}'(w, z)$ to $r\phi(r\sigma) \in \mathscr{W}(w, z)$. It now follows from Theorem 1.10-1 that $g(\alpha\sigma) \in \mathscr{W}'(-\infty, \infty)$. Q.E.D.

As was mentioned before, since $k(\sigma - \tau, t) \in \mathscr{W}(-\infty, \infty)$ for any positive $t \leq 1$ and every fixed σ, it follows that the right-hand side of (7) has a sense for $0 < t \leq 1$ or equivalently for $0 < y < \alpha^2/\kappa$. By analogy to classical analysis, it appears then that the temperature $v(x, y)$ in the rod at any instant of time y in the interval $0 < y < A^2/\kappa \triangleq Y$ can be obtained by choosing α such that $y \leq \alpha^2/\kappa < A^2/\kappa$ and then applying (7) with $t = \kappa y/\alpha^2$. Moreover, with $\alpha = \sqrt{\kappa y}$ we have that $t = 1$, in which case (7) is a generalized Weierstrass transform. According to Theorem 7.3-2 then, for each positive $y < A^2/\kappa$, (7) is a smooth (indeed, analytic) function of x for $-\infty < x < \infty$.

We turn now to the proof of the assertion that (7) provides a solution to our Cauchy problem. We can prove that $v(x, y)$ satisfies the differential equation (1), where $Y \triangleq A^2/\kappa$, by showing that $u(\sigma, t)$ satisfies (5) for $-\infty < \sigma < \infty$ and $0 < t < 1$ (where α is allowed to be any value such that $\sqrt{y\kappa} \leq \alpha < A$).

First, it can be shown that

(9)
$$D_\sigma{}^m u(\sigma, t) = \langle g(\alpha\tau), D_\sigma{}^m k(\sigma - \tau, t)\rangle \qquad m = 1, 2, 3, \ldots; \quad -\infty < \sigma < \infty; \quad 0 < t < 1$$

by temporarily renormalizing the variables to obtain $t = 1$ and then invoking Sec. 7.3, Eq. (3). We omit the details.

Second, we establish that

$$D_t u(\sigma, t) = \langle g(\alpha\tau), D_t k(\sigma - \tau, t)\rangle \qquad -\infty < \sigma < \infty,\ 0 < t < 1 \tag{10}$$

Since

$$D_t k(\sigma - \tau, t) = k(\sigma - \tau, t)\,\frac{(\sigma - \tau)^2 - 2t}{4t^2} \tag{11}$$

it follows that (11), as a function of τ, is a member of $\mathscr{W}(-\infty, \infty)$ when $0 < t \leq 1$; consequently, (10) possesses a sense, Next, consider

$$\frac{u(\sigma, t + \Delta t) - u(\sigma, t)}{\Delta t} - \langle g(\alpha\tau), D_t k(\sigma - \tau, t)\rangle = \langle g(\alpha\tau), \theta_{\Delta t}(\tau)\rangle$$

where $\Delta t \neq 0$ and

$$\theta_{\Delta t}(\tau) = \frac{1}{\Delta t}\,[k(\sigma - \tau, t + \Delta t) - k(\sigma - \tau, t)] - D_t k(\sigma - \tau, t)$$

Since $g(\alpha\tau) \in \mathscr{W}_{a,b}$ for every a and b (Lemma 7.5-1), we will have established (10) when we show that, as $\Delta t \to 0$, $\theta_{\Delta t}(\tau) \to 0$ in $\mathscr{W}_{a,b}$ for any arbitrary choice of a and b. By using the identity:

$$D_\tau{}^2 k(\sigma - \tau, t) = D_t k(\sigma - \tau, t) \tag{12}$$

one can show that

(13)
$$D_\tau{}^p \theta_{\Delta t}(\tau) = \frac{1}{\Delta t}\int_0^{\Delta t} d\xi \int_0^{\xi} D_\tau{}^{p+4} k(\sigma - \tau, t + \zeta)\, d\zeta \qquad p = 0, 1, 2, \ldots$$

Keeping t fixed with $0 < t < 1$, restrict Δt such that $0 < t - |\Delta t| < t + |\Delta t| < 1$. Then, from (13) and Sec. 7.2, Eq. (2), we get

$$|e^{\tau^2/4}\rho_{a,b}(\tau)D_\tau{}^p\theta_{\Delta t}(\tau)|$$

$$\leq \frac{|\Delta t|}{2} \sup_{|\zeta|\leq|\Delta t|} |e^{\tau^2/4}\rho_{a,b}(\tau)D_\tau^{p+4}k(\sigma-\tau, t+\zeta)| \tag{14}$$

$$= \frac{|\Delta t|}{2} \sup_{|\zeta|\leq|\Delta t|} \left| \exp\left[\frac{\tau^2}{4}\left(1-\frac{1}{t+\zeta}\right)+\frac{\sigma\tau-\sigma^2/2}{2(t+\zeta)}\right]\right.$$

$$\times \frac{\rho_{a,b}(\tau)}{\sqrt{4\pi(t+\zeta)}}\, P_{p+4}\left(\sigma-\tau, \frac{1}{t+\zeta}\right)$$

For $|\Delta t| < \frac{1}{2}\min(t, 1-t)$, the last expression is bounded by $M|\Delta t|$ where M is a constant not depending on either τ or Δt. Thus, as $\Delta t \to 0$, $\theta_{\Delta t}(\tau)$ tends to zero in $\mathscr{W}_{a,b}$ whatever be the values of a and b. Consequently (10) is true.

That $u(\sigma, t)$ satisfies (5) for $-\infty < \sigma < \infty$ and $0 < t < 1$ now follows by combining (9) for $m = 2$, (10), and (12).

Next, to show that the initial condition on $v(x, y)$ is satisfied, we need merely establish (6). In Lemma 7.4-1, let us replace x by σ, $\psi(x)$ by $\phi(\sigma)$, and $\theta(\tau, x)$ by $k(\sigma-\tau, t)$. That $k(\sigma-\tau, t)$ satisfies the hypothesis of Lemma 7.4-1 for every a and b, for $0 < t < 1$, and for supp $\phi \subset [A, B]$ is readily seen from Sec. 7.2, Eq. (3). Therefore, we may invoke this Lemma to write

$$\langle u(\sigma, t), \phi(\sigma)\rangle = \langle \phi(\sigma), \langle g(\alpha t), k(\sigma-\tau, t)\rangle\rangle \tag{15}$$

$$= \langle g(\alpha t), \langle \phi(\sigma), k(\sigma-\tau, t)\rangle\rangle$$

We demonstrated in the proof of Theorem 7.4-1 that, as $t \to 0+$,

$$\langle \phi(\sigma), k(\sigma-\tau, t)\rangle \to \phi(\tau)$$

in $\mathscr{W}_{a,b}$ for every a and b. Since $g(\alpha\tau) \in \mathscr{W}'_{a,b}$, (15) tends to $\langle g(\alpha\tau), \phi(\tau)\rangle$ as $t \to 0+$. Thus, the initial condition is verified.

We conclude by pointing out the difference between the way the Weierstrass transformation was used and the way the transformations of the previous chapters were used to solve boundary-value problems. Since the kernel of the Weierstrass transformation is related through certain changes of variables to the Green's function of our present Cauchy problem, we were able to apply the Weierstrass transformation to the initial condition to determine the solution directly. In contrast to this, the previous transformations generated various operational calculi which were used to transform the partial differential equations into differential equations wherein all the derivatives with respect to one of the independent variables were eliminated; the inverse transforms of the solutions to the latter differential equations then provided the solutions to the original problems.

(Nevertheless, the reader should not infer that the Weierstrass transformation does not generate any operational calculus. It does. See Problem 7.3-6).

PROBLEM 7.5-1 Establish (9).

PROBLEM 7.5-2 Establish (12) and (13).

CHAPTER VIII

The Convolution Transformation

8.1. Introduction

Consider the conventional convolution:

$$F(x) = \int_{-\infty}^{\infty} f(t)G(x-t)\,dt \tag{1}$$

of two suitably restricted functions. This expression can be viewed as an integral transformation which has as its kernel $G(x-t)$ and which transforms $f(t)$ into $F(x)$. We shall call (1) the *conventional convolution transformation with kernel G*. The theory of this transformation was pioneered by Hirschman and Widder around 1947 to 1949 (see Hirschman and Widder [1], [2]). By restricting the kernel G to a fairly wide class of functions, they discovered among other things some inversion formulas for (1). Moreover, it turns out that for certain choices of the kernel G, the integral (1) becomes through appropriate changes of variables various standard integral transformations such as the one-sided Laplace, Stieltjes, and K transformations. Indeed, (1) possesses great generality and encompasses quite a variety of integral transformations as special cases (Hirschman and Widder [1], pp. 65–79). In recent years the investigations into the conventional convolution transformation have comprised an active and important part of the research in integral transformations. (See, for example, Blackman and Pollard [1], Cholewinski [1], Dauns and Widder [1], Ditzian [1], [2], Ditzian and Jakimovski [1], Fox [1], Haimo [1], Hirschman [1], Sumner [1], and Tanno [1]–[3]).

In this chapter we indicate how (1) can be extended to generalized functions $f(t)$. As usual, the idea is to construct testing-function spaces which contain various kernels $G(x-t)$ as functions of t, where x is treated as the transform's independent variable. The duals of these testing-function spaces consist of those generalized functions $f(t)$ that possess convolution transforms, and (1) generalizes into

$$F(x) = \langle f(t), G(x-t)\rangle \tag{2}$$

Actually, the testing-function spaces $\mathscr{L}_{c,d}$ defined in Chapter III meet our needs for quite a variety of kernels and allow (2) to have a meaning for a fairly broad class of generalized functions $f(t)$. The $\mathscr{L}_{c,d}$ spaces are precisely the ones we shall use. However, it may be worth mentioning that for certain kernels a meaning can be assigned to (2) for spaces of generalized functions that are broader than the $\mathscr{L}'_{c,d}$ spaces. (In this regard, see Zemanian [11].)

The kernels that we shall consider are required to satisfy a number of conditions, which are listed in the next section. That there truly are kernels which satisfy the stated conditions is proven in the works of Hirschman and Widder [1], [2]. Their results are not easily established; to prove them here would carry us considerably afield from the subject of generalized functions, and we would merely be repeating discussions that already appear in a readily available book. Therefore, we merely assume the properties that are needed and refer the reader to Hirschman and Widder's book for a discussion of certain kernel functions that possess the stated properties.

The generalized convolution transformation is defined in Sec. 8.3, and $F(x)$ as given by (2) is shown to be a smooth function of x for all x. In Sec. 8.4 the real inversion formula of Hirschman and Widder [1], pp. 128, 131, and 132, is shown to invert (2) when the limiting operation in that formula is taken in a distributional sense (i.e., in the space $\mathscr{D}'$). The rest of the chapter is devoted to two special cases, namely, the one-sided Laplace transformation (Sec. 8.5) and the Stieltjes transformation (Sec. 8.6). In these last two sections we freely borrow various results from Hirschman and Widder's book.

8.2. Convolution Kernels

The basic idea in inverting the generalized convolution transformation:

(1) $$F(x) = \langle f(t), G(x-t)\rangle$$

is the following. We assume that corresponding to the given kernel G there exists a sequence of differentiation and shifting operators $\{P_n\}_{n=0}^{\infty}$ which converts $G(x-t)$ into a delta-forming sequence $\{G_n(x-t)\}_{n=0}^{\infty}$, $G_n(x-t) = P_n G(x-t)$. By this we mean that $\{G_n(x-t)\}_{n=0}^{\infty}$ is a sequence of conventional functions that converges in some sense to the delta functional $\delta(x-t)$. This assumption, coupled with the fact that the convolution of a generalized function f with δ yields f, allows us to formally invert (1) as follows. Treating the right-hand side of (1) as a generalized convolution product $f * G$, we write

$$P_n F = P_n(f * G) = f * (P_n G) = f * G_n \to f * \delta = f \qquad n \to \infty$$

Thus, this heuristic argument indicates that an inversion formula for (1) is

$$f = \lim_{n\to\infty} P_n F \tag{2}$$

where the limit is taken in some generalized sense. This is indeed the case, but the proof of (2) is considerably more complicated than the above formal manipulations. In this section we specify some precise conditions on G and P_n, which will be used when we define (1) and prove (2).

Assumptions A: Henceforth, we assume that to every permissible convolution kernel G there exists at least one sequence of operators $\{P_n(D)\}_{n=0}^{\infty}$ where $P_0(D)$ is the identity operator and

$$P_n(D) = e^{B_nD}Q_n(D) \qquad n = 1, 2, 3, \ldots \tag{3}$$

Here D represents conventional differentiation, B_n is a real constant, and Q_n is a polynomial with real coefficients. The expression e^{B_nD} is the standard operational notation for the operation of shifting through the distance B_n; that is, by definition $e^{B_nD}\phi(t) \triangleq \phi(t + B_n)$ for any function $\phi(t)$. We set $G_n(t) \triangleq P_n(D)G(t)$. Thus, $G_0(t) \triangleq P_0(D)G(t) = G(t)$. Furthermore, we assume that the following conditions are satisfied.

*A*1. $G(t)$ is smooth on $-\infty < t < \infty$. (Therefore, the same is true for every $G_n(t)$.)

*A*2. $G_n(t) \geq 0, \quad n = 0, 1, 2, \ldots.$

*A*3. $\int_{-\infty}^{\infty} G_n(t)\,dt = 1, \quad n = 0, 1, 2, \ldots.$

*A*4. $\lim_{n\to\infty} G_n(t) = 0, \quad 0 < |t| < \infty.$

*A*5. For every $\delta > 0$,

$$\lim_{n\to\infty} \int_{|t|>\delta} G_n(t)\,dt = 0$$

*A*6. Given any real numbers c and δ with $\delta > 0$, there exists a positive integer $N(c, \delta)$ such that, for each $n > N(c, \delta)$, $e^{-ct}G_n(t)$ is monotonic increasing on $-\infty < t < -\delta$ and monotonic decreasing on $\delta < t < \infty$.

*A*7. $G(t)$ possesses the following asymptotic properties as $t \to \pm\infty$:

(i) There exists at least one real negative number β such that, for each $k = 0, 1, 2, \ldots,$

$$D^kG(t) = O(e^{\beta t}) \qquad t \to \infty$$

(ii) There exists at least one real positive number α such that, for each $k = 0, 1, 2, \ldots,$

$$D^kG(t) = O(e^{\alpha t}) \qquad t \to -\infty$$

In the following we let α_1 denote the infimum of all β for which (i) holds and α_2 denote the supremum of all α for which (ii) holds. Thus, α_1 is either a real negative number or $-\infty$, and α_2 is either a real positive number or $+\infty$.

This terminates our specification of assumptions A. We shall need no more than these conditions in our discussion of the generalized convolution transformation.

Are there kernels which satisfy all these assumptions? Yes! Hirschman and Widder describe at some length a fairly wide class of such functions. Let us quote some of their results.

Let s be a complex variable, and set

$$G(t) = \frac{1}{2\pi i} \int_{-i\infty}^{i\infty} \frac{e^{st}}{E(s)} \, ds \tag{4}$$

where

$$E(s) = e^{b_0 s} \prod_{k=1}^{\infty} \left(1 - \frac{s}{a_k}\right) e^{s/a_k} \tag{5}$$

Here, b_0 and the a_k are real numbers with $a_k \neq 0$, $|a_k| \to \infty$ as $k \to \infty$, and

$$\sum_{k=1}^{\infty} a_k^{-2} < \infty$$

$E(s)$ is an entire function. Moreover, (4) satisfies conditions $A1$ and $A7$ (Hirschman and Widder [1], pp. 55, 108–109).

Next, let $\{b_n\}_{n=1}^{\infty}$ be any sequence of real numbers such that $\lim_{n\to\infty} b_n = 0$ With $G(t)$ and $E(s)$ being given by (4) and (5), set $E_0(s) = E(s)$ and $G_0(t) = G(t)$. As before, let $P_0(D)$ denote the identity operator, so that $P_0(D)G(t) = G(t)$. Moreover, for $n = 1, 2, 3, \ldots$, set

$$P_n(D) = e^{(b_0 - b_n)D} \prod_{k=1}^{n} \left(1 - \frac{D}{a_k}\right) e^{D/a_k} \tag{6}$$

$$E_n(s) = e^{b_n s} \prod_{k=n+1}^{\infty} \left(1 - \frac{s}{a_k}\right) e^{s/a_k} \tag{7}$$

$$G_n(t) = \frac{1}{2\pi i} \int_{-i\infty}^{i\infty} \frac{e^{st}}{E_n(s)} \, ds \tag{8}$$

Thus, $P_n(D)$ has the form of (3) where

$$B_n = b_0 - b_n + \sum_{k=1}^{n} a_k^{-1}$$

and

$$Q_n(D) = \prod_{k=1}^{n} \left(1 - \frac{D}{a_k}\right)$$

Moreover, upon applying (6) to (4), we obtain $P_n(D)G(t) = G_n(t)$ (Hirschman and Widder [1], p. 57). That (8) satisfies $A2$ through $A6$ is indicated in Hirschman and Widder [1] on the following pages: For $A2$ and $A3$, see p. 53. For $A4$, see p. 126. For $A5$, see p. 58. Finally, for $A6$, combine the results on pp. 80, 81, 125, and 126 as indicated in Zemanian [11], p. 327.

PROBLEM 8.2-1. Prove that, if G satisfies assumptions A, then $\{G_n\}_{n=0}^{\infty}$ is a delta-forming sequence in the sense that $G_n \to \delta$ in $\mathscr{D}'$ as $n \to \infty$.

PROBLEM 8.2-2. Here is a degenerate case of the convolution transformation. Let n be a finite integer no less than 1, let a_k $(k = 1, 2, \ldots, n)$ be real numbers not equal to zero, and set

$$E'(s) = \prod_{k=1}^{n} \left(1 - \frac{s}{a_k}\right)$$

$$G'(t) = \frac{1}{2\pi i} \int_{-i\infty}^{i\infty} \frac{e^{st}}{E'(s)}\, ds$$

Also, let

$$g(t) = \begin{cases} e^t & -\infty < t < 0 \\ \frac{1}{2} & t = 0 \\ 0 & 0 < t < \infty \end{cases}$$

and $g_k(t) = |a_k|\, g(a_k t)$. Finally, let α_1 be the largest negative number in the collection $\{a_k\}$ or $-\infty$ if all a_k are positive, and let α_2 be the smallest positive number in the collection $\{a_k\}$ or $+\infty$ is all a_k are negative.

(a) Show that, in the sense of both conventional convolution and convolution in $\mathscr{L}'(\alpha_1, \alpha_2)$,

$$G' = g_1 * g_2 * \cdots * g_n$$

Also, show that, as a conventional function,

$$G'(t) \geq 0 \qquad -\infty < t < \infty$$

and

$$\int_{-\infty}^{\infty} G'(t)\, dt = 1$$

(b) Show that, in the sense of differentiation in $\mathscr{L}'(\alpha_1, \alpha_2)$,

$$E'(D)G' = \delta$$

(c) Now, let $f \in \mathscr{L}'(\alpha_1, \alpha_2)$, and let F denote the following convolution product in $\mathscr{L}'(\alpha_1, \alpha_2)$:

$$F \triangleq f * G' \tag{9}$$

Show that (9) is inverted by the application of $E'(D)$ to F:

$$f = E'(D)F$$

where again differentiation in $\mathscr{L}'(\alpha_1, \alpha_2)$ is understood.

8.3. The Convolution Transformation

We base our definition of the generalized convolution transformation on the testing-function spaces $\mathscr{L}_{c,d}$ and their dual spaces $\mathscr{L}'_{c,d}$, which were described in Sec. 3.2. Let G be a given kernel, and let α_1 and α_2 be the quantities defined in assumption $A7$. Assumptions $A1$ and $A7$ imply that, for any choice of the real numbers c and d such that $c < \alpha_2$ and $d > \alpha_1$ and for every fixed real number x, the kernel $G(x-t)$ as a function of t is a member of $\mathscr{L}_{c,d}$. Indeed, choose β and α such that $\alpha_1 < \beta < d$ and $c < \alpha < \alpha_2$. Then, with $\kappa_{c,d}(t)$ being defined as in Sec. 3.2, we have

$$\kappa_{c,d}(t) D_t{}^k G(x-t) = O(e^{ct-\alpha t}) = o(1) \qquad t \to \infty$$

and

$$\kappa_{c,d}(t) D_t{}^k G(x-t) = O(e^{dt-\beta t}) = o(1) \qquad t \to -\infty$$

which implies our assertion.

We define the *generalized convolution transformation with kernel* G on any $f \in \mathscr{L}'_{c,d}$, where again $c < \alpha_2$ and $d > \alpha_1$, through the expression:

$$F(x) \triangleq \langle f(t), G(x-t) \rangle \qquad -\infty < x < \infty \tag{1}$$

We refer to $F(x)$ as the *convolution transform of* f; it is a conventional function defined for all x.

Note that we do not make any use of the countable-union spaces $\mathscr{L}(w, z)$, as we did when discussing the Laplace transformation. The reason for this is that, if $G(x-t)$ as a function of t is in $\mathscr{L}_{c,d}$ for one real value of x, it will be in $\mathscr{L}_{c,d}$ for every real value of x. This is in contrast to the kernel e^{-st} of the Laplace transformation; e^{-st} will not remain in $\mathscr{L}_{c,d}$ if Re s ranges outside the interval $c \le \text{Re } s \le d$. This is why we were led in Chapter III to consider a collection of $\mathscr{L}_{c,d}$ spaces, and indeed the $\mathscr{L}(w, z)$ spaces.

THEOREM 8.3-1. *Given a kernel G, let α_1 and α_2 be defined as in assumption $A7$. Also, let $f \in \mathscr{L}'_{c,d}$, where $c < \alpha_2$ and $d > \alpha_1$, and let $F(x)$ be defined by* (1). *Then, $F(x)$ is a smooth function on $-\infty < x < \infty$, and*

$$F^{(k)}(x) = \langle f(t), G^{(k)}(x-t) \rangle \qquad k = 1, 2, 3, \ldots \tag{3}$$

Here, $H^{(k)}(z)$ denotes the conventional kth derivative of H with respect to its argument z.

PROOF. Note that, since $\mathscr{L}_{c,d}$ is closed under differentiation, $G^{(k)}(x-t) \in \mathscr{L}_{c,d}$, so that the right-hand side of (3) has a sense. Thus, we need merely prove (3), which we do through an inductive argument. Assume that (3) is true for k replaced by $k-1$. It is true by definition for $k=0$. Letting x be fixed and $\Delta x \neq 0$, consider

$$\frac{1}{\Delta x}[F^{(k-1)}(x+\Delta x) - F^{(k-1)}(x)] - \langle f(t), G^{(k)}(x-t)\rangle \tag{4}$$

$$= \langle f(t), \theta_{\Delta x}(t)\rangle$$

where

$$\theta_{\Delta x}(t) = \frac{1}{\Delta x}[G^{(k-1)}(x+\Delta x - t) - G^{(k-1)}(x-t)] - G^{(k)}(x-t)$$

For any nonnegative integer m we may write

$$\theta_{\Delta x}^{(m)}(t) = \frac{(-1)^m}{\Delta x}\int_{x-t}^{x-t+\Delta x} G^{(m+k)}(y)\,dy - (-1)^m G^{(m+k)}(x-t)$$

$$= \frac{(-1)^m}{\Delta x}\int_{x-t}^{x-t+\Delta x} dy \int_{x-t}^{y} G^{(m+k+1)}(z)\,dz$$

Next, let Λ denote the interval $x-t-|\Delta x| < z < x-t+|\Delta x|$. Then,

$$|\kappa_{c,d}(t)\theta_{\Delta x}^{(m)}(t)| \leq \frac{|\Delta x|}{2}\kappa_{c,d}(t)\sup_{z\in\Lambda}|G^{(m+k+1)}(z)| \tag{5}$$

Now, for $|\Delta x| < 1$,

$$\kappa_{c,d}(t)\sup_{z\in\Lambda}|G^{(m+k+1)}(z)|$$

is bounded on $-\infty < t < \infty$; this can be seen through an argument similar to that given in the first paragraph of this section. Therefore, we have from (5) that $\theta_{\Delta x}(t)$ converges in $\mathscr{L}_{c,d}$ to zero as $\Delta x \to 0$. Since $f \in \mathscr{L}'_{c,d}$, (4) converges to zero as $\Delta x \to 0$. This completes our inductive proof of (3).

The next theorem states some restrictions on the rates of growth of $F(x)$ and each of its derivatives as $x \to \pm\infty$.

THEOREM 8.3-2. *Under the hypothesis of Theorem* 8.3-1, *$F(x)$ is a member of $\mathscr{L}_{a,b}$, where a and b are any real numbers such that $a < \min(-\alpha_1, -c)$ and $b > \max(-\alpha_2, -d)$.*

The conclusion implies that, for each nonnegative integer k, $F^{(k)}(x) = O(e^{-ax})$ as $x \to \infty$ and $F^{(k)}(x) = O(e^{-bx})$ as $x \to -\infty$.

PROOF. $F(x)$ is smooth according to Theorem 8.3-1. Therefore, we have only to prove that $\kappa_{a,b}(x)F^{(k)}(x)$ is bounded on $-\infty < x < \infty$. By

(3) and the boundedness property stated in Sec. 3.2, note *III*, we may write

$$|\kappa_{a,b}(x)F^{(k)}(x)| = |\langle f(t), \kappa_{a,b}(x)G^{(k)}(x-t)\rangle|$$
$$\leq C \max_{0\leq p\leq r} \sup_t |\kappa_{c,d}(t)\kappa_{a,b}(x)G^{(k+p)}(x-t)|$$

Our proof will be complete when we show that for every nonnegative integer k

$$A_k(t, x) \triangleq \kappa_{c,d}(t)\kappa_{a,b}(x)G^{(k)}(x-t)$$

is bounded on the (t, x) plane.

By the conditions stated in Theorems 8.3-1 and 8.3-2, we have $\alpha_1 < d$, $\alpha_1 < -a$, $c < \alpha_2$, and $-b < \alpha_2$. Therefore, we can choose two real numbers α and β such that $\alpha_1 < \beta < \min(-a, d)$ and $\max(-b, c) < \alpha < \alpha_2$. By Sec. 8.2, assumption $A7$, there exists a constant K for which

$$|G^{(k)}(x-t)| < \begin{cases} Ke^{\alpha(x-t)} & x-t \leq 0 \\ Ke^{\beta(x-t)} & x-t \geq 0 \end{cases}$$

We now consider six different sectors of the (t, x) plane which together cover the (t, x) plane. The first sector is $t \geq 0$, $x \geq t$. On it we have $\exp \beta(x-t) \leq \exp[-a(x-t)]$ and

$$|A_k(t, x)| < Ke^{ct}e^{ax}e^{\beta(x-t)} \leq Ke^{(c+a)t} \leq K$$

since $c + a < 0$ by hypothesis. The second sector is $x \geq 0$, $t \geq x$. Here, we have $\exp \alpha(x-t) \leq \exp c(x-t)$ and

$$|A_k(t, x)| < Ke^{ct}e^{ax}e^{\alpha(x-t)} \leq Ke^{(a+c)x} \leq K$$

again because $a + c < 0$. The third sector is $t \geq 0$, $x \leq 0$. Now, we may write

$$|A_k(t, x)| < Ke^{ct}e^{bx}e^{\alpha(x-t)} = Ke^{(c-\alpha)t}e^{(b+\alpha)x} \leq K$$

since $c - \alpha < 0$ and $b + \alpha < 0$.

We denote the sector $t \leq 0$, $x \leq t$ as the fourth, the sector $x \leq 0$, $t \leq x$ as the fifth, and the sector $t \leq 0$, $x \geq 0$ as the sixth. The boundedness of $A_k(t, x)$ on the fourth, fifth, and sixth sectors is established by arguments which are respectively similar to those of the first, second, and third sectors. This completes the proof.

8.4. Inversion

The generalized convolution transformation defined by Sec. 8.3, Eq. (1), is inverted by the formula:

(1) $$f = \lim_{n\to\infty} P_n(D)F$$

where the $P_n(D)$ are some shifting and differentiation operators corresponding to the kernel G as specified in assumptions A, and the limit in (1) is taken in the sense of (weak) convergence in $\mathscr{D}'$. To prove this result, we shall need

LEMMA 8.4-1. *For a given kernel G let α_1 and α_2 be defined as in assumption A7 of Sec. 8.2. Also, let $\psi \in \mathscr{D}$, and let $f \in \mathscr{L}'_{c,d}$ where $c < \alpha_2$ and $d > \alpha_1$. Then,*

$$\langle f(t), \langle \psi(x), G(x-t)\rangle\rangle = \langle \psi(x), \langle f(t), G(x-t)\rangle\rangle \tag{2}$$

PROOF. By Theorem 8.3-1, $\langle f(t), G(x-t)\rangle$ is a smooth function of x. Therefore, the right-hand side of (2) has a sense as the integral:

$$\int_{-\infty}^{\infty} \psi(x)\langle f(t), G(x-t)\rangle\, dx$$

Since $f(t) \in \mathscr{L}'_{c,d}$, we can show that the left-hand side of (2) has a sense by proving that

$$I(t) \triangleq \langle \psi(x), G(x-t)\rangle \triangleq \int_{-\infty}^{\infty} \psi(x) G(x-t)\, dx$$

is a member of $\mathscr{L}_{c,d}$. Choose the real numbers α and β such that $\alpha_1 < \beta < d$ and $c < \alpha < \alpha_2$. Assume that the support of $\psi(x)$ is contained in the finite interval $A \le x \le B$. Then, by assumptions A1 and A7 of Sec. 8.2, for any nonnegative integer k there exists a constant M_k such that for all x in $A \le x \le B$

$$|\kappa_{c,d}(t) D_t^k G(x-t)| \le \begin{Bmatrix} M_k e^{(d-\beta)t} & t \le 0 \\ M_k e^{(c-\alpha)t} & t \ge 0 \end{Bmatrix} \le M_k \tag{3}$$

We may differentiate under the integral sign to obtain

$$|\kappa_{c,d}(t) D_t^k I(t)| = \left| \int_A^B \psi(x) \kappa_{c,d}(t) D_t^k G(x-t)\, dx \right|$$

$$\le M_k \int_A^B |\psi(x)|\, dx < \infty$$

So, $I(t)$ truly is a member of $\mathscr{L}_{c,d}$.

It remains to prove the equality in (2). Set $X = B - A$, and

$$J(t, m) = \frac{X}{m} \sum_{\nu=1}^{m} \psi\left(A + \frac{\nu X}{m}\right) G\left(A + \frac{\nu X}{m} - t\right)$$

Assume for the moment that $J(t, m) \to I(t)$ in $\mathscr{L}_{c,d}$ as $m \to \infty$. Since $f \in \mathscr{L}_{c,d}$, it follows that

$$\langle f(t), J(t, m)\rangle \to \langle f(t), \langle \psi(x), G(x-t)\rangle\rangle \qquad m \to \infty \tag{4}$$

On the other hand, by applying $f(t)$ to the sum $J(t, m)$ term by term and recalling that $\langle f(t), G(x-t)\rangle$ is a smooth function of x (Theorem 8.3-1), we get

$$\langle f(t), J(t, m)\rangle = \frac{X}{m} \sum_{\nu=1}^{m} \psi\left(A + \frac{\nu X}{m}\right) \left\langle f(t), G\left(A + \frac{\nu X}{m} - t\right)\right\rangle$$

$$(5) \qquad \to \int_A^B \psi(x)\langle f(t), G(x-t)\rangle\, dx \qquad m \to \infty$$

$$\triangleq \langle \psi(x), \langle f(t), G(x-t)\rangle\rangle$$

The equality in (2) is established by (4) and (5).

Thus, we have only to prove that $J(t, m) \to I(t)$ in $\mathscr{L}_{c,d}$ as $m \to \infty$. This is obvious if $\psi(x) \equiv 0$. So, assume that $\psi(x) \not\equiv 0$, and let k be a fixed nonnegative integer. Set

$$H(t, m) = \kappa_{c,d}(t) D_t^k [I(t) - J(t, m)]$$

We have to show that $H(t, m)$ converges uniformly to zero on $-\infty < t < \infty$ as $m \to \infty$. Again, choose α and β such that $\alpha_1 < \beta < d$ and $c < \alpha < \alpha_2$. In view of (3), given an $\varepsilon > 0$ there exists a T such that for $|t| > T$ and $A \le x \le B$

$$|\kappa_{c,d}(t) D_t^k G(x-t)| < \frac{\varepsilon}{3} \left[\int_A^B |\psi(x)|\, dx\right]^{-1}$$

Consequently,

$$\sup_{|t|>T} \kappa_{c,d}(t) D_t^k I(t) < \frac{\varepsilon}{3}$$

Also, for all m,

$$(6) \qquad \sup_{|t|>T} |\kappa_{c,d}(t) D_t^k J(t, m)| < \frac{\varepsilon}{3} \left[\int_A^B |\psi(x)|\, dx\right]^{-1} \frac{X}{m} \sum_{\nu=1}^{m} \left|\psi\left(A + \frac{\nu X}{m}\right)\right|$$

Moreover, there exists an m_0 such that for all $m > m_0$ the right-hand side of (6) is bounded by $2\varepsilon/3$. Thus, for $m > m_0$ and $|t| > T$, we have $|H(t, m)| < \varepsilon$.

Next, set $K = \sup_{|t|<T} \kappa_{c,d}(t)$. Then, for $|t| \le T$,

$$(7) \qquad |H(t, m)| \le K \left| \int_A^B \psi(x) D_t^k G(x-t)\, dx - \frac{X}{m} \sum_{\nu=1}^{m} \psi\left(A + \frac{\nu X}{m}\right) D_t^k G\left(A + \frac{\nu X}{m} - t\right)\right|$$

Since $\psi(x) D_t^k G(x-t)$ is uniformly continuous for all (t, x) such that $-T \le t \le T$ and $A \le x \le B$, there exists an m_1 such that for all $m > m_1$ the right-hand side of (7) is bounded by ε on $-T \le t \le T$.

Thus, we have proven that $H(t, m)$ converges uniformly to zero on $-\infty < t < \infty$ as $m \to \infty$. The proof of Lemma 8.4-1 is therefore complete.

As our inversion theorem, we have

THEOREM 8.4-1. *For a given kernel G, let α_1 and α_2 be defined as in assumption A7 of Sec. 8.2, and let $P_n(D)$, $n = 0, 1, 2, \ldots$, denote some shifting and differentiation operators as specified in assumptions A of Sec. 8.2. Assume that $f \in \mathscr{L}'_{c,d}$ where $c < \alpha_2$ and $d > \alpha_1$. If $F(x) = \langle f(t), G(x-t)\rangle$ then*

$$\lim_{n \to \infty} P_n(D)F = f \tag{8}$$

in the sense of convergence in $\mathscr{D}'$; that is, for every $\psi \in \mathscr{D}$,

$$\lim_{n \to \infty} \langle P_n(D)F, \psi\rangle = \langle f, \psi\rangle$$

PROOF. Since $F(x)$ is a smooth function and since $P_n(D) = e^{B_nD}Q_n(D)$ where B_n is a real number and Q_n is a polynomial, we have

$$P_n(D)F(x) = Q_n(D_x)F(x + B_n)$$

which is also a smooth function. Therefore, $\langle P_n(D)F, \psi\rangle$ is an integral, and for any $\psi \in \mathscr{D}$ we have through repeated integrations by parts and a change of variable

$$\begin{aligned}\langle P_n(D)F, \psi\rangle &= \langle F, P_n(-D)\psi\rangle \\ &= \langle P_n(-D_x)\psi(x), \langle f(t), G(x-t)\rangle\rangle\end{aligned}$$

Clearly, $P_n(-D)\psi$ is also a member of $\mathscr{D}$. Therefore, we may invoke Lemma 8.4-1 to write

$$\begin{aligned}\langle P_n(D)F, \psi\rangle &= \langle f(t), \langle P_n(-D_x)\psi(x), G(x-t)\rangle\rangle \\ &= \langle f(t), \langle \psi(x), P_n(D_x)G(x-t)\rangle\rangle \\ &= \langle f(t), \langle \psi(x), G_n(x-t)\rangle\rangle\end{aligned}$$

Since $f(t) \in \mathscr{L}'_{c,d}$, all that remains to be shown is that

$$\langle \psi(x), G_n(x-t)\rangle \to \psi(t)$$

in $\mathscr{L}_{c,d}$ as $n \to \infty$. In other words, we have to prove that for each nonnegative integer k

$$\kappa_{c,d}(t)D_t^k\left[\int_{-\infty}^{\infty} \psi(x)G_n(x-t)\,dx - \psi(t)\right] \tag{9}$$

converges uniformly to zero on $-\infty < t < \infty$ as $n \to \infty$. Since G is smooth and $\psi \in \mathscr{D}$, we may differentiate under the integral sign to convert (9) into

$$\kappa_{c,\,d}(t)\left[\int_{-\infty}^{\infty}\psi(x)D_t{}^kG_n(x-t)\,dx-\psi^{(k)}(t)\right]$$
$$=\kappa_{c,\,d}(t)\left[\int_{-\infty}^{\infty}\psi(x)(-D_x)^kG_n(x-t)\,dx-\psi^{(k)}(t)\right]$$

By integrating by parts k times and using assumption $A3$, namely,

$$\int_{-\infty}^{\infty}G_n(x-t)\,dt=1 \tag{10}$$

we find that (9) is equal to

$$\kappa_{c,\,d}(t)\int_{-\infty}^{\infty}[\psi^{(k)}(x)-\psi^{(k)}(t)]G_n(x-t)\,dx=I_1(t)+I_2(t)+I_3(t)$$

Here, $I_1(t)$, $I_2(t)$, and $I_3(t)$ denote the terms obtained by integrating over the intervals $-\infty<x<t-\delta$, $t-\delta<x<t+\delta$, and $t+\delta<x<\infty$ respectively, δ being a positive number.

Assumption $A2$ asserts that $G_n(t)\geq 0$. From this and (10) we obtain

$$|I_2(t)|\leq\kappa_{c,\,d}(t)\sup_{t-\delta<x<t+\delta}|\psi^{(k)}(x)-\psi^{(k)}(t)|$$
$$\leq\delta\kappa_{c,\,d}(t)\sup_{t-\delta<\tau<t+\delta}|\psi^{(k+1)}(\tau)|$$

Restrict δ by $0<\delta<1$. Then, since ψ is smooth and of bounded support, the last expression is bounded by δB where B is a constant with respect to t and δ. Thus, given an $\varepsilon>0$, we have that $|I_2(t)|\leq\varepsilon$ for $\delta=\min(1,\varepsilon/B)$ and for all n. Fix δ this way.

Next, consider

$$I_1(t)=\kappa_{c,\,d}(t)\int_{-\infty}^{t-\delta}\psi^{(k)}(x)G_n(x-t)\,dx-\kappa_{c,\,d}(t)\psi^{(k)}(t)\int_{-\infty}^{t-\delta}G_n(x-t)\,dx \tag{11}$$

Since $\psi\in\mathscr{D}$, it follows that $\kappa_{c,\,d}(t)\psi^{(k)}(t)$ is bounded on $-\infty<t<\infty$. Moreover, by assumption $A5$,

$$\int_{-\infty}^{t-\delta}G_n(x-t)\,dx=\int_{-\infty}^{-\delta}G_n(y)\,dy\to 0\qquad n\to\infty$$

Hence, the second term on the right-hand side of (11) converges uniformly to zero on $-\infty<t<\infty$ as $n\to\infty$.

Denote the first term on the right-hand side of (11) by $J_1(t)$. As before, let $A\leq x\leq B$ be a finite interval containing the support of $\psi(x)$. For $-\infty<t-\delta\leq A$, $J_1(t)\equiv 0$. On the other hand, for $A<t-\delta<\infty$, $\kappa_{c,\,d}(t)<Ke^{ct}$ if K is some sufficiently large constant, and therefore

$$|J_1(t)|\leq Ke^{ct}\int_A^{t-\delta}|\psi^{(k)}(x)|e^{c(x-t)}e^{-c(x-t)}G_n(x-t)\,dx$$

By virtue of assumption $A6$, for all sufficiently large n, $e^{-c(x-t)}G_n(x-t)$ is a monotonic increasing function of $x-t$ for $x-t<-\delta$, so that it is dominated by $e^{c\delta}G_n(-\delta)$. Consequently, for $A<t-\delta<\infty$,

$$|J_1(t)| \leq Ke^{c\delta}G_n(-\delta)\int_A^B |\psi^{(k)}(x)|e^{cx}\,dx$$

By assumption $A4$, $G_n(-\delta)\to 0$ as $n\to\infty$. This proves the uniform convergence of $J_1(t)$ (and therefore of $I_1(t)$) to zero on $-\infty<t<\infty$.

A similar argument shows that $I_3(t)$ also converges uniformly to zero on $-\infty<t<\infty$ as $n\to\infty$. Altogether this proves that as $n\to\infty$ the limit superior of (9) is uniformly bounded by ε on $-\infty<t<\infty$. Since $\varepsilon>0$ is arbitrary, our proof is complete.

As a consequence of Theorem 8.4-1, we have

THEOREM 8.4-2 (*The Uniqueness Theorem*). *Let G, c, and d be as specified in Theorem 8.4-1, and let $f\in\mathscr{L}'_{c,d}$ and $h\in\mathscr{L}'_{c,d}$. Also let $F(x)=\langle f(t), G(x-t)\rangle$ and $H(x)=\langle h(t), G(x-t)\rangle$. If $F(x)=H(x)$ for all x, then $f=h$ in the sense of equality in $\mathscr{D}'$.*

PROOF. Let $P_n(D)$ be as specified in Theorem 8.4-1. Then, in the sense of convergence in $\mathscr{D}'$, we have

$$f=\lim_{n\to\infty} P_n(D)F=\lim_{n\to\infty} P_n(D)H=h$$

Q.E.D.

Actually, we can strengthen the conclusion of Theorem 8.4-2. We have from Sec. 3.2 that the restrictions of f and h to the countable-union space $\mathscr{L}(c, d)$ are both members of $\mathscr{L}'(c, d)$. Moreover, $\mathscr{D}$ is dense in $\mathscr{L}(c, d)$ (see Sec. 3.2, note I). Consequently, since f and h coincide on $\mathscr{D}$, they must also coincide on $\mathscr{L}(c, d)$. In other words, $f=h$ in the sense of equality in $\mathscr{L}'(c, d)$.

Before leaving this section, we mention that there is another inversion formula for the generalized convolution transformation (Pandey and Zemanian [1]). It makes use of integration in the complex plane and is therefore called a *complex inversion formula*. This is in contrast to (8) wherein the independent variable for f and F is real, and so one is lead to refer to (8) as a *real inversion formula*. The complex inversion formula does not hold for as wide a class of kernels G as does (8).

8.5. The One-Sided Laplace Transformation

The rest of this chapter is devoted to two standard integral transformations which through changes of variables become special cases of the convolution transformation. In this section we discuss that form of the Laplace transformation wherein the lower limit on the integral is fixed at the origin:

(1) $$J(y) = \int_0^\infty j(\tau)e^{-y\tau}\, d\tau$$

Thus, this is a particular case of the one-sided Laplace transformation discussed in Sec. 3.10 wherein the lower limit was allowed to be arbitrary (but finite). We indicated one means of generalizing (1) in Problem 3.10-13. Our generalized convolution transformation provides another method.

Consider the particular kernel:

(2) $$G(t) = e^{-e^t}e^t$$

It is a fact that this kernel satisfies assumptions A of Sec. 8.2. In this case, $\alpha_1 = -\infty$, $\alpha_2 = 1$, and for the operators $P_n(D)$ we may use

(3) $$P_n(D) = e^{D\log n} \prod_{k=1}^{n} \left(1 - \frac{D}{k}\right) \qquad n = 1, 2, 3, \ldots$$

(See Hirschman and Widder [1], pp. 66–67.) Assume for the moment that we have a conventional function $f(t)$, and apply the change of variables $y = e^x$, $\tau = e^{-t}$ to the convolution transform:

(4) $$F(x) = \int_{-\infty}^{\infty} f(t)G(x-t)\, dt \qquad -\infty < x < \infty$$

We obtain $G(x - t) = y\tau e^{-y\tau}$ and

(5) $$y^{-1}F(\log y) = \int_0^\infty f(-\log \tau)e^{-y\tau}\, d\tau \qquad 0 < y < \infty$$

which can be identified with (1).

The same change-of-variables technique can be used when dealing with a generalized function $f(t)$. First note that the substitution $\tau = e^{-t}$ is precisely the one used in Sec. 4.2 when relating the $\mathscr{L}_{c,d}$ spaces to the $\mathscr{M}_{c,d}$ spaces. In particular, Theorem 4.2-1 asserts that $\tau^{-1}\phi(-\log \tau) \leftrightarrow \phi(t)$ is an isomorphism from $\mathscr{M}_{c,d}$ onto $\mathscr{L}_{c,d}$. If $f(t) \in \mathscr{L}'_{c,d}$, we may define $f(-\log \tau)$ by

(6) $$\langle f(-\log \tau), \tau^{-1}\phi(-\log \tau)\rangle \triangleq \langle f(t), \phi(t)\rangle \qquad \phi \in \mathscr{L}_{c,d}$$

Then, Theorem 4.2-2 asserts that $f(t) \leftrightarrow f(-\log \tau)$ is an isomorphism from $\mathscr{L}'_{c,d}$ onto $\mathscr{M}'_{c,d}$. Thus, if we choose $c < 1$ and d as any real number, we may replace $\phi(t)$ by $G(x - t)$ and set $y = e^x$, $\tau = e^{-t}$ again to obtain

$$\langle f(-\log \tau), ye^{-y\tau}\rangle = \langle f(t), G(x-t)\rangle = F(\log y)$$

Setting $J(y) = y^{-1}F(\log y)$ and $j(\tau) = f(-\log \tau)$, we finally obtain the new definition of the generalized one-sided Laplace transformation corresponding to (1):

$$(7) \qquad J(y) \triangleq \langle j(\tau), e^{-y\tau} \rangle \qquad 0 < y < \infty$$

This has a meaning as the application of $j(\tau) \in \mathscr{M}'_{c,d}$ to $e^{-y\tau} \in \mathscr{M}'_{c,d}$ where $c < 1$ and d is arbitrary. This is a fairly narrow definition of the generalized one-sided Laplace transformation. The one given in Sec. 3.10 as well as the one mentioned in Problem 3.10-13 are considerably broader.

Theorem 8.3-1 implies that $J(y)$ is a smooth function on $0 < y < \infty$.

A real inversion formula for (7) can be obtained from Sec. 8.4, Eq. (8), where now $P_n(D)$ is given by (3). The substitution of $\tau - e^{-t}$ into

$$P_n(D_t)F(t) = P_n(D_t)[e^t J(e^t)] \qquad -\infty < t < \infty$$

yields after some computation (Hirschman and Widder [1], pp. 66–67)

$$\frac{(-1)^n}{n!}\left(\frac{n}{\tau}\right)^{n+1} J^{(n)}\left(\frac{n}{\tau}\right) \qquad 0 < \tau < \infty$$

Moreover, $\tau^{-1}\phi(-\log \tau) \leftrightarrow \phi(t)$ is readily seen to be an isomorphism from $\mathscr{D}(I)$ onto $\mathscr{D}$, where now I denotes the interval $0 < \tau < \infty$. Thus, using the definition (6) again where now $\phi(t)$ is restricted to $\mathscr{D}$, we obtain from theorem 8.4-1 the following result (see Problem 8.5-2): If $j \in \mathscr{M}'_{c,d}$, where $c < 1$ and d is arbitrary, and if J is defined by (7), then

$$(8) \qquad \lim_{n\to\infty} \frac{(-1)^n}{n!}\left(\frac{n}{\tau}\right)^{n+1} J^{(n)}\left(\frac{n}{\tau}\right) = j(\tau)$$

in the sense of convergence in $\mathscr{D}'(I)$ where I is the interval $0 < \tau < \infty$. Formula (8) is the classical Post-Widder inversion formula (Widder [1]; p. 288); it now has a meaning for certain generalized functions $j(\tau)$. Actually, (8) can be used for a still wider class of generalized functions than that described here. (See Zemanian [3].)

PROBLEM 8.5-1. Prove directly that, for any fixed real $y > 0$, $e^{-y\tau}$ is a member of $\mathscr{M}_{c,d}$ for every $c < 1$ and every d.

PROBLEM 8.5-2. Prove that $\tau^{-1}\phi(-\log \tau) \leftrightarrow \phi(t)$, where $t = -\log \tau$, is an isomorphism from $\mathscr{D}(I)$ onto $\mathscr{D}$. As a consequence, the mapping $f(t) \leftrightarrow f(-\log \tau)$, as defined by (6), is an isomorphism from $\mathscr{D}'$ onto $\mathscr{D}'(I)$. It is this result combined with Theorem 8.4-1 that allows us to assert that (8) possesses a sense in terms of convergence in $\mathscr{D}'(I)$.

PROBLEM 8.5-3. Let

$$(9) \qquad G(t) = \int_{-\infty}^{\infty} H(u)H(t-u)\, du \qquad H(u) = e^{-e^u}e^u$$

Show that the change of variables:

$$y = e^x, \quad \tau = e^{-t}, \quad J(y) = y^{-1}F(\log y), \quad j(\tau) = f(-\log \tau)$$

converts the conventional convolution transformation (4) into

$$J(y) = 2 \int_0^\infty j(\tau) K_0(2\sqrt{y\tau})\, d\tau \tag{10}$$

where K_0 denotes the modified Bessel function of third kind and order zero. Note that this is precisely the same change of variables as that used before in this section. (*Hint*: Use the formula:

$$K_0(2w) = \tfrac{1}{2} \int_0^\infty v^{-1} e^{-w(v+4/v)/2}\, dv \qquad w > 0$$

See Erdelyi (Ed.) [1], Vol. 2, Equation 7.12(23).)

The kernel (9) also happens to satisfy assumptions A; here again, $\alpha_1 = -\infty$ and $\alpha_2 = 1$. (See Hirschman and Widder [1], p. 71.) In view of these facts, state how (10) can be extended to certain generalized functions.

8.6. The Stieltjes Transformation

As another special case of the convolution transformation, we consider the Stieltjes transformation which is defined on certain conventional functions $j(\tau)$ by

$$J(y) = \int_0^\infty \frac{j(\tau)}{y + \tau}\, d\tau \tag{1}$$

We start again with the conventional convolution transformation:

$$F(x) = \int_{-\infty}^\infty f(t) G(x - t)\, dt \qquad -\infty < x < \infty \tag{2}$$

and choose for the kernel:

$$G(x - t) = \tfrac{1}{2} \operatorname{sech} \frac{x - t}{2} \tag{3}$$

Now, we have $\alpha_1 = -\frac{1}{2}$ and $\alpha_2 = \frac{1}{2}$; for the operators $P_n(D)$ we can choose

$$P_n(D) = \frac{1}{\pi} \prod_{k=1}^{n} \left[1 - \left(\frac{2D}{2k - 1}\right)^2\right] \tag{4}$$

(See Hirschman and Widder [1], pp. 69–70 and Titchmarsh [1], p. 114.) Upon applying the change of variables $y = e^x$, $\tau = e^t$ to (2), we obtain

$$y^{-1/2}F(\log y) = \int_0^\infty \tau^{-1/2}f(\log \tau)\,\frac{1}{y+\tau}\,d\tau \qquad 0<y<\infty$$

which can be identified with (1).

To extend these results to generalized functions we first construct a testing-function space $\mathscr{I}_{c,d}$ by applying the change of variable $\tau = e^t$ to the definition of $\mathscr{L}_{c,d}$ and setting $\sqrt{\tau}\,\psi(\tau) = \phi(\log \tau)$ in Sec. 3.2, Eq. (1). This yields the following definition: Given any two real numbers c and d, $\mathscr{I}_{c,d}$ is the space of all smooth functions $\psi(\tau)$ on $0 < \tau < \infty$ such that

$$\iota_k(\psi) \triangleq \iota_{c,d,k}(\psi) \triangleq \sup_{0<\tau<\infty} |\kappa_{c,d}(\log \tau)(\tau D_\tau)^k \sqrt{\tau}\,\psi(\tau)| < \infty \qquad k = 0, 1, 2, \ldots$$

where

$$\kappa_{c,d}(\log \tau) = \begin{cases} \tau^c & 1 \le \tau < \infty \\ \tau^d & 0 < \tau < 1 \end{cases}$$

The topology of $\mathscr{I}_{c,d}$ is that generated by the multinorm $\{\iota_{c,d,k}\}_{k=0}^\infty$. As a consequence, $\mathscr{I}_{c,d}$ is a complete countably multinormed space. Moreover, the mapping $\psi(\tau) \triangleq \tau^{-1/2}\phi(\log \tau) \leftrightarrow \phi(t)$ is an isomorphism from $\mathscr{I}_{c,d}$ onto $\mathscr{L}_{c,d}$ since $\iota_{c,d,k}[\psi(\tau)] = \gamma_{c,d,k}[\phi(t)]$.

Next, if $f(t) \in \mathscr{L}'_{c,d}$, we define $\tau^{-1/2}f(\log \tau)$ as a functional on $\mathscr{I}_{c,d}$ by

$$\langle \tau^{-1/2}f(\log \tau), \tau^{-1/2}\phi(\log \tau)\rangle \triangleq \langle f(t), \phi(t)\rangle \qquad f \in \mathscr{L}'_{c,d}, \qquad \phi \in \mathscr{L}_{c,d}$$

(5)

By Theorem 1.10-2, $f(t) \leftrightarrow \tau^{-1/2}f(\log \tau)$ is an isomorphism from $\mathscr{L}'_{c,d}$ onto $\mathscr{I}'_{c,d}$. Now, if $c < \frac{1}{2}$ and $d > -\frac{1}{2}$, (3) as a function of t is a member of $\mathscr{L}_{c,d}$ for each fixed x. So, setting $\phi(t) = G(x - t)$ and $x = \log y$ and using (5), we can convert the convolution transform:

$$(6) \qquad F(x) = \langle f(t), G(x-t)\rangle \qquad -\infty < x < \infty$$

into

$$y^{-1/2}F(\log y) = \left\langle \tau^{-1/2}f(\log \tau), \frac{1}{y+\tau}\right\rangle \qquad 0<y<\infty$$

Upon denoting $y^{-1/2}F(\log y)$ by $J(y)$ and $\tau^{-1/2}f(\log \tau)$ by $j(\tau)$, we arrive at the following generalized Stieltjes transformation: If $j(\tau) \in \mathscr{I}'_{c,d}$ for some $c < \frac{1}{2}$ and some $d > -\frac{1}{2}$, then the Stieltjes transform J of j is defined by

$$(7) \qquad J(y) \triangleq \left\langle j(\tau), \frac{1}{y+\tau}\right\rangle \qquad 0<y<\infty$$

where for each fixed y the right-hand side has a sense as the application of $j(\tau) \in \mathscr{I}'_{c,d}$ to $(y+\tau)^{-1} \in \mathscr{I}_{c,d}$. (Actually, (7) has this meaning even when $c = \frac{1}{2}$ and $d = -\frac{1}{2}$; see Problem 8.6-1.) By virtue of Theorem 8.3-1, $J(y)$ is a smooth function on $0 < y < \infty$.

We turn now to the inversion formula for (7) resulting from Sec. 8.4, Eq. (8). The substitutions $t = \log \tau$ and $F(\log \tau) = \sqrt{\tau}\, J(\tau)$ into $P_n(D_t)F(t)$, where now P_n is given by (4), yield

$$\frac{(-1)^n}{\pi}\left[\frac{n!\,4^n}{(2n)!}\right]^2 \sqrt{\tau}\, D_\tau^{\,n}[\tau^{2n} D_\tau^{\,n} J(\tau)]$$

(See Hirschman and Widder [1], pp. 66, 69–70.) Upon using Stirling's formula (Widder [2], p. 383) and setting $j(\tau) = \tau^{-1/2} f(\log \tau)$, we can convert Sec. 8.4, Eq. (8) into

$$j(\tau) = \lim_{n\to\infty} \frac{(-1)^n}{2\pi}\left(\frac{e}{n}\right)^{2n} D_\tau^{\,n}[\tau^{2n} D_\tau^{\,n} J(\tau)] \tag{8}$$

This has a sense as a limit in $\mathscr{D}'(I)$ where (I) is the interval $0 < \tau < \infty$; indeed, Sec. 8.4, Eq. (8) is a limit in $\mathscr{D}'$, and the change of variable we have used (namely, (5)) defines an isomorphism from $\mathscr{D}'$ onto $\mathscr{D}'(I)$. In summary, if $j \in \mathscr{I}'_{c,d}$ for some $c < \frac{1}{2}$ and some $d > -\frac{1}{2}$ and if J is defined by (7), then (8) holds true in the sense of convergence in $\mathscr{D}'(I)$.

For another development of a generalized Stieltjes transformation and its inversion, see Benedetto [2].

One last remark: Any of the particular integral transforms that are encompassed by the convolution transformation (see Hirschman and Widder [1], pp. 66–79) can be extended to generalized functions by following the method used in this and the preceding section. In particular, by employing an appropriate change of variables, one can convert the convolution transformation (6) into

$$J(y) = \langle j(\tau), K(y, \tau)\rangle \tag{9}$$

where $K(y, \tau)$ is the kernel for the particular transformation under consideration. This entails the formulation of a new testing-function space and its dual by applying the appropriate change of variable to $\mathscr{L}_{c,d}$. (Problem 8.5-3 indicates still another example of this.) Similarly, Sec. 8.4 Eq. (8) can be converted into an inversion formula for (9) by the same means.

PROBLEM 8.6-1. Prove that, for each real positive number y, $(y + \tau)^{-1}$ as a function of τ is a member of $\mathscr{I}_{c,d}$ whenever $c \le \frac{1}{2}$ and $d \ge -\frac{1}{2}$.

PROBLEM 8.6-2. Prove that the mapping $\tau^{-1/2}\phi(\log \tau) \leftrightarrow \phi(t)$, where $t = \log \tau$, is an isomorphism from $\mathscr{D}(I)$ onto $\mathscr{D}$ where I is the interval $0 < \tau < \infty$. This justifies our assertion that $f(t) \leftrightarrow \tau^{-1/2} f(\log \tau)$ as defined by (5) is an isomorphism from $\mathscr{D}'$ onto $\mathscr{D}'(I)$.

CHAPTER IX

Transformations Arising from Orthonormal Series Expansions

9.1. Introduction

This chapter is of a somewhat different character than are the previous ones. The method we shall now use is related to Hilbert-space techniques, and its prototype is the Fourier series expansion of a periodic distribution (Zemanian [1], Chapter 11). A procedure will be developed for expanding a generalized function f into a series of the form:

$$f = \sum_{n=1}^{\infty} F(n)\psi_n \tag{1}$$

where the ψ_n constitute a complete system of orthonormal functions and the $F(n)$ are the corresponding Fourier coefficients of f.

This procedure leads to a whole new class of generalized integral transformations. The basic idea is to view the mapping $f \leftrightarrow F(n)$ as a transformation $\mathfrak{A}$ from a certain class of generalized functions f into the space of functions $F(n)$ mapping the integers into the complex plane. Then, (1) defines the inverse transformation; of course, the convergence of the series (1) must be interpreted in a generalized sense. Moreover, the permissible orthonormal functions ψ_n will be eigenfunctions of a certain type of self-adjoint differential operator $\mathfrak{R}$. As a result, the corresponding transformation $\mathfrak{A}$ will generate an operational calculus for solving differential equations involving the operator $\mathfrak{R}$. Particular generalized integral transformations that are encompassed by this technique are the finite Fourier transformation (i.e., the transformation corresponding to any Fourier series), the Laguerre transformation, the Hermite transformation, the Jacobi transformation with its special cases such as the Legendre, Chebyshev, and Gegenbauer transformations, and finally the finite Hankel transformations.

The method described herein was first discussed in Zemanian [12]. Other works that discuss orthogonal series expansions involving generalized functions are by Bouix [1] Chapter 7, Braga and Schönberg [1], Gelfand and Kostyuchenko [1] (See also Gelfand and Shilov [1], Vol. 3, Chapter 4, Gelfand and Vilenkin [1] pp. 103–127, 372, and the references therein), Giertz [1], and Walter [1]. The procedures of Giertz and Walter are appropriate for the generalized functions of Temple [1], Lighthill [1], and Korevaar [1], whereas the present technique is suitable for generalized functions as defined in Sec. 2.4. Braga and Schönberg exploit the concept of "formal series"; in this regard, see also Korevaar (2). Bouix's method can be used even when the ψ_n are not eigenfunctions of a differential operator; but by the same token, it is not as suitable for the generation of operational calculi for differential equations. The method of Gelfand and Kostyuchenko is actually a technique for the spectral analysis of operators that do not possess eigenfunctions in the usual sense but do possess generalized functions that act like eigenfunctions. It has been useful in a variety of subjects in functional analysis.

There are still other methods that have been devised for particular orthonormal sets such as the Fourier and Hermite sets. See, for example, Korevaar [2], Schwartz [1], Vol. II, pp. 80–87, 116–119, and Widlund [1].

9.2. The Space $L_2(I)$

In this section we survey a number of very well known classical results. We shall have need of them subsequently. As references on these results, see Williamson [1] or Liusternik and Sobolev [1].

Let I denote any open interval $a < x < b$ on the real line. Here, $a = -\infty$ and $b = \infty$ are permitted. A function $f(x)$ is said to be *quadratically integrable on I* if it is a locally integrable function on I such that

$$(1) \qquad \alpha_0(f) \triangleq \left[\int_a^b |f(x)|^2 \, dx \right]^{1/2} < \infty$$

The set of all quadratically integrable functions can be partitioned into equivalence classes of such functions by stipulating that two such functions f and g are in the same class if and only if $\alpha_0(f-g) = 0$. This is the case if and only if $f = g$ almost everywhere on I. The resulting space of equivalence classes is denoted by $L_2(I)$. It is customary to speak of $L_2(I)$ as a space of functions, even though it is actually a space of equivalence classes of functions. Every such class is represented by any one of its members and one speaks of "a function $f(x)$ being a member of $L_2(I)$." Moreover, the functional α_0 is also defined on $L_2(I)$ by (1); that is, the number that α_0 assigns to any equivalence class is defined as the number that α_0 assigns to any one of its members, this number being the same for all members of a given class.

$L_2(I)$ is a linear space whose zero element is the class of all functions that are equal to zero almost everywhere on I. Moreover, α_0 is a norm on $L_2(I)$ and is therefore a special case of a countable multinorm (the multinorm has only one element). $L_2(I)$ is assigned the topology generated by α_0. It turns out that $L_2(I)$ is a complete space.

Note that, if I has a compact closure and if $\{\phi_m\}$ is a sequence of continuous bounded functions that converges uniformly on I, then $\{\phi_m\}$ converges in $L_2(I)$ as well. Every function in that equivalence class in $L_2(I)$ which contains the uniform limit ϕ of $\{\phi_m\}$ will be equal to ϕ almost everywhere on I.

An *inner product*, which is a rule assigning a complex number (f, g) to each ordered pair f, g of elements in $L_2(I)$, is defined by

$$(f, g) \triangleq \int_a^b f(x)\overline{g(x)}\, dx \tag{2}$$

where $\overline{g(x)}$ denotes the complex conjugate of $g(x)$. It possesses the following properties:

$$(f + h, g) = (f, g) + (h, g)$$

$$(\beta f, g) = (f, \bar{\beta} g) = \beta (f, g)$$

Here, β denotes a complex number.

$$(f, g) = \overline{(g, f)}$$

$$(f, f) = [\alpha_0(f)]^2 \geq 0$$

If $(f, f) = 0$, then f is the zero element in $L_2(I)$. Furthermore, the inner product is continuous with respect to each of its arguments; that is, if $f_m \to f$ in $L_2(I)$ as $m \to \infty$, then $(f_m, g) \to (f, g)$ and $(g, f_m) \to (g, f)$. A useful result is the *Schwarz inequality*, which states that

$$|(f, g)| \leq \alpha_0(f)\alpha_0(g) \tag{3}$$

The dual of $L_2(I)$ is $L_2(I)$. That is, for each continuous linear functional H on $L_2(I)$ there exists a unique member h of $L_2(I)$ such that $H(f) = (h, f)$ for every $f \in L_2(I)$; here, $H(f)$ denotes the number that H assigns to f.

Throughout this chapter $\mathfrak{N}$ shall denote a linear differential operator of the form:

$$\mathfrak{N} = \theta_0 D^{n_1} \theta_1 D^{n_2} \cdots D^{n_\nu} \theta_\nu$$

where $D = d/dx$, the n_k are positive integers, and the θ_k are smooth functions on I that are never equal to zero anywhere on I. We also require that the θ_k and n_k be such that

$$\mathfrak{N} = \bar{\theta}_\nu(-D)^{n_\nu} \cdots (-D)^{n_2}\bar{\theta}_1(-D)^{n_1}\bar{\theta}_0 \tag{4}$$

Here, $\bar{\theta}_k$ denotes the complex-conjugate function, that is, $\bar{\theta}_k(x) = \overline{\theta_k(x)}$. Moreover, we assume that there exists a sequence $\{\lambda_n\}_{n=0}^{\infty}$ of real numbers called *eigenvalues of* $\mathfrak{N}$, and a sequence $\{\psi_n\}_{n=0}^{\infty}$ of smooth functions in $L_2(I)$, called *eigenfunctions of* $\mathfrak{N}$, such that $|\lambda_n| \to \infty$ as $n \to \infty$ and

$$\mathfrak{N}\psi_n = \lambda_n \psi_n \qquad n = 0, 1, 2, \ldots$$

The zero function (more precisely, the equivalence class of functions that are equal to zero almost everywhere on I) is not allowed as an eigenfunction. We shall number the ψ_n and λ_n according to $|\lambda_0| \leq |\lambda_1| \leq |\lambda_2| \leq \cdots$. We also assume that the ψ_n form a complete orthonormal sequence in $L_2(I)$. By the *orthonormality* of the sequence $\{\psi_n\}$ one means that

$$(\psi_n, \psi_m) = \begin{cases} 0 & n \neq m \\ 1 & n = m \end{cases}$$

The *completeness* of $\{\psi_n\}$ means that every $f \in L_2(I)$ can be expanded into the series:

$$f = \sum_{n=0}^{\infty} (f, \psi_n)\psi_n \tag{5}$$

which converges in $L_2(I)$, that is,

$$\alpha_0[f - \sum_{n=0}^{N} (f, \psi_n)\psi_n] \to 0 \qquad N \to \infty$$

We call (5) the *orthonormal series expansion of f with respect to* $\{\psi_n\}$. It should be pointed out that for a given operator $\mathfrak{N}$ there may be more than one complete orthonormal system $\{\psi_n\}_{n=0}^{\infty}$ of eigenfunctions.

An important classical result states that $\{\psi_n\}$ is complete if and only if, for every $f \in L_2(I)$, the coefficients (f, ψ_n) satisfy *Parseval's equation*:

$$\sum_{n=0}^{\infty} |(f, \psi_n)|^2 = \int_a^b |f|^2\, dx \triangleq [\alpha_0(f)]^2$$

Furthermore, we have the *Riesz-Fischer theorem*, which in our context can be stated as follows:

THEOREM 9.2-1. *Let* $\{\psi_n\}_{n=0}^{\infty}$ *be a complete orthonormal system as specified above, and let* $\{c_n\}_{n=0}^{\infty}$ *be a sequence of complex numbers such that* $\sum_{n=0}^{\infty}|c_n|^2$ *converges. Then, there exists a unique* $f \in L_2(I)$ *such that* $c_n = (f, \psi_n)$. *Consequently,*

$$f = \sum_{n=0}^{\infty} c_n \psi_n$$

in the sense of convergence in $L_2(I)$.

All the functions in the orthonormal systems arising from the classical orthogonal polynomials as well as in many other standard orthonormal systems are eigenfunctions of differential operators $\mathfrak{N}$ possessing the aforementioned properties. We present just two examples of such systems now and several more in Sec. 9.8.

EXAMPLE 9.2-1. *A Fourier System:* Let I be the open interval $(-\pi, \pi)$ and let

$$\mathfrak{N} = -iD = i^{-1/2}Di^{-1/2}$$

We may choose

$$\psi_n(x) = \frac{e^{inx}}{\sqrt{2\pi}} \qquad n = 0, \pm 1, \pm 2, \ldots$$

Then,

$$\lambda_n = n$$

(Here, we have used a different numbering system than that mentioned above. In this regard, see Problem 9.2-1.) Thus, any $f \in L_2(I)$ can be expanded into

$$f(x) = \frac{1}{2\pi} \sum_{n=-\infty}^{\infty} (f(t), e^{int})e^{inx}$$

which is the exponential form of the Fourier series for f.

EXAMPLE 9.2-2. *The Laguerre System.* Now, we set $I = (0, \infty)$ and

$$\mathfrak{N} = e^{x/2}Dxe^{-x}De^{x/2} = xD^2 + D - \frac{x}{4} + \frac{1}{2}$$

One complete set of eigenfunctions for $\mathfrak{N}$ consists of the *Laguerre functions* (Erdelyi (Ed.) [1], Vol. II, pp. 188–189):

$$\psi_n(x) = e^{-x/2} \sum_{m=0}^{n} \binom{n}{m} \frac{(-x)^m}{m!} \qquad n = 0, 1, 2, \ldots \tag{6}$$

The corresponding eigenvalues are

$$\lambda_n = -n$$

PROBLEM 9.2-1. With respect to convergence in $L_2(I)$, show that the series (5) can be rearranged in any fashion and that the Fourier series given in Example 9.2-1 is the limit of

$$\frac{1}{2\pi} \sum_{n=-M}^{N} (f(t), e^{int})\, e^{inx}$$

as N and M tend to infinity independently.

9.3. The Testing-Function Space $\mathscr{A}$

We construct a testing-function space $\mathscr{A}$, which depends on the choices of the interval I, the differential operator $\mathfrak{N}$, and the complete orthonormal system $\{\psi_n\}_{n=0}^{\infty}$ of eigenfunctions. Its dual $\mathscr{A}'$ is a space of generalized functions, each of which can be expanded into a series of the eigenfunctions ψ_n. In the sequel, n and k always denote nonnegative integers.

$\mathscr{A}$ consists of all functions $\phi(x)$ that possess the following three properties:

(i) $\phi(x)$ is defined, complex-valued, and smooth on I.

(ii) For each k the quantity:

$$\alpha_k(\phi) \triangleq \alpha_0(\mathfrak{N}^k\phi) \triangleq \left[\int_a^b |\mathfrak{N}^k\phi(x)|^2\, dx\right]^{1/2} \tag{1}$$

exists (i.e., is finite).

(iii) For each n and k

$$(\mathfrak{N}^k\phi, \psi_n) = (\phi, \mathfrak{N}^k \psi_n) \tag{2}$$

$\mathscr{A}$ is a linear space. Moreover, $\{\alpha_k\}_{k=0}^{\infty}$ is a multinorm on $\mathscr{A}$. Indeed, for any constant β, $\alpha_k(\beta\phi) = |\beta|\alpha_k(\phi)$ obviously. Also, $\alpha_k(\phi_1 + \phi_2) \leq \alpha_k(\phi_1) + \alpha_k(\phi_2)$, as is readily shown by using Minkowski's inequality. Hence, each α_k is a seminorm, and in addition α_0 is clearly a norm on $\mathscr{A}$. We equip $\mathscr{A}$ with the topology that is generated by $\{\alpha_k\}_{k=0}^{\infty}$, and this makes $\mathscr{A}$ a countably multinormed space. Under this formulation $\mathscr{A}$ turns out to be a testing-function space; this will be proven later on. Furthermore, $\mathscr{A}$ is clearly a subspace of $L_2(I)$ when we identify each function in $\mathscr{A}$ with the corresponding equivalence class in $L_2(I)$; similarly, convergence in $\mathscr{A}$ implies convergence in $L_2(I)$.

Every ψ_n is a member of $\mathscr{A}$. Indeed, ψ_n satisfies condition (i) by assumption. Moreover,

$$[\alpha_k(\psi_n)]^2 = \int_a^b |\mathfrak{N}^k\psi_n|^2\, dx = |\lambda_n|^{2k} \int_a^b |\psi_n|^2\, dx = |\lambda_n|^{2k}$$

so that condition (ii) is satisfied. Finally, if $n \neq m$, we have

$$(\mathfrak{N}^k\psi_n, \psi_m) = \lambda_n{}^k(\psi_n, \psi_m) = 0 = \lambda_m{}^k(\psi_n, \psi_m) = (\psi_n, \mathfrak{N}^k\psi_m)$$

whereas, for $n = m$,

$$(\mathfrak{N}^k\psi_n, \psi_n) = (\lambda_n{}^k\psi_n, \psi_n) = (\psi_n, \lambda_n{}^k\psi_n) = (\psi_n, \mathfrak{N}^k\psi_n)$$

since λ_n is real. Thus, condition (iii) is also satisfied.

We also note that the operator $\mathfrak{N}$ is a continuous linear mapping of $\mathscr{A}$ into itself. This follows directly from the definition of $\mathscr{A}$.

LEMMA 9.3-1. *$\mathscr{A}$ is complete and therefore a Fréchet space.*

PROOF. Let $\{\phi_m\}_{m=1}^{\infty}$ be a Cauchy sequence in $\mathscr{A}$. Then, for each k we have that $\{\mathfrak{N}^k\phi_m\}_m$ is a Cauchy sequence in $L_2(I)$. By the completeness of $L_2(I)$, there exists a function $\chi_k' \in L_2(I)$ which is the limit in $L_2(I)$ of $\{\mathfrak{N}^k\phi_m\}_m$. We shall show that χ_k' is almost everywhere (on I) equal to $\mathfrak{N}^k\chi_0$ where $\chi_0 \in \mathscr{A}$ is independent of k.

Let x_1 be any fixed point of I and x a variable point in I. Also, let D^{-1} be the integration operator:

$$D^{-1} = \int_{x_1}^{x} \cdots \, dt$$

Thus, for any smooth function ζ on I,

$$D^{-1}D\zeta(x) = \zeta(x) \quad \zeta(x_1)$$

Using Schwarz's inequality (Sec. 9.2, Eq. (3)) on the interval whose end points are x and x_1, we may write

$$(3) \quad |D^{-1}\theta_0^{-1}\mathfrak{N}^{k+1}(\phi_m - \phi_n)|^2 = \left|\int_{x_1}^{x} \theta_0^{-1}\mathfrak{N}^{k+1}(\phi_m - \phi_n)\, dt\right|^2$$

$$\leq \int_{x_1}^{x} |\theta_0^{-1}|^2\, dt \int_{x_1}^{x} |\mathfrak{N}^{k+1}(\phi_m - \phi_n)|^2\, dt$$

$$\leq \left|\int_{x_1}^{x} |\theta_0^{-1}|^2\, dt\right| \int_{a}^{b} |\mathfrak{N}^{k+1}(\phi_m - \phi_n)|^2\, dt$$

Now, let Ω denote an arbitrary open interval whose closure is compact in I. The first integral on the right-hand side of (3) is a bounded smooth function on every Ω since $\theta_0 \neq 0$ anywhere on I. The second integral converges to zero as m and n tend to infinity independently. This shows that the left-hand side of (3) converges to zero uniformly on every Ω.

We may remove the differentiations and multiplications by θ_k in the operator $\mathfrak{N}$ step by step to obtain $\mathfrak{N}^{-1}\mathfrak{N}^{k+1}\phi_m$ where

$$\mathfrak{N}^{-1} = \theta_\nu^{-1}D^{-n_\nu}\cdots\theta_1^{-1}D^{-n_1}\theta_0^{-1}$$

and $D^{-n} = (D^{-1})^n$. At each step the resulting quantity will converge uniformly on every Ω as $m \to \infty$.

We now observe that $\mathfrak{N}^{-1}\,\mathfrak{N}^{k+1}\phi_m(x)$ vanishes at $x = x_1$ together with all its derivatives of order less than $s \triangleq n_1 + \cdots + n_\nu$. Thus,

$$(4) \quad \mathfrak{N}^k\phi_m(x) = \mathfrak{N}^{-1}\mathfrak{N}^{k+1}\phi_m(x) + \sum_{j<s} a_{mj}g_j(x)$$

where the sum denotes that solution of the differential equation $\mathfrak{N}y = 0$ whose derivatives at x_1 of order less than s are the same as those of $\mathfrak{N}^k\phi_m(x)$. Let $g_j(x)$ be the solution of the equation $\mathfrak{N}y = 0$ such that $g_j^{(i)}(x_1) = \delta_{ij}$, $i = 0, 1, \ldots, s-1$. ($g_j^{(i)}$ denotes the ith derivative of g_j, and δ_{ij} is the Kronecker delta: $\delta_{ij} = 0$ when $i \neq j$ and $\delta_{ii} = 1$.) The $g_j(x)$ are linearly independent smooth functions on every Ω (Hurewicz [1], p. 46); also, each coefficient a_{mj} will be equal to the jth derivatives of $\mathfrak{N}^k\phi_m$ at x_1. Since $\mathfrak{N}^k\phi_m$ converges in $L_2(I)$ and $\mathfrak{N}^{-1}\mathfrak{N}^{k+1}\phi_m$ converges uniformly on every Ω as $m \to \infty$, we conclude that

$$\sum_{j<s} a_{mj} g_j(x) \tag{5}$$

converges in $L_2(\Omega)$ for every Ω. Because the g_j are linearly independent, it follows that for each j the coefficients a_{mj} converge to a limit a_j. Consequently, (5) and therefore $\mathfrak{N}^k\phi_m$ converge uniformly on every Ω.

The uniform limit of $\mathfrak{N}^k\phi_m$, which we denote by χ_k, is a continuous function on I. Moreover, the preceding argument holds for every k, and therefore we obtain from (4)

$$\chi_k(x) = \mathfrak{N}^{-1}\chi_{k+1}(x) + \sum_{j<s} a_j g_j(x) \tag{6}$$

We can conclude that χ_k is a smooth function and that $\chi_{k+1} = \mathfrak{N}\chi_k$; hence, $\chi_k = \mathfrak{N}^k\chi_0$. Moreover, $\chi_k'(x) = \chi_k(x)$ almost everywhere on I since χ_k is the uniform limit on every Ω of $\{\mathfrak{N}^k\phi_m\}_m$ and χ_k' is the limit in $L_2(I)$ of $\{\mathfrak{N}^k\phi_m\}_m$. Thus, both $\mathfrak{N}^k\chi_0$ and χ_k' are in the same equivalence class in $L_2(I)$. It follows that, for every k, $\alpha_k(\chi_0) \triangleq \alpha_0(\mathfrak{N}^k\chi_0) < \infty$ and

$$\alpha_k(\chi_0 - \phi_m) \triangleq \alpha_0(\mathfrak{N}^k\chi_0 - \mathfrak{N}^k\phi_m) \to 0$$

as $m \to \infty$.

To complete the proof, we have to show that $(\mathfrak{N}^k\chi_0, \psi_n) = (\chi_0, \mathfrak{N}^k\psi_n)$ for every n and k. Since the inner product is continuous with respect to the convergence in $L_2(I)$ of one of its arguments,

$$(\mathfrak{N}^k\chi_0, \psi_n) = \lim_{m\to\infty} (\mathfrak{N}^k\phi_m, \psi_n)$$

$$= \lim_{m\to\infty} (\phi_m, \mathfrak{N}^k\psi_n) = (\chi_0, \mathfrak{N}^k\psi_n)$$

Q.E.D.

We now prove three more lemmas that we shall need in subsequent sections.

LEMMA 9.3-2. *If $\phi \in \mathscr{A}$, then*

$$\phi = \sum_{n=0}^{\infty} (\phi, \psi_n)\psi_n$$

where the series converges in $\mathscr{A}$.

PROOF. By condition (ii) at the beginning of this section, $\mathfrak{N}^k\phi$ is in $L_2(I)$ for each nonnegative integer k. Hence, we may expand $\mathfrak{N}^k\phi$ into a series of the orthonormal functions ψ_n. By using (2) and the fact that $\mathfrak{N}\psi_n = \lambda_n\psi_n$ where λ_n is a real number, we obtain

$$\text{(7)} \qquad \mathfrak{N}^k\phi = \sum_{n=0}^{\infty} (\mathfrak{N}^k\phi, \psi_n)\psi_n = \sum (\phi, \mathfrak{N}^k\psi_n)\psi_n$$
$$= \sum (\phi, \lambda_n{}^k\psi_n)\psi_n = \sum (\phi, \psi_n)\lambda_n{}^k\psi_n = \sum (\phi, \psi_n)\, \mathfrak{N}^k\psi_n$$

These series converge in $L_2(I)$. Consequently, for each k,

$$\alpha_k[\phi - \sum_{n=0}^{N} (\phi, \psi_n)\psi_n] = \alpha_0[\mathfrak{N}^k\phi - \sum_{n=0}^{N} (\phi, \psi_n)\, \mathfrak{N}^k\psi_n] \to 0$$

as $N \to \infty$. This is what we had to show.

Incidentally, the last lemma implies that $\mathfrak{N}$ satisfies the condition:

$$\text{(8)} \qquad (\mathfrak{N}\phi, \chi) = (\phi, \mathfrak{N}\chi)$$

where ϕ and χ are arbitrary members of $\mathscr{A}$. One refers to this fact by saying that $\mathfrak{N}$ is *self-adjoint on* $\mathscr{A}$. To prove (8), we use (2), (7), and the fact that the inner product is continuous with respect of each of its arguments to write

$$(\mathfrak{N}\phi, \chi) = \int_a^b \bar{\chi} \sum_n (\phi, \psi_n)\mathfrak{N}\psi_n \, dx = \sum_n (\phi, \psi_n) \int_a^b \bar{\chi}\mathfrak{N}\psi_n \, dx$$
$$= \sum_n (\phi, \psi_n) \int_a^b \psi_n \overline{\mathfrak{N}\chi} \, dx = \int_a^b \sum_n (\phi, \psi_n)\psi_n \overline{\mathfrak{N}\chi} \, dx = (\phi, \mathfrak{N}\chi)$$

We can characterize orthonormal series that converge in $\mathscr{A}$ in the following way:

LEMMA 9.3-3. *Let a_n denote complex numbers. Then, $\sum_{n=0}^{\infty} a_n\psi_n$ converges in $\mathscr{A}$ if and only if $\sum_{n=0}^{\infty} |\lambda_n|^{2k}|a_n|^2$ converges for every nonnegative integer k.*

PROOF. We employ the fact that the ψ_n form an orthonormal set to write

$$\int_a^b \left| \mathfrak{N}^k \sum_{n=q}^{p} a_n\psi_n \right|^2 dx = \int_a^b \left| \sum_{n=q}^{p} a_n\lambda_n{}^k\psi_n \right|^2 dx$$
$$= \int_a^b \sum_{n=q}^{p} \sum_{m=q}^{p} a_n\overline{a_m}\lambda_n{}^k\lambda_m{}^k\psi_n\overline{\psi}_m \, dx = \sum_{n=q}^{p} |\lambda_n|^{2k}|a_n|^2$$

Our assertion follows directly from this equation.

LEMMA 9.3-4. *$\mathscr{A}$ is a subspace of $\mathscr{E}(I)$. Moreover, if $\{\phi_m\}_{m=1}^{\infty}$ converges in $\mathscr{A}$ to the limit ϕ, then $\{\phi_m\}$ also converges in $\mathscr{E}(I)$ to the same limit ϕ.*

PROOF. We only prove the second statement, the first one being obvious. Again, Ω denotes an arbitrary open interval whose closure is compact in I. We have already demonstrated in the proof of Lemma 9.3-1 that, if $\{\phi_m\}$ converges in $\mathscr{A}$, then it, as well as $\{\mathfrak{N}^k\phi_m\}_m$ for each fixed k, converges uniformly on every Ω. Setting $k = 0$ in (4), we have

$$\phi_m = \mathfrak{N}^{-1}\mathfrak{N}\phi_m + \sum_{j<s} a_{mj}g_j(x)$$

where the coefficients a_{mj} converge as $m \to \infty$. Moreover, we may write

$$D\phi_m = [(D\theta_\nu^{-1})D^{-n_\nu} + \theta_\nu^{-1}D^{1-n_\nu}]\theta_{\nu-1}^{-1}D^{-n_{\nu-1}}\cdots\theta_1^{-1}D^{-n_1}\theta_0^{-1}\mathfrak{N}\phi_m + \sum_{j<s} a_{mj}Dg_j(x)$$

It follows that $D\phi_m$ converges uniformly on Ω. By the same argument, $D^\mu\phi_m$ converges uniformly on Ω for $\mu = 2, \ldots, s \triangleq n_1 + \cdots + n_\nu$. A similar argument based on (4) with $k = 1$ (now the a_{mj} are in general different constants) shows that $D^\mu\mathfrak{N}\phi_m$ converges uniformly on Ω as $m \to \infty$ for each $\mu = 1, \ldots, s$. From this one obtains successively the uniform convergence of $D^{s+1}\phi_m, \ldots, D^{2s}\phi_m$. Continuing in this fashion, we see that, for each k, $\{D^k\phi_m\}_m$ converges uniformly on every Ω. Thus, $\{\phi_m\}$ converges in $\mathscr{E}(I)$ to a limit that is clearly the same as its limit in $\mathscr{A}$. Q.E.D.

We can now state that $\mathscr{A}$ is a testing-function space since the three conditions of Sec. 2.4 have been shown to be fulfilled.

PROBLEM 9.3-1. Prove that $\alpha_k(\phi_1 + \phi_2) \leq \alpha_k(\phi_1) + \alpha_k(\phi_2)$ for every $\phi_1, \phi_2 \in \mathscr{A}$.

PROBLEM 9.3-2. Prove that the convergence of (5) in $L_2(\Omega)$ as $m \to \infty$ and the linear independence of the smooth functions $g_j(x)$ imply that the coefficients a_{mj} converge as $m \to \infty$.

PROBLEM 9.3-3. Let $\mathscr{A}$ denote the testing-function space corresponding to the Fourier system described in Example 9.2-1. Show that every $\phi \in \mathscr{A}$ can be extended into a smooth periodic function $\tilde{\phi}(x)$ on $-\infty < x < \infty$ by means of the equations:

$$\tilde{\phi}(x) = \phi(x) \qquad -\pi < x < \pi$$

$$\tilde{\phi}(\pi) = \lim_{x\to\pi-} \phi(x), \qquad \tilde{\phi}(-\pi) = \lim_{x\to-\pi+} \phi(x)$$

$$\tilde{\phi}(x + 2\pi) = \tilde{\phi}(x) \qquad -\infty < x < \infty$$

Then, prove that a sequence $\{\phi_m\}_{m=0}^{\infty}$ converges in $\mathscr{A}$ if and only if $\{\tilde{\phi}_m\}$ converges uniformly on $-\infty < x < \infty$. (This shows that the testing-function space $\mathscr{A}$ for the Fourier system of Example 9.2-1 can be identified through this extension with the periodic testing-function space $\mathscr{P}_{2\pi}$. Consequently, the members of the dual $\mathscr{A}'$ of $\mathscr{A}$ can be identified with the periodic distributions. For the definition of $\mathscr{P}_{2\pi}$, see Zemanian[1], Sec. 11.2.)

9.4. The Generalized-Function Space $\mathscr{A}'$

$\mathscr{A}'$ denotes the dual of $\mathscr{A}$. It is a space of generalized functions that depends on the choices of I, $\mathfrak{N}$, and the ψ_n, as does $\mathscr{A}$. Instead of working with the number $\langle f, \phi \rangle$ that $f \in \mathscr{A}'$ assigns to $\phi \in \mathscr{A}$, it is more convenient now to work with the number that f assigns to the complex conjugate of ϕ. We denote the latter number by (f, ϕ), that is,

$$(1) \qquad (f, \phi) = \langle f, \overline{\phi} \rangle$$

This use of the notation $(\cdot, \cdot)$ both as an inner product in $L_2(I)$ and for the number that $f \in \mathscr{A}'$ assigns to $\phi \in \mathscr{A}$ does not lead to any inconsistency, as we shall see in note *III* below. The rule for the multiplication by a complex number a now takes on the form:

$$(af, \phi) = (f, \bar{a}\phi) = a(f, \phi)$$

Since $\mathscr{A}$ is complete, so also is $\mathscr{A}'$ according to Theorem 1.8-3.

We define a generalized differential operator $\overline{\mathfrak{N}}'$ on $\mathscr{A}'$ through the relationship:

$$(f, \mathfrak{N}\phi) = \langle f, \overline{\mathfrak{N}\phi} \rangle \triangleq \langle \overline{\mathfrak{N}}'f, \overline{\phi} \rangle = (\overline{\mathfrak{N}}'f, \phi)$$

According to the convention stated in Sec. 2.5, $\overline{\mathfrak{N}}'$ is denoted by the differential expression obtained by reversing the order in which the differentiations and multiplications by θ occurr in $\mathfrak{N}$, replacing each D by $-D$, and then taking the complex conjugate of the result. But, this is precisely the expression for $\mathfrak{N}$ according to Sec. 9.2, Eq. (4). Thus, $\mathfrak{N} = \overline{\mathfrak{N}}'$ is defined as a generalized differential operator on $\mathscr{A}'$ through the equation:

$$(2) \qquad (\mathfrak{N}f, \phi) \triangleq (f, \mathfrak{N}\phi) \qquad f \in \mathscr{A}', \qquad \phi \in \mathscr{A}$$

Since $\mathfrak{N}$ is a continuous linear mapping of $\mathscr{A}$ into $\mathscr{A}$, it is also a continuous linear mapping of $\mathscr{A}'$ into $\mathscr{A}'$.

We now list some other properties of $\mathscr{A}'$.

I. $\mathscr{D}(I)$ is obviously a subspace of $\mathscr{A}$, and convergence in $\mathscr{D}(I)$ implies convergence in $\mathscr{A}$. Consequently, the restriction of any $f \in \mathscr{A}'$ to $\mathscr{D}(I)$ is a member of $\mathscr{D}'(I)$. Moreover, convergence in $\mathscr{A}'$ implies convergence in $\mathscr{D}'(I)$.

II. Since $\mathscr{D}(I) \subset \mathscr{A} \subset \mathscr{E}(I)$ and since $\mathscr{D}(I)$ is dense in $\mathscr{E}(I)$, $\mathscr{A}$ is also dense in $\mathscr{E}(I)$. It now follows from Corollary 1.8-2*a* and Lemma 9.3-4 that $\mathscr{E}'(I)$ is a subspace of $\mathscr{A}'$.

III. We imbed $L_2(I)$ (and, therefore, $\mathscr{A}$ since $\mathscr{A} \subset L_2(I)$) into $\mathscr{A}'$ by defining the number that $f \in L_2(I)$ assigns to any $\phi \in \mathscr{A}$ as

$$(f, \phi) \triangleq \int_a^b f(x)\overline{\phi(x)}\, dx \tag{3}$$

f is clearly linear on $\mathscr{A}$. Its continuity on $\mathscr{A}$ follows from the fact that, if $\{\phi_m\}_{m=1}^{\infty}$ converges in $\mathscr{A}$ to zero, then by the Schwarz inequality

$$|(f, \phi_m)| \le \alpha_0(f)\alpha_0(\phi_m) \to 0 \qquad m \to \infty$$

The definition (3) is consistent with the facts, pointed out in Sec. 9.2, that the dual of $L_2(I)$ is $L_2(I)$ and that the inner product (h, χ) is precisely the number that an arbitrary continuous linear functional h on $L_2(I)$ assigns to an arbitrary $\chi \in L_2(I)$.

Note that this imbedding of $L_2(I)$ into $\mathscr{A}'$ is one-to-one. Indeed, if two members f and g of $L_2(I)$ become imbedded as the same element of $\mathscr{A}'$, then $(f, \phi) = (g, \phi)$ for every $\phi \in \mathscr{D}(I)$. But, through an argument similar to the one given in the paragraph following Sec. 3.10, Eq. (3), this implies that $f = g$ almost everywhere on I. Hence, f and g are in the same equivalence class in $L_2(I)$.

A useful result is that, if $f = \mathfrak{N}^k g$ for some $g \in L_2(I)$ and some k, then $f \in \mathscr{A}'$. This follows directly from the facts that $L_2(I) \subset \mathscr{A}'$ and that $\mathfrak{N}$ maps $\mathscr{A}'$ into $\mathscr{A}'$.

It is also worth noting that the conventional operator $\mathfrak{N}$ and the generalized operator $\mathfrak{N} = \overline{\mathfrak{N}}'$ agree on $\mathscr{A}$ since the conventional operator is self-adjoint on $\mathscr{A}$.

IV. Theorem 1.8-1 asserts the following: For each $f \in \mathscr{A}'$ there exists a nonnegative integer r and a positive constant C such that

$$|(f, \phi)| \le C \max_{0 \le k \le r} \alpha_k(\phi)$$

for every $\phi \in \mathscr{A}$. Here, r and C depend on f but not on ϕ.

PROBLEM 9.4-1. As in Problem 9.3-3, let $\mathscr{A}$ be the testing-function space corresponding to the Fourier system described in Example 9.2-1. Let $f \in L_2(I)$ be defined by $f(x) = x$ for $-\pi < x < \pi$. In the sense of generalized differentiation, show that

$$D^m f(x) = -\pi\delta^{(m-1)}(x - \pi) - \pi\delta^{(m-1)}(x + \pi) \qquad m = 1, 2, 3, \ldots \tag{4}$$

where

$$(\delta^{(m-1)}(x - \pi), \phi(x)) \triangleq (-1)^m \lim_{x \to \pi -} D^{m-1}\overline{\phi(x)}$$

with a similar definition holding for π and $x \to \pi-$ replaced by $-\pi$ and $x \to -\pi+$ respectively. As was implied in Problem 9.3-3, the right-hand side of (4) can be extended into the periodic distribution:

$$-2\pi \sum_{\nu=-\infty}^{\infty} \delta^{(m-1)}(x - 2\nu\pi - \pi)$$

9.5. Orthonormal Series Expansions and Generalized Integral Transformations

The fundamental theorem of this chapter is the next one. It states that any generalized function in $\mathscr{A}'$ possesses an orthonormal series expansion with respect to the $\{\psi_n\}$ used in the construction of $\mathscr{A}$.

THEOREM 9.5-1. *If $f \in \mathscr{A}'$, then*

$$f = \sum_{n=0}^{\infty} (f, \psi_n)\psi_n \tag{1}$$

where the series converges in $\mathscr{A}'$.

PROOF. To show this we need merely invoke Lemma 9.3-2 and write, for any $\phi \in \mathscr{A}$,

$$\begin{aligned}(f, \phi) = (f, \textstyle\sum(\phi, \psi_n)\psi_n) &= \textstyle\sum (f, \psi_n)\overline{(\phi, \psi_n)} \\ &= \textstyle\sum (f, \psi_n)(\psi_n, \phi)\end{aligned} \tag{2}$$

Thus, the right-hand side of (2) truly converges for every $\phi \in \mathscr{A}$, which means that the series in (1) converges in $\mathscr{A}'$. Q.E.D.

We can view the orthonormal series expansion (1) as the inversion formula for a certain generalized integral transformation $\mathfrak{A}$, defined by

$$\mathfrak{A}f \triangleq F(n) \triangleq (f, \psi_n) \qquad f \in \mathscr{A}'; \qquad n = 0, 1, 2, \ldots \tag{3}$$

Thus, $\mathfrak{A}$ is a mapping of $\mathscr{A}'$ into the space of complex-valued functions defined on n. The inverse mapping $\mathfrak{A}^{-1}$ is given by (1), which may be rewritten as

$$\mathfrak{A}^{-1}F(n) = \sum_{n=0}^{\infty} F(n)\psi_n = f \tag{4}$$

(Henceforth, the transform function (3) will be denoted by the capital letter corresponding to the lower case letter used for the original generalized function in $\mathscr{A}'$.) Clearly, $\mathfrak{A}$ is a linear mapping and is continuous in the sense that, if $\{f_\nu\}_{\nu=1}^{\infty}$ converges in $\mathscr{A}'$ to f, then $\{F_\nu(n)\}_{\nu=1}^{\infty}$ converges to $F(n)$ for each n.

THEOREM 9.5-2 (*The Uniqueness Theorem*). *If $f, g \in \mathscr{A}'$ and if their transforms satisfy $F(n) = G(n)$ for every n, then $f = g$ in the sense of equality in $\mathscr{A}'$.*

PROOF.

$$f - g = \sum (f - g, \psi_n)\psi_n = \sum [(f, \psi_n) - (g, \psi_n)]\psi_n = 0$$

PROBLEM 9.5-1. Show that, if $f \in \mathscr{A}'$ and $F(n) = (f, \psi_n)$, then there exists an integer r such that $F(n) = O(|\lambda_n|^r)$ as $n \to \infty$.

PROBLEM 9.5-2. Let $f \in \mathscr{E}'(I)$ where I is the interval $0 < x < \infty$. For the Laguerre transformation corresponding to the system of Example 9.2-2, establish the following operation-transform formulas where $F(n) = \mathfrak{A}[f(x)]$.

a. $\mathfrak{A}[e^{-x/2}De^{x/2}f(x)] = \sum_{k=0}^{n} F(k)$

b. $\mathfrak{A}[e^{-x/2}xDe^{x/2}f(x)] = nF(n) - (n+1)F(n+1)$

c. $\mathfrak{A}[e^{x/2}Dxe^{-x/2}f(x) + xf(x)] = (n+1)[F(n) - F(n+1)]$

d. $\mathfrak{A}[e^{-x/2}DxDe^{x/2}f(x)] = -(n+1)F(n+1)$

e. $\mathfrak{A}[e^{-3x/2}Dxe^{x}De^{x/2}f(x)] = -2(n+1)F(n+1) + nF(n)$

f. $\mathfrak{A}[xf(x)] = -(n+1)F(n+1) + (2n+1)F(n) - nF(n-1)$

$$F(-1) = 0$$

Hint: Use the formulas {Erdelyi (Ed.) [1], Vol. II, Eqs. 10.12(8), 10.12(12), and 10.12(16)} for the Laguerre polynomials $L_n(x) = L_n^0(x)$.

$$xL_n(x) = -(n+1)L_{n+1}(x) + (2n+1)L_n(x) - nL_{n-1}(x); \; L_{-1}(x) \equiv 0$$

$$xDL_n(x) = (n+1)L_{n+1}(x) - (n+1-x)L_n(x)$$

$$D[L_n(x) - L_{n+1}(x)] = L_n(x)$$

(For a variety of properties of the conventional Laguerre transformation, see Debnath [1], Hirschman [2], and McCully [1].)

9.6. Characterizations of the Generalized Functions in $\mathscr{A}'$ and Their Transforms

We now turn to the problem of precisely characterizing the functions $F(n)$ that are generated by the transformation $\mathfrak{A}$. In particular, we shall prove that a sequence $\{b_n\}_{n=0}^{\infty}$ of complex numbers is the transform of some member f of $\mathscr{A}'$ (i.e., $b_n = (f, \psi_n)$) if and only if the sequence satisfies the growth condition (2) stated below. This in turn will lead to a characterization of the members of $\mathscr{A}'$ in terms of certain combinations of generalized derivatives of the elements of $L_2(I)$.

THEOREM 9.6-1. *Let b_n denote complex numbers. Then,*

$$\sum_{n=0}^{\infty} b_n \psi_n \tag{1}$$

converges in $\mathscr{A}'$ if and only if there exists a nonnegative integer q such that

$$\sum_{\lambda_n \neq 0} |\lambda_n|^{-2q} |b_n|^2 \tag{2}$$

converges. Furthermore, if f denotes the sum in $\mathscr{A}'$ of (1), *then $b_n = (f, \psi_n)$.* (The symbol $\sum_{\lambda_n \neq 0}$ denotes a summation over all n for which $\lambda_n \neq 0$.)

PROOF. First, assume that (2) converges for some $q \geq 0$. We wish to show that, for each $\phi \in \mathscr{A}$,

$$\sum_{n=0}^{\infty} (b_n \psi_n, \phi) \tag{3}$$

converges. This will confirm the convergence in $\mathscr{A}'$ of (1). Using the Schwarz inequality for sums of real numbers (Widder [2], p. 313), we may write

$$\begin{aligned} \sum_{\lambda_n \neq 0} |(b_n \psi_n, \phi)| &= \sum |b_n(\psi_n, \phi)| \\ &= \sum |\lambda_n^{-q} b_n| \, |\lambda_n^{q}(\phi, \psi_n)| \\ &\leq [\sum |\lambda_n^{-q} b_n|^2 \sum |\lambda_n^{q}(\phi, \psi_n)|^2]^{1/2} \end{aligned}$$

The first series in the right-hand side converges by assumption, whereas the second series converges by Lemmas 9.3-2 and 9.3-3. Thus, (3) truly does converge.

Next, assume that (1) converges in $\mathscr{A}'$, and let f be its sum in $\mathscr{A}'$. Since $\psi_m \in \mathscr{A}$, we may write

$$(f, \psi_m) = \sum_{n=0}^{\infty} b_n(\psi_n, \psi_m)$$

and by the orthonormality of the ψ_n we get $(f, \psi_m) = b_m$. This verifies the last statement of the theorem.

Still assuming that (1) converges in $\mathscr{A}'$, we finally wish to show that (2) converges for some q. We do this by an argument suggested by J. Korevaar. We know from the proof of theorem 9.5-1 (in particular, from Sec. 9.5, Eq. (2)) that, for every $\phi = \sum a_n \psi_n \in \mathscr{A}$, the series $\sum \overline{a_n} b_n$ converges. Call this result condition A. From here on, it will be assumed that the arguments of the a_n are always chosen in such a way that $\overline{a_n} b_n = |a_n b_n|$.

We show first that the sequence $\{\lambda_n^{-q} b_n\}$, where n traverses those integers for which $\lambda_n \neq 0$, is bounded for some value of q, say, q_0. Indeed, if not, the sequence is unbounded for every $q = 1, 2, 3, \ldots$; hence, there exists an increasing sequence of indices n_q such that $\lambda_{n_q} \neq 0$ and

$$|\lambda_{n_q}^{-q} b_{n_q}| \geq 1 \qquad q = 1, 2, 3, \ldots$$

Now, choose

$$|a_{n_q}| = |\lambda_{n_q}^{q} q|^{-1} \qquad q = 1, 2, 3, \ldots$$

and $a_m = 0$ if $m \neq n_q$ for every q. Now, for any fixed nonnegative integer k,

$$\sum_{m=0}^{\infty} |\lambda_m{}^k a_m|^2 = \sum_{q=1}^{\infty} q^{-2} |\lambda_{n_q}|^{2k-2q}$$

The right-hand side converges because $|\lambda_{n_q}|^{2k-2q}$ is bounded by 1 for all sufficiently large q. Lemma 9.3-3 now implies that $\phi \triangleq \sum a_n \psi_n$ is a member of $\mathscr{A}$. Thus, we have found a member of $\mathscr{A}$ for which

$$\sum_{n=0}^{\infty} |a_n b_n| \geq \sum_{q=1}^{\infty} q^{-1} = \infty$$

This contradicts condition A.

We may thus assume that, as $n \to \infty$, $|\lambda_n^{-q} b_n| \to 0$ for each $q > q_0$ because $|\lambda_n| \to \infty$. We now show that (2) converges for some $q > q_0$. If not, then (2) diverges for every $q > q_0$. Therefore, there exists an increasing sequence of integers m_q such that

$$1 \leq \sum_{n=m_{q-1}}^{m_q - 1} |\lambda_n^{-q} b_n|^2 < 2 \qquad q = q_0 + 1, q_0 + 2, \ldots$$

This time choose

$$|a_n| = |b_n \lambda_n^{-2q} q^{-1}| \qquad m_{q-1} \leq n < m_q, \qquad q > q_0$$

Then, for any fixed nonnegative integer k,

$$(4) \qquad \sum_{n=m_{q-1}}^{m_q - 1} |\lambda_n{}^k a_n|^2 = \sum_{n=m_{q-1}}^{m_q - 1} |\lambda_n|^{2k-2q} |\lambda_n{}^{-q} b_n|^2 q^{-2}$$

But, $|\lambda_n| \to \infty$ as $n \to \infty$; therefore, (4) is less than $2q^{-2}$ for all sufficiently large q. This implies that

$$\sum_{n=0}^{\infty} |\lambda_n{}^k a_n|^2$$

converges for every k. By Lemma 9.3-3, $\phi \triangleq \sum a_n \psi_n$ is a member of $\mathscr{A}$. On the other hand, $\sum_{n=0}^{\infty} |a_n b_n|$ diverges because

$$\sum_{n=m_{q-1}}^{m_q - 1} |a_n b_n| = \sum_{n=m_{q-1}}^{m_q - 1} |b_n{}^2 \lambda_n^{-2q} q^{-1}| \geq q^{-1}$$

Again, we have contradicted condition A. The proof is complete.

The next theorem presents a precise characterization of the members of $\mathscr{A}'$.

THEOREM 9.6-2. *A necessary and sufficient condition for f to be a member of $\mathscr{A}'$ is that there be some nonnegative integer q and a $g \in L_2(I)$ such that*

$$(5) \qquad f = \mathfrak{N}^q g + \sum_{\lambda_n = 0} c_n \psi_n$$

where the c_n denote complex constants.

(Here, of course, $\mathfrak{N}^q$ is understood to be a generalized differential operator on $\mathscr{A}'$. Also, $\sum_{\lambda_n=0}$ denotes a summation on those n for which $\lambda_n = 0$; there are only a finite number of such n.)

PROOF. *Sufficiency:* We have already shown in Sec. 9.4, note *III* that $\mathfrak{N}^q g \in \mathscr{A}'$ and that $\mathscr{A} \subset \mathscr{A}'$. Since $\psi_n \in \mathscr{A} \subset \mathscr{A}'$, it follows that $f \in \mathscr{A}'$.

Necessity: Let $f = \sum F(n)\psi_n \in \mathscr{A}'$. Set $G(n) = \lambda_n^{-q}F(n)$ whenever $\lambda_n \neq 0$, where $q \geq 0$ is such that

$$\sum_{\lambda_n \neq 0} |\lambda_n^{-q}F(n)|^2$$

converges. Also, set $G(n) = 0$ when $\lambda_n = 0$. Hence, $\sum_{n=0}^{\infty}|G(n)|^2$ converges, and by the Riesz-Fischer theorem (Theorem 9.2-1), there exists a $g \in L_2(I)$ such that $G(n) = (g, \psi_n)$. Moreover, since $\psi_n \in \mathscr{A}$, the definition of $\mathfrak{N}$ on $\mathscr{A}'$ yields

$$(g, \lambda_n^q\psi_n) = (g, \mathfrak{N}^q\psi_n) = (\mathfrak{N}^q g, \psi_n)$$

Altogether then, we may write

$$\begin{aligned} f = \sum_{n=0}^{\infty} F(n)\psi_n &= \sum_{\lambda_n \neq 0} \lambda_n^q G(n)\psi_n + \sum_{\lambda_n=0} F(n)\psi_n \\ &= \sum_{n=0}^{\infty} (g, \lambda_n^q\psi_n)\psi_n + \sum_{\lambda_n=0} F(n)\psi_n \\ &= \sum_{n=0}^{\infty} (\mathfrak{N}^q g, \psi_n)\psi_n + \sum_{\lambda_n=0} F(n)\psi_n \\ &= \mathfrak{N}^q g + \sum_{\lambda_n=0} F(n)\psi_n \end{aligned}$$

Q.E.D.

PROBLEM 9.6-1. Let $\mathscr{A}$ be the testing-function space corresponding to the Laguerre system of Example 9.2-2. Show that $e^{-sx} \in \mathscr{A}$ for each fixed $s \in \mathscr{C}^1$ having a positive real part. This allows us to define the one-sided Laplace transform of any $f \in \mathscr{A}'$ by

(6) $$F(s) \triangleq (f(x), e^{-\bar{s}x}) \qquad \operatorname{Re} s > 0$$

Next, set

(7) $$s = \frac{1+z}{2(1-z)} \qquad \operatorname{Re} s > 0, \qquad |z| < 1$$

and construct the power series expansion:

(8) $$\frac{F[s(z)]}{1-z} = \sum_{n=0}^{\infty} b_n z^n \qquad |z| < 1$$

Prove the following two assertions, which imply a means of inverting (6). A function $F(s)$ is the Laplace transform of an $f \in \mathscr{A}'$ according to the definition (6) if and only if it possesses the expansion (8) where the sequence $\{b_n\}_{n=0}^{\infty}$ of complex numbers satisfies condition (2) of Theorem 9.6-1. In this case,

$$f(t) = \sum_{n=0}^{\infty} b_n \psi_n(t)$$

where the series converges in $\mathscr{A}'$ and the $\psi_n(t)$ are the Laguerre functions Sec. 9.2, Eq. (6).

Hint: Use the transform Sec. 3.4, Eq. (10).

PROBLEM 9.6-2. With respect to convergence in $\mathscr{A}'$, show that the series Sec. 9.5, Eq. (1) can be rearranged in any fashion.

9.7. An Operational Calculus for $\mathfrak{N}$

We have already indicated that the differential operator $\mathfrak{N}$ is a continuous linear mapping of $\mathscr{A}'$ into $\mathscr{A}'$. Therefore, we may write for every $f \in \mathscr{A}'$

$$\mathfrak{N}^k f = \sum_{n=0}^{\infty} (f, \psi_n) \mathfrak{N}^k \psi_n = \sum_{n=0}^{\infty} (f, \psi_n) \lambda_n^{\,k} \psi_n$$

We can use this fact to solve the differential equation:

$$P(\mathfrak{N})u = g \tag{1}$$

where P is a polynomial and the given g and unknown u are required to be in $\mathscr{A}'$. Upon applying the transformation $\mathfrak{A}$, we obtain

$$P(\lambda_n)U(n) = G(n) \qquad U = \mathfrak{A}u, \qquad G = \mathfrak{A}g$$

If $P(\lambda_n) \neq 0$ for every n, we can divide by $P(\lambda_n)$ and then apply $\mathfrak{A}^{-1}$ to get

$$u = \sum_{n=0}^{\infty} \frac{G(n)}{P(\lambda_n)} \psi_n \tag{2}$$

By Theorems 9.5-2 and 9.6-1 this solution exists and is unique in $\mathscr{A}'$. If $P(\lambda_n) = 0$ for some λ_n, say, for λ_{n_k} $(k = 1, \ldots, m)$, then a solution exists in $\mathscr{A}'$ if and only if $G(n_k) = 0$ for $k = 1, \ldots, m$. In this case a solution to (1) is

$$u = \sum_{p(\lambda_n) \neq 0} \frac{G(n)}{P(\lambda_n)} \psi_n \tag{3}$$

But, it is no longer unique in $\mathscr{A}'$, and we may add to (3) any complementary solution:

$$u_c = \sum_{k=1}^{m} a_k \psi_{n_k}$$

where the a_k are arbitrary numbers.

PROBLEM 9.7-1. State a set of conditions under which the operational calculus for $\mathfrak{N}$ can be extended to simultaneous differential equations.

PROBLEM 9.7-2. Let Q be a complex-valued function on the set $\{\lambda_n\}$ corresponding to a given system: $\mathfrak{N}$, $\{\psi_n\}$, $\{\lambda_n\}$. Assume that Q possesses the following properties: $Q(\lambda_n) \neq 0$ for every n; for some fixed integer k, $Q(\lambda_n) = O(\lambda_n{}^k)$ and $[Q(\lambda_n)]^{-1} = O(\lambda_n{}^k)$ as $|\lambda_n| \to \infty$. Then, define the operator $Q(\mathfrak{N})$ on $\mathscr{A}'$ by

$$Q(\mathfrak{N})f \triangleq \sum_{n=0}^{\infty} Q(\lambda_n)\, F(n)\psi_n \qquad F = \mathfrak{A}f$$

Show that $Q(\mathfrak{N})$ is a linear mapping of $\mathscr{A}'$ into $\mathscr{A}'$. Also, show that a solution to (1) is (2) where P is replaced by Q. Discuss how this formulation must be changed when, for an infinite number of the λ_n, we have $Q(\lambda_n) = 0$.

9.8. Particular Cases

As was mentioned in Sec. 9.2, all the generalized integral transformations arising from the classical orthogonal polynomials, as well as many others, are special cases of our formulation. We list here a number of these and and also state the interval I, the differential operator $\mathfrak{N}$, the eigenfunctions ψ_n, and the eigenvalues λ_n. The ψ_n corresponding to the indicated values of n form a complete orthonormal set in $L_2(I)$. In every case the assumptions stated in Sec. 9.2 are fulfilled. Thus, any particular direct transformation and its inverse can be written down simply by substituting the appropriate quantities into the expressions:

(1) $$F(n) = (f(x), \psi_n(x))$$

(2) $$f(x) = \sum_n (f, \psi_n)\psi_n(x)$$

respectively. In most cases the name of a transformation listed below conforms with the name of the corresponding orthonormal functions ψ_n. A source for the information listed here is Erdelyi (Ed.) [1], Vol. II, wherein still other bibliography can be found.

1. *The Finite Fourier Transformation:* The name of these transformations is due to the facts that their kernels are the exponential functions whereas the interval I is finite.

1*a*. *First Form.* This case was presented in Example 9.2-1. We repeat it here for the sake of completeness.

$$I = (-\pi, \pi)$$

$$\mathfrak{N} = -iD = i^{-1/2} D i^{-1/2}$$

$$\psi_n(x) = \frac{e^{inx}}{\sqrt{2\pi}} \qquad n = 0, \pm 1, \pm 2, \ldots$$

$$\lambda_n = n$$

(The numbering system employed here differs from that used in our general formulation. See Problem 9.6-2.)

1*b*. *Second Form:*

$$I = (0, \pi)$$

$$\mathfrak{N} = D^2$$

$$\psi_0(x) = \pi^{-1/2}$$

$$\psi_n(x) = \sqrt{\frac{2}{\pi}} \cos nx \qquad n = 1, 2, 3, \ldots$$

$$\lambda_n = -n^2 \qquad n = 0, 1, 2, \ldots$$

1*c*. *Third Form:*

$$I = (0, \pi)$$

$$\mathfrak{N} = D^2$$

$$\psi_n(x) = \sqrt{\frac{2}{\pi}} \sin nx \qquad n = 1, 2, 3, \ldots$$

$$\lambda_n = -n^2$$

2. *The Laguerre Transformation:*

$$I = (0, \infty)$$

$$\mathfrak{N} = x^{-\alpha/2} e^{x/2} D x^{\alpha+1} e^{-x} D x^{-\alpha/2} e^{x/2}$$

$$= xD^2 + D - \frac{x}{4} - \frac{\alpha^2}{4x} + \frac{\alpha + 1}{2}$$

Here, α is a real number greater than -1.

$$\psi_n(x) = \left[\frac{\Gamma(n+1)}{\Gamma(\alpha+n+1)}\right]^{1/2} x^{\alpha/2}e^{-x/2}L_n{}^{\alpha}(x) \qquad n=0,1,2,\ldots$$

where the $L_n{}^{\alpha}(x)$ are the generalized Laguerre ploynomials:

$$L_n{}^{\alpha}(x) = \sum_{m=0}^{n} \binom{n+\alpha}{n-m} \frac{(-x)^m}{m!}$$

As always,

$$\binom{y}{v} \triangleq \frac{\Gamma(y+1)}{\Gamma(v+1)\Gamma(y-v+1)}$$

The functions $\psi_n(x)$ are called the generalized Laguerre functions (this is a different use of the adjective "generalized"), and we shall refer to α as their order.

Finally,

$$\lambda_n = -n$$

The system presented in Example 9.2-2 is a particular case of this one and is obtained by setting $\alpha = 0$; $L_n{}^0(x)$ are the customary Laguerre polynomials.

3. *The Hermite Transformation:*

$$I = (-\infty, \infty)$$

$$\mathfrak{N} = e^{x^2/2}De^{-x^2}De^{x^2/2}$$

$$= D^2 - x^2 + 1$$

$$\psi_n(x) = \frac{e^{-x^2/2}H_n(x)}{[2^n(n!)\sqrt{\pi}]^{1/2}} \qquad n=0,1,2,\ldots$$

where the $H_n(x)$ are the Hermite polynomials:

$$H_n(x) = n! \sum_{m=0}^{[n/2]} \frac{(-1)^m(2x)^{n-2m}}{m!(n-2m)!}$$

Here, as well as later on, $[n/2]$ denotes either $n/2$ or $(n-1)/2$ according to the evenness or oddness of n respectively.

$$\lambda_n = -2n$$

In this particular case $\mathscr{A}$ happens to be the space $\mathscr{S}$ of testing functions of rapid descent and $\mathscr{A}'$ the space $\mathscr{S}'$ of distributions of slow growth. A proof of this is given in the appendix of the technical report cited in Zemanian [12].

4. *The Jacobi Transformation:*

$$I = (-1, 1)$$

$$\mathfrak{N} = [w(x)]^{-1/2} D(x^2 - 1) w(x) D[w(x)]^{-1/2}$$

where $w(x) = (1 - x)^\alpha (1 + x)^\beta$ and α and β are real numbers with $\alpha > -1$, $\beta > -1$.

$$\psi_n(x) = \left[\frac{w(x)}{h_n}\right]^{1/2} P_n^{(\alpha,\beta)}(x) \qquad n = 0, 1, 2, \ldots$$

where

$$h_n = \frac{2^{\alpha+\beta+1}\Gamma(n + \alpha + 1)\Gamma(n + \beta + 1)}{n!(2n + \alpha + \beta + 1)\Gamma(n + \alpha + \beta + 1)}$$

and the $P_n^{(\alpha,\beta)}(x)$ are the Jacobi polynomials:

$$P_n^{(\alpha,\beta)}(x) = 2^{-n} \sum_{m=0}^{n} \binom{n+\alpha}{m}\binom{n+\beta}{n-m} (x-1)^{n-m}(x+1)^m$$

Finally,

$$\lambda_n = n(n + \alpha + \beta + 1)$$

The next three transformations are special cases of this one wherein α and β are restricted in certain ways.

4*a*. *The Legendre Transformation:*

$$I = (-1, 1)$$

$$\mathfrak{N} = D(x^2 - 1)D$$

$$\psi_n(x) = \sqrt{n + \tfrac{1}{2}}\, P_n(x) \qquad n = 0, 1, 2, \ldots$$

where the $P_n(x)$ are the Legendre polynomials:

$$P_n(x) = 2^{-n} \sum_{m=0}^{[n/2]} (-1)^m \binom{n}{m}\binom{2n-2m}{n} x^{n-2m}$$

$$\lambda_n = n(n + 1)$$

This is a special case of the Jacobi transformation obtained by setting $\alpha = \beta = 0$.

4*b*. *The Chebyshev Transformation:*

$$I = (-1, 1)$$

$$\mathfrak{N} = (1 - x^2)^{1/4} D(1 - x^2)^{1/2} D(1 - x^2)^{1/4}$$

$$\psi_0(x) = \pi^{-1/2}(1 - x^2)^{-1/4}$$

$$\psi_n(x) = \sqrt{\frac{2}{\pi}}\,(1 - x^2)^{-1/4} T_n(x) \qquad n = 1, 2, 3, \ldots$$

where the $T_n(x)$ are the Chebyshev polynomials:

$$T_n(x) = \frac{n}{2} \sum_{m=0}^{[n/2]} \frac{(-1)^m (n-m-1)!}{m!(n-2m)!} (2x)^{n-2m}$$

$$\lambda_n = -n^2 \qquad n = 0, 1, 2, \ldots$$

One obtains this system from that of the Jacobi transformation by setting $\alpha = \beta = -\frac{1}{2}$.

4c. *The Gegenbauer Transformation:*

$$I = (-1, 1)$$

$$\mathfrak{N} = [w(x)]^{-1/2} D(1 - x^2)^{\rho+1/2} D[w(x)]^{-1/2}$$

where $w(x) = (1 - x^2)^{\rho-1/2}$ and ρ is a real number with $\rho > -\frac{1}{2}$.

$$\psi_n(x) = \left[\frac{w(x)}{h_n}\right]^{1/2} C_n^{\rho}(x) \qquad n = 0, 1, 2, \ldots$$

where

$$h_n = \frac{\sqrt{\pi}\Gamma(\rho + \frac{1}{2})\Gamma(2\rho + n)}{n!(n + \rho)\Gamma(\rho)\Gamma(2\rho)}$$

and the $C_n^{\rho}(x)$ are the Gegenbauer polynomials (also called the ultraspherical polynomials):

$$C_n^{\rho}(x) = \sum_{m=0}^{[n/2]} \frac{(-1)^m \Gamma(\rho + n - m)}{m!(n - 2m)!\Gamma(\rho)} (2x)^{n-2m}$$

$$\lambda_n = -n(n + 2\rho)$$

To get this system set $\alpha = \beta = \rho - \frac{1}{2}$ in the Jacobi system.

5. *The Finite Hankel Transformations:* The name of these transformations is explained by the facts that their kernels involve Bessel functions whereas I is a finite interval. The order μ of the Bessel functions is restricted to the real numbers and is also called the order of the transformation. Some other references for the following information, besides Erdelyi (Ed.) [1], Vol. II, pp. 70–72, are Boas and Pollard [1], MacRobert [1], Titchmarsh [2], and Watson [1] Chapter 18.

5a. *First Form:*

$$I = (0, 1)$$

$$\mathfrak{N} = S_\mu = x^{-\mu-1/2} D x^{2\mu+1} D x^{-\mu-1/2} \qquad \mu \geq -\frac{1}{2}$$

$$\psi_n(x) = \frac{\sqrt{2x}\, J_\mu(y_{\mu,n} x)}{J_{\mu+1}(y_{\mu,n})} \qquad n = 1, 2, 3, \ldots$$

where J_μ is the μth order Bessel function of first kind and the $y_{\mu,n}$ denote all the positive roots of $J_\mu(y) = 0$ with $0 < y_{\mu,1} < y_{\mu,2} < y_{\mu,3} < \ldots$

$$\lambda_n = -y_{\mu,n}^2$$

In this case the expansion (2) is called the *μth-order Fourier-Bessel series for f.*

5b. *Second Form:*

$$I = (0, 1)$$

$$\mathfrak{N} = S_\mu = x^{-\mu-1/2} D x^{2\mu+1} D x^{-\mu-1/2} \qquad \mu \geq -\tfrac{1}{2}$$

$$\psi_n(x) = \sqrt{\frac{2x}{h_n}}\, J_\mu(z_{\mu,n} x) \qquad n = 1, 2, 3, \ldots$$

where the $z_{\mu,n}$ denote all the positive roots of

$$zJ_\mu^{(1)}(z) + aJ_\mu(z) = 0 \tag{3}$$

with $0 < z_{\mu,1} < z_{\mu,2} < z_{\mu,3} \ldots$. Here, a is any fixed real number, and $J_\mu^{(1)}(z) \triangleq D_z J_\mu(z)$. Also,

$$h_n = [J_\mu^{(1)}(z_{\mu,n})]^2 + (1 - \mu^2 z_{\mu,n}^{-2})[J_\mu(z_{\mu,n})]^2$$

Finally,

$$\lambda_n = -z_{\mu,n}^2$$

Now, the expansion (2) is called the *μth-order Dini series for f.*

5c. *Third Form:*

$$I = (a, b) \qquad 0 < a < b < \infty$$

$$\mathfrak{N} = S_\mu = x^{-\mu-1/2} D x^{2\mu+1} D x^{-\mu-1/2}$$

Here, μ is any real number.

$$\psi_n(x) = \sqrt{\frac{\pi x}{2h_n}}\, [J_\mu(w_{\mu,n} x) Y_\mu(w_{\mu,n} b) - Y_\mu(w_{\mu,n} x) J_\mu(w_{\mu,n} b)]$$

where Y_μ is the μth order Bessel function of second kind and the $w_{\mu,n}$ denote all the positive roots of

$$J_\mu(aw) Y_\mu(bw) - Y_\mu(aw) J_\mu(bw) = 0$$

with $0 < w_{\mu,1} < w_{\mu,2} < w_{\mu,3} < \ldots$ Also,

$$h_n = w_{\mu,n}^{-2} \left\{1 - \left[\frac{J_\mu(w_{\mu,n} b)}{J_\mu(w_{\mu,n} a)}\right]^2\right\}$$

Finally,

$$\lambda_n = -w_{\mu,n}^2$$

PROBLEM 9.8-1. By using the first form of the finite Fourier transformation, find all the solutions in the corresponding space $\mathscr{A}'$ of the following differential equations:

a. $(D^2 + 2D + 1)u(x) = D\delta(x) \qquad -\pi < x < \pi$
b. $(D^2 + 4)u(x) = e^{ix} \qquad -\pi < x < \pi$

PROBLEM 9.8-2. By using the Laguerre transformation with $\alpha = 0$, solve the differential equation:

$$\left(xD^2 + D - \frac{x}{4}\right)u(x) = e^{-ax}$$

where $0 < x < \infty$ and a is a complex number with $\operatorname{Re} a > 0$. *Hint:* See Problem 3.4-3.

PROBLEM 9.8-3. For the Hermite transformation, establish the following operation-transform formulas, where $f \in \mathscr{E}'$, $F(n) = \mathfrak{A}[f(x)]$, and $F(-1) = 0$.

$$a. \quad \mathfrak{A}[xf(x)] = \sqrt{\frac{n+1}{2}}\, F(n+1) + \sqrt{\frac{n}{2}}\, F(n-1)$$

$$b. \quad \mathfrak{A}[Df(x)] = \sqrt{\frac{n+1}{2}}\, F(n+1) - \sqrt{\frac{n}{2}}\, F(n-1)$$

$$c. \quad \mathfrak{A}[e^{-x^2/2} D e^{x^2/2} f(x)] = \sqrt{2n+2}\, F(n+1)$$

$$d. \quad \mathfrak{A}[e^{x^2/2} D e^{-x^2/2} f(x)] = -\sqrt{2n}\, F(n-1)$$

Hint: Use the formulas {Erdelyi (Ed.) [1], Vol. II, Eqs. 10.13(10) and 10.13(14)}:

$$H_{n+1}(x) - 2xH_n(x) + 2nH_{n-1}(x) = 0$$

$$DH_n(x) = 2nH_{n-1}(x)$$

$$H_{-1}(x) \equiv 0$$

(Other results of this sort for the conventional Hermite transformation are given by Debnath [2].)

PROBLEM 9.8-4. In terms of the system for the Hermite transformation, consider the differential equation:

$$(4) \qquad (\mathfrak{N} + a)u = \delta$$

where $a \in \mathscr{C}^1$. For what values of a does each one of the following statements hold true?

a. (4) possesses no solution u in $\mathscr{A}'$.
b. (4) possesses a unique solution u in $\mathscr{A}'$.
c. (4) possesses more than one solution u in $\mathscr{A}'$.
Find all solutions in $\mathscr{A}'$ whenever they exist.

PROBLEM 9.8-5. Let $f \in \mathscr{E}'(I)$ where $I = (-1, 1)$. For the Jacobi transformation determine an operation-transform formula for the mapping

$$f(x) \leftrightarrow [w(x)]^{-1/2} D(x^2 - 1)[w(x)]^{1/2} f(x)$$

by using the identities {Erdelyi (Ed.) [1], Vol. II, Eqs. 10.8(11) and 10.8(15)}:

$$\gamma(\gamma + 1)(\gamma + 2)xP_n^{(\alpha,\beta)}(x) = (\beta^2 - \alpha^2)(\gamma + 1)P_n^{(\alpha,\beta)}(x)$$
$$+ 2\gamma(n + 1)(\gamma - n + 1)P_{n+1}^{(\alpha,\beta)}(x) + 2(n + \alpha)(n + \beta)(\gamma + 2)P_{n-1}^{(\alpha,\beta)}(x)$$
$$\gamma(1 - x^2)DP_n^{(\alpha,\beta)}(x) = n(\alpha - \beta - \gamma x)P_n^{(\alpha,\beta)}(x) + 2(n + \alpha)(n + \beta)P_{n-1}^{(\alpha,\beta)}(x)$$

where $\gamma = 2n + \alpha + \beta$ and $P_{-1}^{(\alpha,\beta)}(x) \equiv 0$. Then, state the particular cases of the operation-transform formula for the Legendre, Chebyshev, and Gegenbauer transformations.

(A number of other properties for the corresponding conventional transformations appear in the literature as follows. For the Legendre transformation, see Churchill [3], Churchill and Dolph [1], and Tranter [1]. For the Gegenbauer transformation, see Conte [1], Lakshmanarao [1], and Srivastava [1]. For the Jacobi transformation, see Debnath [3] and Scott [1].)

PROBLEM 9.8-6. Show that, under the first form of the finite Hankel transformation,

$$\mathfrak{A}(x^{\mu+1/2}) = \frac{\sqrt{2}}{y_{\mu,n}} \qquad 0 < x < 1, \qquad \mu \geq -\tfrac{1}{2}, \qquad n = 1, 2, 3, \ldots$$

According to the operational calculus of Sec. 9.7, this implies that

$$\mathfrak{A}(S_\mu x^{\mu+1/2}) = -\sqrt{2}\, y_{\mu,n} \tag{5}$$

But, a conventional differentiation shows that $S_\mu x^{\mu+1/2} = 0$ for $0 < x < 1$. This seems to violate the uniqueness of the finite Hankel transformation (Theorem 9.5-2). Explain this discrepancy, and show that (5) is correct by properly evaluating its left-hand side.

PROBLEM 9.8-7. Using the first form of the finite Hankel transformation, where μ is now required to satisfy $-\frac{1}{2} \leq \mu < 1$, find a solution u to the following differential equation:

$$(S_\mu{}^2 + S_\mu + 2)u(x) = x^{(1/2)-\mu} \qquad 0 < x < 1$$

Is the solution unique in the corresponding space $\mathscr{A}'$?

9.9. Application of the Finite Fourier Transformation: A Dirichlet Problem for a Semi-infinite Channel

The rest of this chapter is devoted to a number of applications of some of the particular transformations listed in the preceding section. We first apply the finite Fourier transformation to solve a Dirichlet problem for the interior of a semi-infinite channel R in the (x, y) plane. In particular, let

$$R \triangleq \{(x, y): 0 < x < \pi,\ 0 < y < \infty\}$$

We wish to find a conventional function $u = u(x, y)$ that satisfies Laplace's equation in R:

$$(1) \qquad D_x{}^2 u + D_y{}^2 u = 0 \qquad (x, y) \in R,\ D_z{}^2 = \frac{\partial^2}{\partial z^2}$$

and the following boundary conditions:

(i) As $y \to 0+$, $u(x, y) \to f(x) \in \mathscr{A}'$ in the sense of convergence in $\mathscr{A}'$. Here, $\mathscr{A}'$ is the generalized-function space corresponding to the system 1c of Sec. 9.8.

(ii) As $x \to 0+$ or $x \to \pi-$, $u(x, y)$ converges uniformly to zero on $Y \le y < \infty$ for each $Y > 0$.

(iii) As $y \to \infty$, $u(x, y)$ converges uniformly to zero on $0 < x < \pi$.

As usual, we develop the solution formally and prove subsequently that we truly have a solution. In view of boundary conditions (i) and (ii). we choose the third form of the finite Fourier transformation; note that the corresponding eigenfunctions $\sqrt{2/\pi}\sin nx$ satisfy boundary condition (ii). Set

$$U(n, y) \triangleq \left(u(x, y), \sqrt{\frac{2}{\pi}}\sin nx\right) \qquad n = 1, 2, 3, \ldots$$

Then, the differential equation is formally transformed into

$$(D_y{}^2 - n^2)U(n, y) = 0$$

Hence,

$$(2) \qquad U(n, y) = A(n)e^{-ny} + B(n)e^{ny}$$

where $A(n)$ and $B(n)$ do not depend on y. In view of boundary condition (iii), it is reasonable to choose $B(n) = 0$. Then, upon matching the transform of the boundary condition (i), we obtain

$$A(n) = \lim_{y \to 0+} U(n, y) = \left(f(t), \sqrt{\frac{2}{\pi}}\sin nt\right) \triangleq F(n)$$

Thus,

$$U(n, y) = F(n)e^{-ny}$$

Upon applying the inverse transformation, we are lead to the possible solution:

(3) $$u(x, y) = \sqrt{\frac{2}{\pi}} \sum_{n=1}^{\infty} F(n) e^{-ny} \sin nx$$

where

(4) $$F(n) = \left(f(t), \sqrt{\frac{2}{\pi}} \sin nt \right)$$

Now to prove that (3) truly is a solution. First, note that, by Theorem 9.6-1 and the fact that $\lambda_n = -n^2$, $F(n)$ is of slow growth (i.e., as $n \to \infty$, $F(n) = O(n^q)$ for some integer q). The factor e^{-ny} therefore insures that (3) converges uniformly on every half-plane in the (x, y) plane of the form $Y \leq y < \infty$ $(Y > 0)$. The same is true after any number of term-by-term differentiations of (3) with respect to either x or y. Thus, we may apply the operator $D_x^2 + D_y^2$ under the summation sign in (3). Since $e^{-ny} \sin nx$ satisfies Laplace's equation, so does u. Thus, the differential equation (1) is satisfied in the conventional sense everywhere in R.

To verify the boundary condition (i) we have to show that, for each $\phi \in \mathscr{A}$,

(5) $$(u(x, y), \phi(x)) \to (f, \phi) \qquad y \to 0+$$

Now, for any fixed $y > 0$, (3) converges in $\mathscr{A}$ and therefore in $L_2(I)$, according to Lemma 9.3-3. Consequently, we can take its inner product with ϕ term-by-term to write

(6) $$(u(x, y), \phi(x)) = \sqrt{\frac{2}{\pi}} \sum_{n=1}^{\infty} F(n) e^{-ny} (\sin nx, \phi(x))$$

But, by Lemmas 9.3-2 and 9.3-3, $(\sin nx, \phi(x))$ is of rapid descent; that is, as $n \to \infty$,

$$|(\sin nx, \phi(x))| = o(n^{-q})$$

for every integer q. It follows that the series in (6) converges uniformly on $0 \leq y < \infty$, so that we may pass to the limit as $y \to 0+$ under the summation sign to get

(7) $$(u(x, y), \phi(x)) \to \sqrt{\frac{2}{\pi}} \sum_{n=1}^{\infty} F(n)(\sin nx, \phi(x))$$

Since $f \in \mathscr{A}'$, the right-hand side is equal to (f, ϕ); see Sec. 9.5, Eq. (2). This proves (5).

Finally, for $Y \leq y < \infty$ $(Y > 0)$, we have from (3) that

(8) $$|u(x, y)| \leq \sqrt{\frac{2}{\pi}} \sum_{n=1}^{\infty} |F(n)| e^{-nY} |\sin nx|$$

By our previous comments, the series herein converges uniformly on $-\infty < x < \infty$. So, we may take the limit as $x \to 0+$ or $x \to \pi-$ under the summation sign in (8) to verify boundary condition (ii). In the same way,

$$|u(x, y)| \leq \sqrt{\frac{2}{\pi}} \sum_{n=1}^{\infty} |F(n)| e^{-ny} \to 0 \qquad y \to \infty$$

which verifies boundary condition (iii).

PROBLEM 9.9-1. State why the solution (3) is not unique.

PROBLEM 9.9-2. In the rectangular region:

$$R = \{(x, y): \quad 0 \leq x \leq a, 0 \leq y \leq b\}$$

solve Laplace's equation $(D_x{}^2 + D_y{}^2)u(x, y) = 0$ under the following boundary conditions:

(i) As $y \to 0+$, $u(x, y)$ converges in $\mathscr{D}'(I)$ to $f_1(x) \in \mathscr{E}'(I)$, where $I = (0, a)$.
(ii) As $x \to 0+$, $u(x, y)$ converges in $\mathscr{D}'(J)$ to $f_2(y) \in \mathscr{E}'(J)$, where $J = (0, b)$.
(iii) As $y \to b-$, $u(x, y)$ converges in $\mathscr{D}'(I)$ to $f_3(x) \in \mathscr{E}'(I)$.
(iv) As $x \to a-$, $u(x, y)$ converges in $\mathscr{D}'(J)$ to $f_4(y) \in \mathscr{E}'(J)$.

Hint: First, solve this problem when $f_2 = f_3 = f_4 = 0$. After appropriately normalizing the variables x and y, choose the $A(n)$ and $B(n)$ in (2) to make $U(n, b) = 0$ for each n. Repeat this procedure for each of the other three sides of R. Then, use the superposition principle: The solution to a sum of boundary conditions is the sum of the solutions to each of the boundary conditions. Justify your answer.

9.10. Applications of the Laguerre and Jacobi Transformations: The Time-Domain Synthesis of Signals

A classical problem in electrical network theory is the so-called "time-domain synthesis problem." One version of it can be stated as follows: Given an electrical signal described as a real-valued conventional function $f(t)$ on $0 < t < \infty$, construct an electrical network consisting of only a finite number of real resistors, capacitors, and inductors, which are all fixed, linear, and positive, such that its output $f_N(t)$, resulting from a delta-functional input $\delta(t)$, approximates $f(t)$ on $0 < t < \infty$ in some sense.

A standard technique for solving this problem is to expand $f(t)$ into a convergent series:

$$f(t) = \sum_{n=0}^{\infty} g_n(t) \tag{1}$$

of real-valued functions $g_n(t)$ such that every partial sum:

$$f_N(t) = \sum_{n=0}^{N} g_n(t) \qquad N = 0, 1, 2, \ldots \tag{2}$$

possesses the following two properties:

(i) $f_N(t) \equiv 0$ for $-\infty < t < 0$.

(ii) The (two-sided) Laplace transform $F_N(s)$ of $f_N(t)$ is a rational function having a zero at $s = \infty$ and all its poles in the left-half s plane ($\operatorname{Re} s < 0$) except possibly for a simple pole at the origin. (This condition is more restrictive than it need be since simple imaginary poles satisfying certain restrictions on their residues can also be allowed.)

After choosing N in (2) sufficiently large to satisfy whatever approximation criterion is being used, one can then employ a variety of standard synthesis techniques to construct the desired electrical network from a knowledge of $F_N(s)$. Thus, an essential part of this procedure is to find an expression (2) having the aforementioned properties. This is quite often accomplished, when f is quadratically integrable on $0 < t < \infty$, by using an appropriate orthonormal series expansion; see, for example, Armstrong [1], Huggins [1], Kautz [1], Lee [1], Lerner [1], and Young and Huggins [1].

We shall now expand this classical problem by allowing the given signal f to be a suitably restricted generalized function and will demand as a solution a sequence $\{f_N(t)\}_{N=0}^{\infty}$ such that each f_N is a real-valued conventional function possessing the properties (i) and (ii) stated above and also such that $f_N \to f$ on $0 < t < \infty$ in some generalized sense.

The Laguerre transformation yields an immediate solution if $f \in \mathscr{A}'$, where $\mathscr{A}'$ now corresponds to the system 2 of Sec. 9.8. Choose the order α of the generalized Laguerre functions ψ_n equal to an even nonnegative integer. Then, truncate the orthonormal series expansion of f at $n = N$ and multiply the result by the unit step function $1_+(t)$. This yields

$$f_N(t) = 1_+(t) \sum_{n=0}^{N} (f, \psi_n)\psi_n(t) \tag{3}$$

It is readily shown that the Laplace transform $F_N(s)$ of $f_N(t)$ is a rational function with a zero at $s = \infty$ and only one discrete pole of multiplicity $N + 1 + \alpha/2$ at $s = -\frac{1}{2}$. Thus, properties (i) and (ii) are satisfied. Moreover, we also know that $f_N(t)$ converges in $\mathscr{A}'$ to $f(t)$ as $N \to \infty$. Therefore, the sequence whose elements are (3) is a solution.

Another solution can be obtained by using the Jacobi transformation and a change of variable. Assume now that $f \in \mathscr{E}'(I)$ where I is the open interval $(0, \infty)$. Let $x(t)$ be a real continuous nondecreasing function identically equal to -1 for $-\infty < t \le 0$, with $dx/dt > 0$ for $0 < t < \infty$, and $\lim_{t\to\infty} x(t) = 1$. Furthermore, let the Laplace transform of $x(t) + 1$ be a rational function with a zero at infinity and all its poles in the left-half s plane (Re $s < 0$) except for a simple pole at the origin. There are a variety of criteria with which one can construct such an $x(t) + 1$ by appropriately choosing the pole and zero locations of its Laplace transform (Zemanian [13], [14]). Let $t(x)$ denote the inverse function corresponding to $x(t)$. By the change of variable formula for distributions on the interval $0 < t < \infty$ (Zemanian [1], p. 30), we can construct the distribution $f[t(x)]$. On the x axis the support of $f[t(x)]$ is a compact subset of $J \triangleq (-1, 1)$; that is, $f[t(x)] \in \mathscr{E}'(J) \subset \mathscr{A}'$, where now $\mathscr{A}'$ is the generalized-function space corresponding to the Jacobi transformation with α and β being chosen as even nonnegative integers. This choice of α and β insures that the Jacobi functions $\psi_n(x)$ are polynomials. We expand $f[t(x)]$ into

$$f[t(x)] = \sum_{n=0}^{\infty} a_n \psi_n(x)$$

where

$$a_n = (f[t(x)], \psi_n(x)) = \left(f(t), \psi_n[x(t)] \frac{dx}{dt} \right)$$

Hence,

$$f(t) = \sum_{n=0}^{\infty} a_n \psi_n[x(t)]$$

Since the ψ_n are polynomials and, for $t > 0$, $x(t)$ is a finite linear combination of a real constant and real-valued damped exponential functions multiplied by nonnegative powers of t, it follows that

$$f_N(t) \triangleq 1_+(t) \sum_{n=0}^{N} a_n \psi_n[x(t)] \tag{4}$$

satisfies conditions (i) and (ii). By Sec. 9.4, note I, convergence in $\mathscr{A}'$ implies convergence in $\mathscr{D}'(J)$. It immediately follows that $f_N(t) \to f(t)$ in $\mathscr{D}'(I)$ as $N \to \infty$, where $I = (0, \infty)$. Thus, the sequence $\{f_N\}$, where each f_N is given by (4), is also a solution to our problem.

Note that, if we choose

$$x(t) = 1 - 2e^{-\rho t} \qquad 0 < t < \infty$$

where ρ is a real positive number, then the Laplace transform $F_N(s)$ of $f_N(t)$ will have only real simple poles situated at points of the form $-m\rho$ where m is a nonnegative integer. More generally, if the terms of $x(t)$ have only real exponents, $F_N(s)$ will have only real poles. On the other hand, by allowing $x(t)$ to have damped sinusoidal terms in it, we find that $F_N(s)$ can have complex poles as well as real ones.

PROBLEM 9.10-1. Show that the two-sided Laplace transform of (3) is a rational function with a zero at $s = \infty$ and having only one discrete pole of multiplicity $N + 1 + \alpha/2$ located at $s = -\frac{1}{2}$. As before, take the order α of the generalized Laguerre functions ψ_n equal to a nonnegative even integer.

9.11. Application of the Legendre Transformation: A Dirichlet Problem for the Interior of the Unit Sphere

The Legendre transformation provides a means for solving the following Dirichlet problem for the interior of the unit sphere. We choose a spherical coordinate system:

$$\{(r, \theta, \alpha): 0 \leq r < 1, 0 \leq \theta \leq \pi, 0 \leq \alpha < 2\pi\}$$

and assume that the desired potential function v does not depend upon the α coordinate (i.e., $v = v(r, \theta)$). Upon setting $\mu = \cos\theta$, we may write Laplace's equation as

(1) $$rD_r^2 rv + D_\mu(1 - \mu^2)D_\mu v = 0$$

$$v = v(r, \cos^{-1}\mu), \qquad 0 < r < 1, \qquad -1 < \mu < 1$$

Furthermore, we require that $v(r, \theta)$ remain bounded in some neighborhood of the origin $r = 0$. Finally, let $\mathscr{A}'$ be the generalized-function space corresponding to the system $4a$ of Sec. 9.8 for the Legendre transformation, and let $f(\mu) \in \mathscr{A}'$. We impose the boundary condition:

(2) $$v(r, \cos^{-1}\mu) \to f(\mu) \text{ in } \mathscr{A}' \text{ as } r \to 1-$$

To formally derive a solution, we first apply the Legendre transformation $\mathfrak{A}$ with respect to μ to the differential equation (1). By reversing the order of application of $\mathfrak{A}$ and $rD_r^2 r$ and setting

$$V = V(r, n) = \mathfrak{A}v(r, \cos^{-1}\mu) \qquad n = 0, 1, 2, \ldots$$

we obtain

(3) $$rD_r^2 rV - n(n + 1)V = 0$$

The solution to (3) is

(4) $$V(r, n) = A(n)r^n + B(n)r^{-n-1}$$

Since $v(r, \theta)$ is finite in a neighborhood of $r = 0$, we choose $B(n) = 0$ to avoid terms that diverge as $r \to 0$. The $A(n)$ are determined by matching the transform of the boundary condition (2); here, it is assumed that $\mathfrak{A}$ and the passage to the limit as $r \to 1-$ can be interchanged. Thus,

$$A(n) = \lim_{r \to 1-} V(r, n) = (\mathfrak{A} f)(n) \tag{5}$$

This yields

$$V(r, n) = (f(\mu), \psi_n(\mu)) r^n$$

where

$$\psi_n(\mu) = \sqrt{n + \tfrac{1}{2}}\, P_n(\mu) \qquad n = 0, 1, 2, \ldots$$

The $P_n(\mu)$ are the Legendre polynomials. Upon applying the inverse Legendre transformation to $V(r, n)$, we obtain our formal solution:

$$v(r, \theta) = v(r, \cos^{-1}\mu) = \sum_{n=0}^{\infty} (f, \psi_n) r^n \psi_n(\cos\theta) \tag{6}$$

We now prove that (6) is a conventional function that satisfies (1) in the conventional sense. Formally applying the operator $D_r^p D_\mu^q$ $(p, q = 0, 1, 2, \ldots)$ under the summation sign in (6), we obtain

(7) $D_r^p D_\mu^q v =$

$$\sum_{n=p}^{\infty} (f, \psi_n) n(n-1) \cdots (n-p+1) r^{n-p} \sqrt{n + \tfrac{1}{2}}\, D_\mu^q P_n(\mu)$$

Theorem 9.6-1 implies that (f, ψ_n) is of slow growth; that is, there exists a constant C and an integer k such that

$$|(f, \psi_n)| \le C n^k \qquad n = 1, 2, 3, \ldots$$

Moreover, it is a fact that

$$|D_\mu^q P_n(\mu)| \le n^{2q} \qquad -1 \le \mu \le 1, \qquad n = 1, 2, 3, \ldots$$

(See Churchill [2], pp. 211–212.) It follows that (7) converges uniformly on the domain $0 < r \le a < 1$, $-1 < \mu < 1$ because a^{n-p} decreases exponentially as $n \to \infty$ whereas all the other factors are bounded by a polynomial in n. From this we see that (6) is a conventional, indeed, smooth function and that the Laplacian operator may be applied to (6) term by term. Since $r^n \psi_n(\mu)$ satisfies (1) in the conventional sense, as is easily shown, it follows that (6) does too.

Next, by what we have shown, we may take the limit as $r \to 0$ under the summation sign to conclude that (6) tends to $(f, \psi_0)/\sqrt{2}$ uniformly on $0 \le \theta \le \pi$. Thus, (6) certainly remains bounded in a neighborhood of the origin.

Finally, we verify (2). Let ϕ be an arbitrary member of $\mathscr{A}$. According to Lemma 9.3-3, (6) converges in $\mathscr{A}$ and therefore in $L_2(I)$ for each fixed

$r < 1$. Consequently, we can take its inner product with ϕ term by term to write

$$(8) \qquad (v(r, \cos^{-1}\mu), \phi(\mu)) = \sum_{n=0}^{\infty} (f, \psi_n) r^n (\psi_n, \phi) \qquad r < 1$$

Now, (f, ψ_n) is of slow growth as $n \to \infty$ (Theorems 9.5-1 and 9.6-1), whereas (ψ_n, ϕ) is of rapid descent (Lemmas 9.3-2 and 9.3-3). Consequently, the right-hand side of (8) converges uniformly on $0 < r \leq 1$, and we may take the limit as $r \to 1-$ under the summation sign. Thus,

$$\lim_{r \to 1-} (v(r, \cos^{-1}\mu), \phi(\mu)) = \sum_{n=0}^{\infty} (f, \psi_n)(\psi_n, \phi) = (f, \phi)$$

which verifies (2).

PROBLEM 9.11-1. Find a conventional function $v(r, \cos^{-1}\mu)$ that satisfies the differential equation in (1) on the outside of the unit sphere (i.e., for $1 < r < \infty$, $-1 < \mu < 1$) such that $v(r, \cos^{-1}\mu)$ converges pointwise to zero as $r \to \infty$ and converges in $\mathscr{A}'$ to $f(\mu) \in \mathscr{A}'$ as $r \to 1+$. Here too, $\mathscr{A}'$ is the generalized-function space for the system 4a of Sec. 9.8. Show that your solution satisfies the required conditions.

PROBLEM 9.11-2. Consider the problem of one-dimensional heat flow is a nonuniform rod described by the differential equation:

$$(9) \qquad [aD_x(1 - x^2)D_x + (\mu x + \nu)D_x]v(x, t) = cD_t v(x, t) \qquad -1 < x < 1, \quad 0 < t < \infty$$

where a, c, μ, and ν are real constants with $a > 0$, $c > 0$. Here, $v(x, t)$ is the temperature in the rod. All the surfaces of the rod are perfectly insulated. Moreover, $a(1 - x^2)$ represents the thermal conductivity of the rod, c is a constant depending on the rod's density and specific heat, and the term $(\mu x + \nu)D_x v$ is due to a distributed heat source within the rod. Show that the differential equation (9) can be converted into

$$-a\mathfrak{N}_x u(x, t) = cD_t u(x, t)$$

where $\mathfrak{N}_x$ is the differential operator (with respect to x) in system 4 of Sec. 9.8 for the Jacobi transformation, and $u = \sqrt{w}\, v$, $w(x) \triangleq (1 - x)^\alpha (1 + x)^\beta$. What restrictions on μ and ν do the conditions $\alpha > -1$ and $\beta > -1$ impose?

Next, assume that, as $t \to 0+$, $v(x, t)$ converges in $\mathscr{D}'(I)$, $I = (-1, 1)$, to $g(x)/\sqrt{w(x)}$ where $g \in \mathscr{A}'$, $\mathscr{A}'$ being the appropriate generalized-function space for the Jacobi transformation. Formally, find a solution to this Cauchy problem. (This example is taken from Debnath [3].)

9.12. Application of the Finite Hankel Transformation of the First Form: A Dirichlet Problem for a Semi-Infinite Cylinder

The finite Hankel transformation is useful for solving various boundary-value problems with cylindrical boundaries. In this and the next section we give two examples of this, the present one being devoted to a Dirichlet problem.

We wish to solve Laplace's equation in the interior R of a semi-infinite cylinder of unit radius. In terms of cylindrical coordinates R can be described by

$$R = \{(r, \theta, z): 0 < r < 1, 0 \leq \theta < 2\pi, 0 < z < \infty\}$$

Assuming that the desired potential function v does not depend upon θ (i.e., $v = v(r, z)$) and setting $u(r, z) = \sqrt{r}\, v(r, z)$, we convert Laplace's equation into

$$S_{0,r} u + D_z^2 u = 0 \tag{1}$$

where, as always,

$$S_{0,r} = r^{-1/2} D_r r D_r r^{-1/2}$$

The boundary conditions we impose are

(i) As $z \to 0+$, $u(r, z)$ converges in $\mathscr{A}'$ to $g(r) \in \mathscr{A}'$. Here, $\mathscr{A}'$ is the generalized-function space corresponding to the system $5a$ of Sec. 9.8.

(ii) As $z \to \infty$, $u(r, z)$ converges uniformly to zero on $0 < r < 1$.

(iii) As $r \to 1-$, $u(r, z)$ converges uniformly to zero on $Z \leq z < \infty$ for each $Z > 0$.

(iv) As $r \to 0+$, $u(r, z) = O(\sqrt{r})$ uniformly on $Z \leq z < \infty$ for each $Z > 0$.

Note that this problem is quite similar to that of Sec. 5.8. There, r ranged through the semi-infinite interval $(0, \infty)$, and we used the zero-order (nonfinite) Hankel transformation $\mathfrak{H}_0$. On the other hand, r is now restricted to the finite interval $(0, 1)$, and we use the zero-order finite Hankel transformation $\mathfrak{A}$ of the first form.

The notation of system $5a$ of Sec. 9.8 is employed here with $x = r$ and $y_{0,n} = y_n$. The transformation $\mathfrak{A}$ formally converts (1) into

$$-y_n^2 U + D_z^2 U = 0 \tag{2}$$

where

$$U = U(n, z) = (u(r, z), \psi_n(r))$$

In view of the boundary condition (ii), we choose as a solution to (2):

(3) $$U(n, z) = A(n)e^{-y_n z}$$

where $A(n)$ does not depend on z. Matching the transform of the boundary condition (i), we obtain

$$A(n) = G(n) \triangleq (g, \psi_n)$$

Finally, taking the inverse transform of (3), we obtain the tentative solution:

(4) $$u(r, z) = \sum_{n=1}^{\infty} G(n)e^{-y_n z}\psi_n(r)$$

where

$$\psi_n(r) = \frac{\sqrt{2r}\, J_0(y_n r)}{J_1(y_n)}$$

We now verify that (4) is truly a solution. First of all, it is a fact that

(5) $$y_n \sim \pi(n - \tfrac{1}{4}) \qquad n \to \infty$$

and

$$|J_1(y_n)|^{-1} \leq K\sqrt{n} \qquad n = 1, 2, 3, \ldots$$

for some constant K. (See Jahnke, Emde, and Losch [1], pp. 147 and 153.) Moreover, by (5) and Theorems 9.5-1 and 9.6-1 we have that, as $n \to \infty$, $G(n) = O(n^p)$ for some integer p, whereas $e^{-y_n z} = O(e^{-\pi n Z})$ uniformly on $Z \leq z < \infty$ for each fixed $Z > 0$. Finally, we note that

$$D_r J_0(y_n r) = -y_n J_1(y_n r)$$
$$D_r r J_1(y_n r) = y_n r J_0(y_n r)$$

and that $J_0(y_n r)$ and $J_1(y_n r)$ are uniformly bounded for $0 < r < 1$, $n = 1, 2, 3, \ldots$. These facts imply that (4) converges uniformly on $0 < r < 1$, $Z \leq z < \infty$ $(Z > 0)$. This remains true after any of the operators $D_r r^{-1/2}$, $D_r r D_r r^{-1/2}$, D_z, D_z^2 are applied under the summation sign in (4). Thus, it is valid to apply $S_{0,r} + D_z^2$ term by term to (4), whereupon all terms vanish. We conclude that (4) satisfies the differential equation (1) in the conventional sense.

The stated uniform convergence of (4) allows us to take the limit under its summation sign as either $z \to \infty$, $r \to 1-$, or $r \to 0+$; indeed, this can be done even after absolute values have been taken term by term. This leads directly to a verification of boundary conditions (*i*), (*iii*), or (*iv*), respectively.

Finally, by virtue of Lemma 9.3-3 and our previous comments, the right-hand side of (4) converges in $\mathscr{A}$ and therefore in $L_2(I)$ for each fixed $z > 0$. Consequently we can form the inner product of $u(r, z)$ with any $\phi(r) \in \mathscr{A}$ term by term to get

$$(u(r, z), \phi(r)) = \sum_{n=1}^{\infty} G(n)e^{-y_n z}(\psi_n, \phi) \tag{6}$$

But, $G(n)$ is of slow growth, while (ψ_n, ϕ) is of rapid descent as $n \to \infty$. So, (6) converges uniformly on $0 \leq z < \infty$. and we may pass to the limit under the summation sign as $z \to 0+$ to get

$$(u(r, z), \phi(r)) \to \sum_{n=1}^{\infty} G(n)(\psi_n, \phi) = (g, \phi)$$

Thus, boundary condition (i) is also satisfied. This completes our verification of (4) as a solution.

Problem 9.12-1. Solve the differential equation (1) inside the finite domain:

$$\{(r, z): 0 < r < 1, 0 < z < c\}$$

under the following boundary conditions:

(i) As $z \to 0+$, $u(r, z)$ converges in $\mathscr{A}'$ to $g_1(r) \in \mathscr{A}'$.

(ii) As $z \to c-$, $u(r, z)$ converges in $\mathscr{A}'$ to $g_2(r) \in \mathscr{A}'$.

(iii) As $r \to 1-$, $u(r, z)$ converges to zero uniformly on $Z_1 \leq z \leq Z_2$ whenever $0 < Z_1 < Z_2 < c$.

(iv) As $r \to 0+$, $u(r, z) = O(\sqrt{r})$ uniformly on $Z_1 \leq z \leq Z_2$ whenever $0 < Z_1 < Z_2 < c$.

Here, $\mathscr{A}'$ is the same generalized-function space as before in this section. Justify your answer.

PROBLEM 9.12-2. Derive formally a solution to the following problem: Let R denote the interior of two semi-infinite concentric cylinders of radii a and b: more specifically, in terms of cylindrical coordinates,

$$R \triangleq \{(r, \theta, z): \quad a < r < b, 0 < a < b < \infty, 0 \leq \theta < 2\pi, \ 0 < z < \infty\}$$

Find a conventional function $u = u(r, z)$ that does not depend on θ and satisfies the differential equation (1) inside R. Also, let $\mathscr{A}'$ be the generalized-function space of system 5c for the third form of the finite Hankel transformation where $\mu = 0$. It is also required that u satisfy the boundary conditions:

(i) As $z \to 0+$, $u(r, z)$ converges in $\mathscr{A}'$ to $g(r) \in \mathscr{A}'$.

(ii) As $z \to \infty$, $u(r, z)$ converges uniformly to zero on $a < r < b$.

(iii) As $r \to b-$ or $r \to a+$, $u(r, z)$ converges uniformly to zero on $Z \leq z < \infty$ for every $Z > 0$.

9.13. Application of the Finite Hankel Transformation of the Second Form: Heat Flow in an Infinite Cylinder with a Radiation Condition

Our objective is to solve the heat equation in cylindrical coordinates inside an infinitely long cylinder of radius 1. We seek a conventional function v that depends on the radial dimension r and time t, but not on θ or z, and that satisfies the differential equation:

$$(1) \qquad D_r^2 v + r^{-1} D_r v = D_t v \qquad v = v(r, t), \qquad 0 < r < 1, \qquad 0 < t < \infty$$

As an initial condition we require that, as $t \to 0+$, $v(r, t)$ converge in some sense to a certain generalized function $f(r)$, whereas for a boundary condition we require that for each fixed $t > 0$

$$(2) \qquad \lim_{r \to 1-} [D_r v + av] = 0$$

where a is a real positive constant. When v denotes the temperature within the cylinder, (2) signifies that heat is being radiated away from the surface of the cylinder.

We can use the zero-order finite Hankel transformation of the second form to solve this problem if we first set $u = \sqrt{r}\, v$, $g = \sqrt{r}\, f$, and choose the constant a in Sec. 9.8, Eq. (3) equal to the a appearing in (2). We specify g as a member of $\mathscr{A}'$, the generalized-function space corresponding to the system 5b of Sec. 9.8. The differential equation now becomes

$$(3) \qquad S_{0,r} u = D_t u \qquad u = u(r, t), \quad 0 < r < 1, \quad 0 < t < \infty$$

whereas the initial condition is taken to be

$$(4) \qquad u(r, t) \to g(r) \text{ in } \mathscr{A}' \text{ as } t \to 0+$$

The boundary condition is converted into

$$(5) \qquad \lim_{r \to 1-} \left[D_r u + \left(a - \frac{1}{2r} \right) u \right] = 0$$

Here, it is understood that t is any fixed positive value, so that the limit is taken in the pointwise sense on $0 < t < \infty$.

Now, let $\mathfrak{A}$ denote the transformation corresponding to the system 5b of Sec. 9.8, and set

$$U(n, t) = \mathfrak{A} u(r, t) \qquad n = 1, 2, 3, \ldots.$$

$\mathfrak{A}$ transforms the differential equation (3) into

$$(6) \qquad -z_n^2 U(n, t) = D_t U(n, t) \qquad 0 < t < \infty$$

where the z_n are the successive positive roots of

$$(7) \qquad zDJ_0(z) + aJ_0(z) = 0$$

By solving (6) and matching the transformed initial condition, we obtain

$$U(n, t) = G(n)e^{-z_n^2 t}$$

where

$$G(n) = (g, \psi_n)$$

The inverse transformation yields thereby the following formal solution:

$$(8) \qquad u(r, t) = \sum_{n=1}^{\infty} G(n)e^{-z_n^2 t}\psi_n(r)$$

where

$$\psi_n(r) = \sqrt{\frac{2r}{h_n}}\, J_0(z_n r)$$

and

$$h_n = [J_1(z_n)]^2 + [J_0(z_n)]^2$$

That (8) truly is a conventional function that satisfies the differential equation (3) in the conventional sense, as well as (4) and (5), can be shown as in the preceding section by making use of the following facts: As $n \to \infty$, the roots z_n of (7) are asymptotic to those of $DJ_0(z) = -J_1(z)$, which in turn are asymptotic to $\pi(n + \frac{1}{4})$. Moreover, the zeros of $J_0(z)$ and $J_1(z)$ never coincide, and h_n is asymptotic to $2/\pi z_n$ as $n \to \infty$. (See Jahnke, Emde, and Losch [1], pp. 147, 152–153.) This implies that $1/h_n = O(n)$ as $n \to \infty$.

PROBLEM 9.13-1. Prove that (8) is a conventional function that satisfies (3) in the conventional sense. Also, show that (4) and (5) are satisfied.

PROBLEM 9.13-2. Set up and solve a problem similar to that described in the first paragraph of this section where now the radius of the cylinder is equal to $R \neq 1$, and the left-hand side of (1) is multiplied by a real positive constant $\varkappa$ representing the diffusivity of the cylinder's material.

Bibliography

Apostol, T. M.

[1] Mathematical Analysis, Addison-Wesley, Reading, Mass., 1957.

Armstrong, H. L.

[1] On the Representation of Transients by Series of Orthogonal Functions, IRE Trans. on Circuit Theory, Vol. CT-6, pp. 351–354, 1959.

Aseltine, J. A.

[1] Transforms for Linear Time-Varying Systems, Department of Electrical Engineering, University of California, Los Angeles, Rept. No. 52-1, 1952.

Beltrami, E. J., and M. R. Wohlers

[1] Distributional Boundary Value Theorems and Hilbert Transforms, Arch. Rational Mech. Anal., Vol. 18, pp. 304-309, 1965.

[2] Distributions and the Boundary Values of Analytic Functions, Academic Press, New York, 1966.

Benedetto, J. J.

[1] The Laplace Transform of Generalized Functions, Canadian J. Math., 18, pp. 357–374, 1966.

[2] Analytic Representation of Generalized Functions, Math. Zeitschrift, Vol. 97, pp. 303–319, 1967.

Blackman, J., and H. Pollard

[1] The Finite Convolution Transform, Trans. Amer. Math. Soc., Vol. 91, pp. 399–409, 1959.

Boas, R. P., Jr.

[1] Inversion of a Generalized Laplace Integral, Proc. Natl. Acad. Sci., USA, Vol. 28, pp. 21–24, 1942.

[2] Generalized Laplace Integrals, Bull. Amer. Math. Soc. Vol. 48, pp. 286–294, 1942.

Boas, R. P., Jr., and H. Pollard

[1] Complete Sets of Bessel and Legendre Functions, Annals of Math., Vol. 48, pp. 366–384, 1947.

Bochner, S.

[1] Lectures on Fourier Integrals, Annals of Math. Studies, Vol. 42, Princeton University Press, Princeton, N.J., 1959.

Bochner, S., and W. T. Martin

[1] Several Complex Variables, Princeton University Press, Princeton, N.J., 1948.

Bouix, M.

[1] Les Fonctions Généralisees ou Distributions, Masson, Paris, 1964.

Braga, C. L. R., and M. Schönberg

[1] Formal Series and Distributions, An. da Acad. Brasileira de Ciências, Vol. 31, pp. 333–360, 1959.

Bremermann, H. J.

[1] Distributions, Complex Variables, and Fourier Transforms, Addison-Wesley, Reading, Mass., 1965.

Bremermann, H. J., and L. Durand

[1] On Analytic Continuation, Multiplication, and Fourier Transformations of Schwartz Distributions, J. Math. Phys., Vol. 2, pp. 240–258, 1961.

Cholewinski, F. M.

[1] A Hankel Convolution Complex Inversion Theory, Memoirs Amer. Math. Soc., No. 58, 1965.

Churchill, R. V.

[1] Complex Variables and Applications, 2nd ed., McGraw-Hill, New York, 1960.

[2] Fourier Series and Boundary Value Problems, 2nd ed., McGraw-Hill, New York, 1963.

[3] The Operational Calculus of Legendre Transforms, J. Math. Phys., Vol. 33, pp. 165–177, 1954.

Churchill, R. V., and C. L. Dolph

[1] Inverse Transforms of Products of Legendre Transforms, Proc. Amer. Math. Soc., Vol. 5, pp. 93–100, 1954.

Conte, S. D.

[1] Gegenbauer Transforms, Quart. J. Math. Oxford (2), Vol. 6, pp. 48–52, 1955.

Cooper, J. L. B.

[1] Laplace Transformations of Distributions, Canadian J. Math., Vol. 18, pp. 1325–1332, 1966.

Dauns, J., and D. V. Widder

[1] Convolution Transforms Whose Inversion Functions Have Complex Roots, Pacific J. Math., Vol. 15 (2), pp. 427–442, 1965.

Debnath, L.

[1] On Laguerre Transform, Bull. Calcutta Math. Soc., Vol. 52, pp. 69–77, 1960.

[2] On Hermite Transform, Mat. Vesnik, Vol. 1 (16), pp. 285–292, 1964.

[3] On Jacobi Transform, Bull. Calcutta Math. Soc., Vol. 55, pp. 113–120, 1963.

Dimovski, I.

[1] Operational Calculus for a Class of Differential Operators, Compt. Rend. l'Académie Bulgare Sci., Vol. 19, pp. 1111–1114, 1966.

Ditkin, V. A., and A. P. Prudinkov

[1] Integral Transforms and Operational Calculus, Pergamon Press, New York, 1965.

Ditzian, Z.

[1] On a Class of Convolution Transforms, Pacific J. Math. (to appear)

[2] On Asymptotic Estimates for Kernels of Convolution Transforms, Pacific J. Math., Vol. 21, pp. 249–254, 1967.

Ditzian, Z., and A. Jakimovski

[1] Inversion and Jump Formulas for Variation Diminishing Transforms, Ann. Mat. Pura Appl. (to appear)

Dolezal, V.

[1] Dynamics of Linear Systems, Czech. Acad. Sci., Prague, 1964.

Dudley, R. M.

[1] On Sequential Convergence, Trans. Amer. Math. Soc., Vol. 112, pp. 483–507, 1964.

Ehrenpreis, L.

[1] Analytic Functions and the Fourier Transform of Distributions, I, Annals of Math., Vol. 63, pp. 129–159, 1956.

Erdelyi, A.

[1] On Some Functional Transformations, Rend. Sem. Mat. Univ. Torino, Vol. 10, pp. 217–234, 1950–1951.

Erdelyi, A. (Ed.)

[1] Higher Transcendental Functions, Vols. I, II, and III, McGraw-Hill, New York, 1953.

Fenyo, I.

[1] Hankel-Transformation Verallgemeinerter Funktionen, Mathematica, Vol. 8 (31), 2, pp. 235–242, 1966.

Fox, C.

[1] The Inversion of Convolution Transforms by Differential Operators, Proc. Amer. Math. Soc., Vol. 4, pp. 880–887, 1953.

Friedman, A.

[1] Generalized Functions and Partial Differential Equations, Prentice-Hall, Englewood Cliffs, N.J., 1963.

Fung Kang

[1] Generalized Mellin Transforms I, Sci. Sinica 7, pp. 582–605, 1958.

Garnir, H. G., and M. Munster

[1] Transformation de Laplace des Distributions de L. Schwartz, Bull. Soc. Royale Sciences Liège, 33e Année, pp. 615–631, 1964.

Gelfand, I. M., and A. G. Kostyuchenko

[1] On Eigenfunction Expansions of Differential and Other Operators, Dokl. Akad. Nauk, SSSR, Vol. 103, pp. 349–352, 1955.

Gelfand, I. M., and G. E. Shilov

[1] Generalized Functions, Vols. 1 and 3, Academic Press, New York, 1964, 1967.

[2] Generalized Functions, Vol. 2, Moscow, 1958.

Gelfand, I. M., and N. Ya. Vilenkin

[1] Generalized Functions, Vol. 4, Academic Press, New York, 1964.

Gerardi, F. R.

[1] Application of Mellin and Hankel Transforms to Networks with Time-Varying Parameters, IRE Trans. on Circuit Theory, Vol. CT-6, pp. 197–208, 1959.

Giertz, M.

[1] On the Expansion of Certain Generalized Functions in Series of Orthogonal Functions, Proc. London Math. Soc., 3rd Ser., Vol. 14, pp. 45–52, 1964.

Gonzalez Dominguez, A.

[1] Contribución a la teoria de las funciones de Hille, Ciencia y Técnica, No. 475, pp. 475–521, 1941.

Griffith, J. L.

[1] Hankel Transforms of Functions Zero Outside a Finite Interval, J. Proc. Roy. Soc. New South Wales, Vol. 89, pp. 109—115, 1955.

[2] An Iteration of Fourier Transforms, J. Australian Math. Soc. (to appear)

Grobner, W., and H. Hofreiter

[1] Integraltafel, Vols. 1 and 2, Springer, Innsbruk, Austria, 1958.

Guttinger, W.

[1] Generalized Functions and Dispersion Relations in Physics, Fortschr. Physic, Vol. 14, pp. 483–602, 1966.

Haimo, D. T.

[1] Integral Equations Associated with Hankel Convolutions, Trans. Amer. Math. Soc., Vol. 116, pp. 330–375, 1965.

Hille, E., and R. S. Phillips
[1] Functional Analysis and Semi-groups, Amer. Math. Soc. Colloq. Publ., Vol. 31, revised ed., Providence, R. I., 1957.
Hirschman, I. I., Jr.
[1] Variation Diminishing Hankel Transforms, J. Analyse Math., Vol. 8, pp. 307–336, 1960, 1961.
[2] Laguerre Transforms, Duke Math. J., Vol. 30, pp. 495–510, 1963.
Hirschman, I. I., and D. V. Widder
[1] The Convolution Transform, Princeton University Press, Princeton, N.J., 1955.
[2] The Inversion of a General Class of Convolution Transforms, Trans. Amer. Math. Soc., Vol. 66, pp. 135–201, 1949.
Horváth, J.
[1] Transformadas de Hilbert de Distributions, Segundo Symposium de Matematicas, Buenos Aires, Imprenta y Casa Editora "Coni," 1954.
[2] Singular Integral Operators and Spherical Harmonics, Trans. Amer. Math. Soc., Vol. 82, pp. 52–63, 1956.
[3] Topological Vector Spaces and Distributions, Vol. I, Addison-Wesley. Reading, Mass., 1966.
Huggins, W. H.
[1] Signal Theory, IRE Trans. on Circuit Theory, Vol. CT-3, pp. 210–216, 1956.
Hurewicz, W.
[1] Lectures on Ordinary Differential Equations, John Wiley, New York, 1958.
Ishihara, T.
[1] On Generalized Laplace Transforms, Proc. Japan Acad., Vol. 37, pp. 556–561, 1961.
Jahnke, E., F. Emde, and F. Losch
[1] Tables of Higher Functions, McGraw-Hill, New York, 1960.
Jones, D. S.
[1] Generalized Functions, McGraw-Hill, New York, 1966.
[2] Some Remarks on Hilbert Transforms, J. Inst. Math. Applics., Vol. 1, pp. 226–240, 1965.
Kantorovich, L. V., and G. P. Akilov
[1] Functional Analysis in Normed Spaces, Macmillan, New York, 1963.
Kautz, W. H.
[1] Transient Synthesis in the Time Domain, IRE Trans. on Circuit Theory, Vol. CT-3, pp. 210–216, 1956.
Koh, E., and A. H. Zemanian
[1] The Complex Hankel and I Transformations of Generalized Functions, SIAM J. Appl. Math. (to appear)
Korevaar, J.
[1] Distributions Defined from the Point of View of Applied Mathematics, Koninkl. Ned. Akad. Wetenschap., Ser. A., Vol, 58, pp. 368–389, 483–503, 663–674, 1955.
[2] Pansions and the Theory of Fourier Transforms, Trans. Amer. Math. Soc., Vol. 91, pp. 53–101, 1959.
Lakshmanarao, S. K.
[1] Gegenbauer Transforms, Math. Student, Vol. 22, pp. 161–165, 1954.
Laughlin, T. A.
[1] A Table of Distributional Mellin Transforms, College of Engineering Tech., Rept. 40, State University, Stony Brook, New York, June 15, 1965.

Lauwerier, H. A.

[1] The Hilbert Problem for Generalized Functions, Arch. Rational Mech. Anal., Vol. 13, pp. 157–166, 1963.

Lavoine, J.

[1] Calcul Symbolique, Distributions et Pseudo-Fonctions, Centre Nat. Rech. Sci., Paris, 1959.

Lee, Y. W.

[1] Synthesis of Electrical Networks by Means of the Fourier Transforms of Laguerre Functions, J. Math, Phys., Vol. 11, pp. 83–113, 1931–1932.

Lerner, R. M.

[1] The Representation of Signals, IRE Trans. on Circuit Theory, Vol. CT-6, Special Supplement, pp. 197–216, May, 1959.

Lighthill, M. J.

[1] Fourier Analysis and Generalized Functions, Cambridge University Press, London, 1958.

Lions J. L.

[1] Opérateurs de transmutation singuliers et équations d'Euler Poisson Darboux généralisées, Rend. Seminario Math. Fis. Milano, Vol. 28, pp. 3–16, 1959.

Liusternik, L. A., and V. J. Sobolev

[1] Elements of Functional Analysis, Frederick Ungar, New York, 1961.

Liverman, T. P. G.

[1] Generalized Functions and Direct Operational Methods, Vol. 1, Prentice-Hall, Englewood Cliffs, N.J., 1964.

Macauley-Owen, P.

[1] Parseval's Theorem for Hankel Transforms, Proc. London Math. Soc., Vol. 45, pp. 458–474, 1939.

MacRobert, T. M.

[1] Fourier Integrals, Proc. Edinburgh Math. Soc., Vol. 51, pp. 116–126, 1931.

McCully, J.

[1] The Laguerre Transform, SIAM Review, Vol. 2, pp. 185–191, 1960.

Meijer, C. S.

[1] Asympototische Entwicklungen von Besselschen, Hankelschen und verwandten Funktionen, Proc. Kon. Akad. Wetensch., Amsterdam, Vol. 35, pp. 656–667, 852–866, 948–958, 1079–1090, 1932.

[2] Uber eine Erweiterung der Laplace-Transformation, Proc. Amsterdam Akad. Wet., Vol. 43, pp. 599–608, 702–711, 1940.

Meller, N. A.

[1] On an Operational Calculus for the Operator $B_\alpha = t^{-\alpha}(d/dt)t^{\alpha+1}(d/dt)$ Vichislitelnaya Matematika, pp. 161–168, 1960.

Miller, J. B.

[1] Generalized Function Calculi for the Laplace Transform, Arch. Rational Mech. Anal., Vol. 12, pp. 409–419, 1963.

Myers, D. E.

[1] An Imbedding Space for Schwartz Distributions, Pacific, J. Math., Vol. 11, pp. 1467–1477, 1961.

Paige, L. J., and J. D. Swift

[1] Elements of Linear Algebra, Ginn, Boston, 1961.

Pandey, J. N., and A. H. Zemanian

[1] Complex Inversion for the Generalized Convolution Transformation, Pacific J. Math. (to appear)

Queen, W. C.,

[1] The N-dimensional Generalized Weierstrass Transformation, Doctoral Thesis, State University, Stony Brook, New York, 1967.

Rehberg, C. F.

[1] The Theory of Generalized Functions for Electrical Engineers, Dept. Elect. Eng., New York University, Tech. Rept. 400–42, August, 1961.

Robertson, A. P., and W. Robertson

[1] Topological Vector Spaces, Cambridge University Press, London, 1964.

Rooney, P. G.

[1] On the Inversion of the Gauss Transformation, Canadian J. Math., Vol. 9, pp. 459–464, 1957.

[2] A Generalization of an Inversion Formula for the Gauss Transformation, Canadian Math. Bull., Vol. 6, pp. 45–53, 1963.

Schwartz, L.

[1] Théorie des Distributions, Vols. I, II, Hermann, Paris, 1957, 1959.

[2] Transformation de Laplace des Distributions, Seminaire Mathematique de l'Université de Lund, Tome Supplémentaire, dedié á M. Riesz, pp. 196–206, 1952.

[3] Causalité et analyticité, Anais da Academia Brasileira de Ciências, Vol. 34, pp. 13–21, 1962.

Scott, E. J.

[1] Jacobi Transforms, Quart. J. Math., Oxford (2), Vol. 4, pp. 36–40, 1953.

Soboleff, S. L.

[1] Méthode nouvelle à résoudre le probléme de Cauchy pour les équations linéaires hyperboliques normales, Mat. Sbornik, Vol. 1, pp. 39–72, 1936.

Srivastav, R. P., and K. S. Parihar

[1] Dual Integral Equations—A Distributional Approach (to appear).

Srivastava, K. N.

[1] On Gegenbauer Transforms, Math. Student, Vol. 33, pp. 129–132, 1965.

Sumner, D. B.

[1] A Convolution Transform Admitting an Inversion Formula of Integro-Differential Type, Canadian J. Math., Vol. 5, pp. 114–117, 1953.

Tanno, Y.

[1] On the Convolution Transform, Kodai Math. Sem. Rept., Vol. 11, pp. 40–50, 1959.

[2] On the Convolution Transform II, III, Sci. Rept. Hirosaki University, Vol. 9, pp. 5–20, 1962.

[3] On a Class of Convolution Transforms, Tôhoku Math. J., Vol. 18, pp. 156–173, 1966.

Temple, G.

[1] The Theory of Generalized Functions, Proc. Roy. Soc., Ser. A, Vol. 228, pp. 175–190, 1955.

Tillman, H. G.

[1] Randverteilungen Analytischer Funktionen und Distributionen, Math. Zeitschrift, Vol. 59, pp. 61–83, 1953.

[2] Distributionen als Randverteilungen Analytischer Funktionen II, Math. Zeitschrift, Vol. 76, pp. 5–21, 1961.

Titchmarsh, E. C.

[1] The Theory of Functions, 2nd ed, Oxford University Press, London, 1939.

[2] A Class of Expansions in Series of Bessel Functions, Proc. London Math. Soc., Vol. 22, pp. 13–16, 1923.

Tranter, C. J.
[1] Legendre Transforms, Quart. J. Math., Oxford (2), Vol. 1, pp. 1–8, 1950.
Walter, G. G.
[1] Expansions of Distributions, Trans. Amer. Math. Soc., Vol. 116, pp.492–510, 1965
Watson, G. N.
[1] Theory of Bessel Functions, 2nd ed., Canbridge University Press, Cambridge, 1958.
Weston, J. D.
[1] Positive Perfect Operators, Proc. London Math. Soc., (3), Vol. 10, pp. 545–565 1960.
Widder, D. V.
[1] The Laplace Transform, Princeton University Press, Princeton, N. J., 1946.
[2] Advanced Calculus, 2nd ed., Prentice-Hall, Englewood Cliffs, N. J., 1961.
Widlund, O.
[1] On the Expansion of Generalized Functions in Series of Hermite Functions, Kgl. Tekn. Hogsk. Handl., Stockholm, No. 173, 1961.
Wilansky, A.
[1] Functional Analysis, Blaisdell, New York, 1964.
Williamson, J. H.
[1] Lebesgue Integration, Holt, Rinehart, and Winston, New York, 1962.
Young, T. Y., and W. H. Huggins
[1] Complementary Signals and Orthogonalized Exponentials, IRE Trans. on Circuit Theory, Vol. CT-9, pp. 362–370, 1962.
Zemanian, A. H.
[1] Distribution Theory and Transform Analysis, McGraw-Hill, New York, 1965.
[2] The Distributional Laplace and Mellin Transformations, SIAM J. Appl. Math., Vol. 14, pp. 41–59, 1966.
[3] Inversion Formulas for the Distributional Laplace Transformation, SIAM J. Appl. Math., Vol 14, pp. 159–166, 1966.
[4] A Distributional Hankel Transformation, SIAM J. Appl. Math., Vol. 14, pp. 561–576, 1966.
[5] The Hankel Transformation of Certain Distributions of Rapid Growth, SIAM J. Appl. Math., Vol. 14, pp. 678–690, 1966.
[6] Some Abelian Theorems for the Distributional Hankel and K Transformations, SIAM J. Appl. Math., Vol. 14, pp. 1255–1265, 1966.
[7] A Solution of a Division Problem Arising from Bessel-Type Differential Equations, SIAM J. Appl. Math., Vol. 15, pp. 1106–1111, 1967.
[8] Hankel Transforms of Arbitrary Order, Duke Math. J., Vol. 34, pp. 761–769, 1967.
[9] A Distributional K. Transformation, SIAM J. Appl. Math., Vol. 14, pp. 1350–1365, 1966, and Vol. 15, p. 765, 1967.
[10] A Generalized Weierstrass Transformation, SIAM J. Appl. Math., Vol. 15, pp. 1088–1105, 1967.
[11] A Generalized Convolution Transformation, SIAM J. Appl. Math., Vol. 15, pp. 324–346, 1967.
[12] Orthonormal Series Expansions of Certain Distributions and Distributional Transform Calculus, J. Math. Anal. and Appl., Vol. 14, pp. 263–275, 1966. See also: Tech. Rept. No. 22, College of Engineering, State University of New York at Stony Brook, Nov. 15, 1964.

[13] The Properties of Pole and Zero Locations for Nondecreasing Step Responses, AIEE Trans., Communications and Electronics, No. 50, pp. 421–426, Sept., 1960.

[14] On the Pole and Zero Locations of Rational Laplace Transformations of Nonnegative Functions: I and II, Proc. Amer. Math. Soc., Vol. 10, pp. 868–872, 1959, and Vol. 12, pp. 870–874, 1961.

Index of Symbols

Some additional special symbols

Index

A CATALOG OF SELECTED

DOVER BOOKS

IN SCIENCE AND MATHEMATICS